Finite Elements

Finite element method (FEM) is a numerical technique for approximation of the solution of partial differential equations. This book on FEM and related simulation tools starts with a brief introduction to Sobolev spaces and elliptic scalar problems and continues with explanation of finite element spaces and estimates for the interpolation error. Construction of finite elements on simplices, quadrilaterals, hexahedral is discussed by the authors in self-contained manner. The last chapter focuses on object-oriented finite element algorithms and efficient implementation techniques. Readers can also find the concepts of scalar parabolic problems and high dimensional parabolic problems in different chapters. Besides, the book includes some recent advances in FEM, including nonconforming finite elements for boundary value problems of higher order and approaches for solving differential equations in high-dimensional domains. There are plenty of solved examples and mathematical theorems interspersed throughout the text for better understanding.

Sashikumaar Ganesan is Assistant Professor at the Indian Institute of Science (IISc). He heads the research group on Computational Mathematics at the Department of Computational and Data Sciences at IISc. His areas of interest include numerical analysis, finite elements in fluid dynamics, moving boundary value problems, numerical methods for turbulent flows, variational multiscale methods, population balance modelling, scalable algorithms and high performance computing.

Lutz Tobiska is Professor at the Institute for Analysis and Computational Mathematics, Otto-von-Guericke University Magdeburg, Germany. His research interests include finite elements in fluid dynamics, parallel algorithms, multigrid methods and adaptive methods for convection diffusion equations.

CAMBRIDGE–IISc SERIES

Cambridge–IISc Series aims to publish the best research and scholarly work on different areas of science and technology with emphasis on cutting-edge research.

The books will be aimed at a wide audience including students, researchers, academicians and professionals and will be published under three categories: research monographs, centenary lectures and lecture notes.

The editorial board has been constituted with experts from a range of disciplines in diverse fields of engineering, science and technology from the Indian Institute of Science, Bangalore.

IISc Press Editorial Board:
G. K. Ananthasuresh, *Professor, Department of Mechanical Engineering*
K. Kesava Rao, *Professor, Department of Chemical Engineering*
Gadadhar Misra, *Professor, Department of Mathematics*
T. A. Abinandanan, *Professor, Department of Materials Engineering*
Diptiman Sen, *Professor, Centre for High Energy Physics*

Titles in print in this series:

- *Continuum Mechanics: Foundations and Applications of Mechanics* by C. S. Jog
- *Fluid Mechanics: Foundations and Applications of Mechanics* by C. S. Jog
- *Noncommutative Mathematics for Quantum Systems* by Uwe Franz and Adam Skalski
- *Mechanics, Waves and Thermodynamics* by Sudhir Ranjan Jain

Cambridge - IISc Series

FINITE ELEMENTS

Theory and Algorithms

Sashikumaar Ganesan

Lutz Tobiska

CAMBRIDGE
UNIVERSITY PRESS

CAMBRIDGE
UNIVERSITY PRESS

Shaftesbury Road, Cambridge CB2 8EA, United Kingdom

One Liberty Plaza, 20th Floor, New York, NY 10006, USA

477 Williamstown Road, Port Melbourne, VIC 3207, Australia

314–321, 3rd Floor, Plot 3, Splendor Forum, Jasola District Centre, New Delhi – 110025, India

103 Penang Road, #05–06/07, Visioncrest Commercial, Singapore 238467

Cambridge University Press is part of Cambridge University Press & Assessment, a department of the University of Cambridge.

We share the University's mission to contribute to society through the pursuit of education, learning and research at the highest international levels of excellence.

www.cambridge.org
Information on this title: www.cambridge.org/9781108415705

First published 2017

A catalogue record for this publication is available from the British Library

ISBN 978-1-108-41570-5 Hardback

Additional resources for this publication at www.cambridge.org/9781108415705

Contents

Preface

The purpose of this book is to present—in a coherent and lucid way—the mathematical theory and algorithms of the finite element method, which is the most widely-used method for the solution of partial differential equations in the field of Computational Science. We believe that the full potential of the finite element method can be realised only when the theoretical background and the implemented algorithms are considered as a unit.

The selection of the basic mathematical theory of finite elements in this book is based on lectures given in the "Finite Element I and II" courses offered for several years at Otto-von-Guericke University, Magdeburg and in the "Finite Element I" course offered at the Indian Institute of Science, Bangalore. Furthermore, the finite element algorithms presented are based on the knowledge and experience gained through the development of our in-house finite element package for more than 10 years. The theory and algorithms of finite elements that we describe here are self-contained; our aim is that beginners will find our book to be both readable and useful.

We start in Chapter 1 with a brief introduction to Sobolev spaces and the necessary basics of functional analysis. This will help those readers who are unfamiliar with functional analysis. The goal of Chapter 2 is to explain the finite element method to beginners in the simplest possible way. The concepts of weak solutions, variational formulation of second-order elliptic boundary value problems, incorporation of different boundary conditions in a variational form, and the standard Galerkin approach are all introduced in this chapter. Moreover, existence and uniqueness theory (the Lax-Milgram Theorem) and an abstract error analysis (quasi-optimality of the method) are presented here.

The next two Chapters give the basic theory of the finite element method. In Chapter 3, the construction of finite elements on simplices, quadrilaterals, and hexahedrals is discussed in detail. Furthermore, linear, bilinear and isoparametric transformations are explained, and mapped finite elements are considered. Chapter 4 deals with the interpolation theory of affine equivalent finite elements in Sobolev spaces. We also discuss the interpolation of functions that are less smooth; in particular, Scott-Zhang interpolation is described in this Chapter.

Finite elements for more advanced scalar problems are presented in Chapters 5 and 6. As an example of a finite element method for elliptic higher-order problems, conforming and nonconforming finite element methods for the biharmonic equation are studied in Chapter 5. Next, the finite element method for scalar parabolic problems is presented in Chapter 6. This includes a discussion of the standard θ-methods and discontinuous Galerkin methods for temporal discretisation. In addition, splitting schemes in the context of finite element methods for high-dimensional scalar parabolic problems are examined.

Finite element methods for systems of equations in solid mechanics and fluid mechanics are presented in Chapters 7 and 8. In particular, finite element methods for systems of equations in linear elasticity and for Mindlin-Reissner plate problems are discussed in Chapter 7. Chapter 8 deals in detail with finite element methods for the Stokes and Navier-Stokes equations. Starting from the derivation of conservation of mass and momentum equations, we discuss the implementation of different boundary conditions, a mixed variational formulation, the necessary condition for the stability of the finite element scheme, and finite elements that satisfy that stability condition. Furthermore, the derivation of a priori error estimates for Stokes problems, conforming and nonconforming finite element discretizations, and an array of inf-sup finite element pairs in two and three dimensions are presented. As an alternative to mixed finite element methods, stabilized finite elements with equal-order interpolations that circumvent the inf-sup stability are discussed. Hints on the choice of pressure spaces and on linearization strategies for the Navier-Stokes problem are given.

Finally, in Chapter 9, finite element algorithms and implementations based on object-oriented concepts are provided. All finite element algorithms and computing tools that are needed in an efficient and robust object-oriented finite element package are presented in detail.

We believe that the theoretical material of Chapters 1 to 4 and the implementation matters of Chapter 9 should be presented in any course on finite elements. The selection of the other Chapters is a matter of taste and depends on which application is in the foreground. We have made an effort to write each chapter between Chapters 5 and 8 as a self-contained unit, so—for example—a reader who is mainly interested in partial differential equations of higher order can concentrate on our discussion of the biharmonic equation.

We wish to thank all our colleagues and friends, including D. Braess, V. John, P. Knobloch, Q. Lin, G. Matthies, H.-G. Roos, F. Schieweck, M. Stynes, R. Verfürth, Z. Zhang and A. Zhou, for their help and constructive discussions on several of these topics. Also, we would like to thank our wives Sangeetha and Inge for their patient and continuous encouragement. More importantly, we are grateful to all our funding agencies (in particular the Alexander von Humboldt (AvH) foundation and German Academic Exchange Service (DAAD)) for their generous support. Finally, we would like to express our appreciation to Cambridge University Press and Indian Institute of Science (IISc) Press for their cooperation in the production and publishing of this book.

CHAPTER 1

INTRODUCTION TO SOBOLEV SPACES

In this chapter we recall some basics on functional analysis and provide a brief introduction to Sobolev spaces. For a more detailed and comprehensive study, we refer to Adams (1975).

1.1. Banach and Hilbert spaces

Definition 1.1. Let X be a real linear space. A mapping $\|\cdot\| : X \to \mathbb{R}$ is called a norm on X if

- $\|x\| = 0 \quad \Leftrightarrow \quad x = 0$ for all $x \in X$,
- $\|\lambda x\| = |\lambda| \, \|x\|$ for all $x \in X$, $\lambda \in \mathbb{R}$,
- $\|x + y\| \leq \|x\| + \|y\|$ for all $x, y \in X$.

The pair $(X, \|\cdot\|)$ is called a normed space.

Setting $y = -x$ in the inequality and using the other two properties we get $\|x\| \geq 0$ for all $x \in X$.

Definition 1.2. A sequence $(x_n)_{n\in\mathbb{N}}$ is called Cauchy sequence, if for all $\varepsilon > 0$ there exists an index $n_0(\varepsilon)$ such that for all $m, n > n_0(\varepsilon)$ it holds $\|x_m - x_n\| < \varepsilon$.

Definition 1.3. A sequence $(x_n)_{n\in\mathbb{N}}$ converges to $x \in X$ if for all $\varepsilon > 0$ there is an index $n_0(\varepsilon)$ such that for all $n > n_0(\varepsilon)$ it holds $\|x_n - x\| < \varepsilon$.

Definition 1.4. A normed space $(X, \|\cdot\|)$ is called complete if every Cauchy sequence in X converges in X.

Definition 1.5. A complete normed space is called Banach space.

Example 1.1. Let $\Omega \subset \mathbb{R}^d$, d is the dimension, be an open and bounded domain. We denote by $L^p(\Omega)$, $1 \le p < \infty$, the set of all measurable functions $f : \Omega \to \mathbb{R}$ for which

$$\int_\Omega |f(x)|^p \,\mathrm{d}x < \infty.$$

Similarly, the set of all measurable functions $f : \Omega \to \mathbb{R}$ satisfying

$$\operatorname{ess\,sup}\{\, |f(x)| \, : \, x \in \Omega \} < \infty$$

will be denoted by $L^\infty(\Omega)$. Then, $L^p(\Omega)$, $p \in [1,\infty]$ is a real linear space. Setting for $1 \le p < \infty$

$$\|f\|_{L^p(\Omega)} := \left(\int_\Omega |f(x)|^p \,\mathrm{d}x \right)^{1/p} \quad \text{and} \quad \|f\|_{L^\infty(\Omega)} := \operatorname{ess\,sup}\{\, |f(x)| \, : \, x \in \Omega \},$$

respectively, $L^p(\Omega)$ becomes a normed space for any $p \in [1,\infty]$ if we identify functions which are equal up to a set of measure zero. This identification is necessary because we only have for a set $M \subset \Omega$ of measure zero

$$\int_\Omega |f(x)|^p \,\mathrm{d}x = 0 \qquad \Rightarrow \qquad f(x) = 0 \text{ for all } x \in \Omega \backslash M.$$

Strictly speaking, the elements in $L^p(\Omega)$, $p \in [1,\infty]$ are classes of equivalent functions (equal up to a set of measure zero). Then, the first condition in Definition 1.1 becomes true. The second condition follows directly from the definition of $\|\cdot\|_{L^p(\Omega)}$ and the third condition is just the inequality stated in Lemma 1.1.

Lemma 1.1 (Minkowski's inequality). *For $f, g \in L^p(\Omega)$, $p \in [1,\infty]$, we have*

$$\|f + g\|_{L^p(\Omega)} \le \|f\|_{L^p(\Omega)} + \|g\|_{L^p(\Omega)}.$$

Theorem 1.1. *$L^p(\Omega)$, $p \in [1,\infty]$, is a Banach space.*

Lemma 1.2 (Hölders's inequality). *For $f \in L^p(\Omega)$ and $g \in L^q(\Omega)$ with $p, q \in [1,\infty]$, $1/p + 1/q = 1$, we have $fg \in L^1(\Omega)$ and*

$$\|fg\|_{L^1(\Omega)} \le \|f\|_{L^p(\Omega)} \|g\|_{L^q(\Omega)}.$$

Definition 1.6. Two norms $\|\cdot\|_1$ and $\|\cdot\|_2$ on a normed space X are called equivalent if there exist positive constants C_1, C_2 such that

$$C_1\|x\|_1 \le \|x\|_2 \le C_2\|x\|_1 \quad \text{for all } x \in X.$$

Note that topological properties do not change when switching to an equivalent norm, for example, a sequence is convergent in $(X, \|\cdot\|_1)$ iff it is convergent in $(X, \|\cdot\|_2)$.

Definition 1.7. Let $(X, \|\cdot\|_X)$ be a normed space. A mapping $g : X \to \mathbb{R}$ is called linear if

$$g(\alpha x + \beta y) = \alpha g(x) + \beta g(y) \quad \text{for all } \alpha, \beta \in \mathbb{R}, \ x, y \in X.$$

A linear mapping $g : X \to \mathbb{R}$ is continuous if there is a constant C such that

$$|g(x)| \le C\|x\|_X \quad \text{for all } x \in X.$$

Definition 1.8. We define the sum of two continuous linear functionals, g_1 and g_2, and the multiplication of a continuous linear functional g with a real number α by

$$(g_1+g_2)(x) := g_1(x)+g_2(x) \quad \text{and} \quad (\alpha g)(x) := \alpha g(x), \quad \alpha \in \mathbb{R}, x \in X.$$

Then, it can be shown that the set of all continuous linear functionals create a real linear space, the dual space X^*. In the following, for the elements $g \in X^*$ we use the notation $\langle g,x\rangle := g(x)$.

Lemma 1.3. *The set X^* of continuous linear functionals $x \mapsto \langle g,x\rangle$ on X is a Banach space with respect to the norm*

$$\|g\|_{X^*} := \sup_{0\neq x\in X} \frac{|\langle g,x\rangle|}{\|x\|_X}.$$

Proof. Using the definition of the dual norm $\|\cdot\|_{X^*}$, we see

$$\|g\|_{X^*} = 0 \quad \Leftrightarrow \quad \langle g,x\rangle = 0 \text{ for all } x \in X \quad \Leftrightarrow \quad g = 0.$$

Further, we have

$$\|\lambda g\|_{X^*} = \sup_{0\neq x\in X} \frac{|\langle \lambda g,x\rangle|}{\|x\|_X} = \sup_{0\neq x\in X} \frac{|\lambda\langle g,x\rangle|}{\|x\|_X} = |\lambda| \sup_{0\neq x\in X} \frac{|\langle g,x\rangle|}{\|x\|_X} = |\lambda|\,\|g\|_{X^*}$$

and for the sum of two functionals it holds

$$\begin{aligned}\|g_1+g_2\|_{X^*} &= \sup_{0\neq x\in X} \frac{|\langle g_1+g_2,x\rangle|}{\|x\|_X} = \sup_{0\neq x\in X} \frac{|\langle g_1,x\rangle + \langle g_2,x\rangle|}{\|x\|_X} \\ &\le \sup_{0\neq x\in X} \frac{|\langle g_1,x\rangle|}{\|x\|_X} + \sup_{0\neq x\in X} \frac{|\langle g_2,x\rangle|}{\|x\|_X} = \|g_1\|_{X^*} + \|g_2\|_{X^*}\end{aligned}$$

It remains to show that $(X^*, \|\cdot\|_{X^*})$ is complete. Let $(g_n)_{n\in\mathbb{N}}$ be a Cauchy sequence in X^*, that is, for all $\varepsilon > 0$ there is an index $n_0(\varepsilon)$ such that for all $m,n > n_0(\varepsilon)$ it holds $\|g_m - g_n\|_{X^*} < \varepsilon$. Then, for a fixed $x \in X$, the sequence of real numbers $(\langle g_n,x\rangle)_{n\in\mathbb{N}}$ is a Cauchy sequence in $\mathbb{R}$ due to

$$|\langle g_m,x\rangle - \langle g_n,x\rangle| = |\langle g_m - g_n,x\rangle| \le \|g_m - g_n\|_{X^*}\,\|x\|_X < \varepsilon\|x\|_X$$

for all $m,n > n_0(\varepsilon)$. Any Cauchy sequence of real numbers is convergent in $\mathbb{R}$, that means $\lim_{n\to\infty}\langle g_n,x\rangle = g(x)$ for all $x \in X$. The linearity of g follows from

$$\begin{aligned} g(\alpha x+\beta y) &= \lim_{n\to\infty} \langle g_n, \alpha x+\beta y\rangle = \lim_{n\to\infty} \left(\alpha\langle g_n,x\rangle + \beta\langle g_n,y\rangle\right) \\ &= \alpha \lim_{n\to\infty}\langle g_n,x\rangle + \beta\lim_{n\to\infty}\langle g_n,y\rangle = \alpha g(x)+\beta g(y).\end{aligned}$$

Concerning the continuity of g we recall that Cauchy sequences are bounded, thus

$$|\langle g_n,x\rangle| \le \|g_n\|_{X^*}\,\|x\|_X \le C\|x\|_X \quad \text{for all } x \in X.$$

The limit $n \to \infty$ shows that the linear mapping $g : X \to \mathbb{R}$ is continuous. ■

Definition 1.9. Let V be a linear space. A mapping $(\cdot,\cdot): V \times V \to \mathbb{R}$ is called an inner product on V if

- $(u,v) = (v,u)$ for all $u,v \in V$,
- $(\alpha u + \beta v, w) = \alpha(u,w) + \beta(v,w)$ for all $u,v,w \in V$, $\alpha,\beta \in \mathbb{R}$,
- $(u,u) > 0$ for $u \neq 0$.

A norm on V is induced by setting $\|u\| := \sqrt{(u,u)}$. If $(V, \|\cdot\|_V)$ is complete we call V a Hilbert space.

Lemma 1.4 (Schwarz inequality). *Let V be a Hilbert space. Then,*

$$|(u,v)| \leq \|u\|_V \, \|v\|_V \quad \text{for all } u,v \in V.$$

Example 1.2. $V = L^2(\Omega)$ equipped with the inner product

$$(f,g) := \int_\Omega f(x)g(x)\,\mathrm{d}x$$

is a Hilbert space.

Theorem 1.2 (Riesz representation theorem). *Let V be a Hilbert space. There is a unique linear mapping $R: V^* \to V$ from the dual space V^* into the Hilbert space V such that for all $g \in V^*$ and $v \in V$*

$$(Rg, v) = \langle g, v\rangle \quad \text{and} \quad \|Rg\|_V = \|g\|_{V^*}.$$

Fixed point theorems are often used to establish the unique solvability of an abstract operator equation. Let us consider the following problem in a Banach space V with an operator $P: V \to V$:

$$\text{Find } u \in V \text{ such that } u = Pu. \tag{1.1}$$

Solutions $u \in V$ of Problem (1.1) are called fixed points of the mapping $P: V \to V$.

Definition 1.10. The mapping $P: V \to V$ is called contractive or to be a contraction if there is a positive constant $\rho < 1$ such that

$$\|Pv_1 - Pv_2\|_V \leq \rho\, \|v_1 - v_2\|_V \quad \text{for all } v_1, v_2 \in V.$$

Theorem 1.3 (Banach's fixed point theorem). *Let V be a Banach space and $P: V \to V$ be a contraction. Then, we have the following statements:*

(1) *There is a unique fixed point $u^* \in V$ solving Problem* (1.1).
(2) *The sequence $u_n = Pu_{n-1}$, $n \geq 1$, converges to the fixed point $u^* \in V$ for any initial guess $u_0 \in V$.*
(3) *It holds the error estimate*

$$\|u_n - u^*\|_V \leq \frac{\rho^n}{1-\rho}\|Pu_0 - u_0\|_V \quad \text{for all } n \in \mathbb{N}.$$

Proof. We start by estimating the distance of $u_i \in V$ to $u_{i-1} \in V$

$$\begin{aligned}
\|u_i - u_{i-1}\|_V &= \|Pu_{i-1} - Pu_{i-2}\|_V \leq \rho \|u_{i-1} - u_{i-2}\|_V \leq \rho^2 \|u_{i-2} - u_{i-3}\|_V \\
\vdots \quad &\leq \quad \vdots \\
\|u_i - u_{i-1}\|_V &\leq \rho^{i-1} \|Pu_0 - u_0\|_V \qquad \text{for all } i \in \mathbb{N}.
\end{aligned}$$

From that we get for $m > n \geq 1$

$$\begin{aligned}
\|u_m - u_n\|_V &= \left\| \sum_{i=n+1}^{m} (u_i - u_{i-1}) \right\|_V \leq \sum_{i=n+1}^{m} \|u_i - u_{i-1}\|_V \\
&\leq \sum_{i=n+1}^{m} \rho^{i-1} \|Pu_0 - u_0\|_V \leq \rho^n \frac{1-\rho^{m-n}}{1-\rho} \|Pu_0 - u_0\|_V \\
&\leq \frac{\rho^n}{1-\rho} \|Pu_0 - u_0\|_V \to 0 \quad \text{for } n \to \infty.
\end{aligned}$$

Therefore, for any $\varepsilon > 0$, there exists an index bound $n_0(\varepsilon)$ such that for all $m > n \geq n_0(\varepsilon)$

$$\|u_m - u_n\|_V \leq \frac{\rho^n}{1-\rho} \|Pu_0 - u_0\|_V < \varepsilon,$$

consequently, (u_n) is a Cauchy sequence and converges to some $u^* \in V$. We show that u^* is a fixed point of P. Indeed,

$$\begin{aligned}
\|u^* - Pu^*\|_V &\leq \|u^* - u_n\|_V + \|Pu_{n-1} - Pu^*\|_V \\
&\leq \|u^* - u_n\|_V + \rho \, \|u_{n-1} - u^*\|_V \to 0 \quad \text{for } n \to \infty.
\end{aligned}$$

Next we prove that there is at most one fixed point. Assume that there are two fixed points $u_1^* \neq u_2^*$. Then, we conclude

$$\|u_1^* - u_2^*\|_V = \|Pu_1^* - Pu_2^*\|_V \leq \rho \|u_1^* - u_2^*\|_V \quad \Rightarrow \quad (1-\rho) \, \|u_1^* - u_2^*\|_V \leq 0.$$

Since $\rho < 1$ we get $u_1^* = u_2^*$. Finally, we recall the estimate for $m > n \geq 1$

$$\|u_n - u^*\|_V \leq \|u_n - u_m\|_V + \|u_m - u^*\|_V \leq \frac{\rho^n}{1-\rho} \|Pu_0 - u_0\|_V + \|u_m - u^*\|_V$$

from which the last statement follows by $m \to \infty$. ■

1.2. Weak derivatives

First we introduce a generalization of the integration by parts formula known for smooth functions $u, v : [a,b] \to \mathbb{R}$:

$$\int_a^b u' v \, \mathrm{d}x = uv|_a^b - \int_a^b uv' \, \mathrm{d}x.$$

We start with a characterization of geometric properties of a domain $\Omega \subset \mathbb{R}^d$.

Definition 1.11. The domain $\Omega \subset \mathbb{R}^d$ is said to have a Lipschitz continuous boundary, if for every point of the boundary $\Gamma = \partial\Omega$ there exists a local coordinate system in which the boundary corresponds to some hypersurface with the domain Ω lying on one side of that surface and $\partial\Omega$ can locally be represented by the graph of a Lipschitz continuous mapping.

Theorem 1.4 (Gauss theorem). *Let $\Omega \subset \mathbb{R}^d$ be a bounded domain with Lipschitz continuous boundary Γ and outer normal $\mathbf{n}$. Then, for $\mathbf{w} \in C^1(\overline{\Omega})^d$ we have*

$$\int_\Omega \operatorname{div} \mathbf{w} \, \mathrm{d}x = \int_\Omega \sum_{i=1}^d \frac{\partial w_i}{\partial x_i} \, \mathrm{d}x = \int_\Gamma \mathbf{w} \cdot \mathbf{n} \, \mathrm{d}\gamma \, .$$

Let $j \in \{1, 2, \dots, d\}$ be a fixed index. The formula of integration by parts follows by setting $w_i = uv$ for $i = j$ and $w_i = 0$ for $i \neq j$, and using the product rule.

Lemma 1.5 (Integration by parts). *For $u, v \in C^1(\overline{\Omega})$ we have*

$$\int_\Omega \frac{\partial u}{\partial x_j} v \, \mathrm{d}x = \int_\Gamma u v n_j \, \mathrm{d}\gamma - \int_\Omega u \frac{\partial v}{\partial x_j} \, \mathrm{d}x, \quad j = 1, \dots, d.$$

Now we have the tools to generalize the concept of differentiability.

Definition 1.12. The support of a function $f : \Omega \to \mathbb{R}$ is defined by

$$\operatorname{supp} f = \overline{\{x \in \Omega : f(x) \neq 0\}}.$$

The set of all functions that are differentiable of any order with compact support in Ω will be denoted by $C_0^\infty(\Omega)$.

For $\Omega = (0,1)$, the function $x \mapsto x(1-x)$ does not belong to $C_0^\infty(\Omega)$ because its support is $[0,1] \not\subset (0,1)$. Indeed, functions in $C_0^\infty(\Omega)$ are zero in the neighbourhood of the boundary $\partial\Omega$. As a consequence, a function $f \in C_0^\infty(\Omega)$ can be extended to a smooth function defined on $\mathbb{R}^d$ by setting $f = 0$ outside of Ω.

Definition 1.13. We define

$$L^1_{\mathrm{loc}}(\Omega) := \{f : \Omega \to \mathbb{R} : f \in L^1(A) \text{ for all compact } A \subset \Omega\}.$$

For bounded domains Ω, we have $L^2(\Omega) \subset L^1(\Omega) \subset L^1_{\mathrm{loc}}(\Omega)$.

Example 1.3. Consider $\Omega = (0,1)$. The mapping $x \mapsto 1/x$ belongs to $L^1_{\mathrm{loc}}(\Omega)$ but not to $L^1(\Omega)$. The mapping $x \mapsto 1/\sqrt{x}$ belongs to $L^1(\Omega)$ but not to $L^2(\Omega)$.

Definition 1.14. Let $\alpha = (\alpha_1, \dots, \alpha_d)$ be a multi-index, that is, α_i are non-negative integers and $|\alpha| = \alpha_1 + \cdots + \alpha_d$. A function, $f \in L^1_{\mathrm{loc}}(\Omega)$ has a weak derivative $v = D^\alpha f \in L^1_{\mathrm{loc}}(\Omega)$ if for any $\varphi \in C_0^\infty(\Omega)$

$$\int_\Omega v \varphi \, \mathrm{d}x = (-1)^{|\alpha|} \int_\Omega f D^\alpha \varphi \, \mathrm{d}x.$$

The notion of a weak derivative generalizes the classical partial derivatives. Indeed, a partial derivative in the classical sense satisfies the above condition by partial integration.

Example 1.4. Let $f:(-1,+1)\to\mathbb{R}$ with $f(x)=2-\sqrt{|x|}$. Then, the generalized derivative $Df\in L^1_{\text{loc}}(-1,+1)$ is given by

$$Df(x)=-\frac{\operatorname{sgn} x}{2\sqrt{|x|}}\qquad x\in(-1,0)\cup(0,+1).$$

Proof. Applying integration by parts separately in the two subintervals, we get for $\varphi\in C_0^\infty(-1,+1)$

$$\begin{aligned}
-\int_{-1}^{+1}\frac{\operatorname{sgn} x}{2\sqrt{|x|}}\varphi\,\mathrm{d}x &= \int_{-1}^{0}\frac{1}{2\sqrt{|x|}}\varphi\,\mathrm{d}x-\int_{0}^{+1}\frac{1}{2\sqrt{|x|}}\varphi\,\mathrm{d}x\\
&=\left(2-\sqrt{|x|}\right)\varphi|_{x=-1}^{0}-\int_{-1}^{0}\left(2-\sqrt{|x|}\right)\varphi'\,\mathrm{d}x\\
&\quad+\left(2-\sqrt{|x|}\right)\varphi|_{x=0}^{+1}-\int_{0}^{+1}\left(2-\sqrt{|x|}\right)\varphi'\,\mathrm{d}x\\
&=-\int_{-1}^{+1}\left(2-\sqrt{|x|}\right)\varphi'\,\mathrm{d}x,
\end{aligned}$$

where we used the continuity of φ at $x=0$ and $\varphi(\pm 1)=0$. ■

1.3. SOBOLEV SPACES

Definition 1.15. We denote by $W^{m,p}(\Omega)$, $1\le p\le\infty$, $m\ge 0$, the set of all functions, $f\in L^p(\Omega)$ with weak derivatives $D^\alpha f\in L^p(\Omega)$ up to the order $|\alpha|\le m$. The sets $W^{m,p}(\Omega)$ are called Sobolev spaces.

Lemma 1.6. *The Sobolev space $W^{m,p}(\Omega)$ equipped with the norm*

$$\|f\|_{W^{m,p}(\Omega)}:=\left(\int_\Omega\sum_{|\alpha|\le m}|D^\alpha f(x)|^p\,\mathrm{d}x\right)^{1/p}\qquad \textit{if } 1\le p<\infty$$

and

$$\|f\|_{W^{m,p}(\Omega)}:=\max_{|\alpha|\le m}\left(\operatorname{ess\,sup}_{x\in\Omega}|D^\alpha f(x)|\right)\qquad \textit{if } p=\infty$$

is a Banach space. In particular, the Sobolev space $H^m(\Omega):=W^{m,2}(\Omega)$ is a Hilbert space with respect to the inner product

$$(u,v)_{H^m(\Omega)}:=\sum_{|\alpha|\le m}(D^\alpha u,D^\alpha v)=\sum_{|\alpha|\le m}\int_\Omega D^\alpha u\,D^\alpha v\,\mathrm{d}x\quad \textit{for all } u,v\in H^m(\Omega).$$

Proof. Here, we give the arguments for $1 \leq p < \infty$. For the case $p = \infty$, we refer to Adams (1975). The sets $W^{m,p}(\Omega)$ are real linear spaces (below we show that the sum $(f+g)$ of two elements $f, g \in W^{m,p}(\Omega)$ is an element of $W^{m,p}(\Omega)$). Next we have to show that the mapping $f \mapsto \|f\|_{W^{m,p}(\Omega)}$ is a norm, that is, the three conditions in Definition 1.1 hold. If $f = 0$, then $\|f\|_{W^{m,p}(\Omega)} = 0$ by definition of $\|\cdot\|_{W^{m,p}(\Omega)}$. Further, since $0 = \|f\|_{W^{m,p}(\Omega)} \geq \|f\|_{L^p(\Omega)}$ we conclude $f = 0$ in the sense of $L^p(\Omega)$, that is, functions which are equal up to a set of measure zero are identified. Thus, the first condition in Definition 1.1 holds. The second condition follows directly from the definition

$$\|\lambda f\|_{W^{m,p}(\Omega)} = \left(\int_\Omega \sum_{|\alpha|\leq m} |D^\alpha \lambda f(x)|^p \,\mathrm{d}x\right)^{1/p} = \left(\int_\Omega |\lambda|^p \sum_{|\alpha|\leq m} |D^\alpha f(x)|^p \,\mathrm{d}x\right)^{1/p}$$

$$= |\lambda| \left(\int_\Omega \sum_{|\alpha|\leq m} |D^\alpha f(x)|^p \,\mathrm{d}x\right)^{1/p} = |\lambda| \, \|f\|_{W^{m,p}(\Omega)}.$$

Let $f, g \in W^{m,p}(\Omega)$ and $|\alpha| \leq m$. The Minkowski's inequality, Lemma 1.1, implies that $D^\alpha(f+g) = D^\alpha f + D^\alpha g \in L^p(\Omega)$ and

$$\|D^\alpha(f+g)\|_{L^p(\Omega)} \leq \|D^\alpha f\|_{L^p(\Omega)} + \|D^\alpha g\|_{L^p(\Omega)}.$$

Using the Minkowski's inequality for sums, we get

$$\|f+g\|^p_{W^{m,p}(\Omega)} = \sum_{|\alpha|\leq m} \|D^\alpha(f+g)\|^p_{L^p(\Omega)} \leq \sum_{|\alpha|\leq m} \left(\|D^\alpha f\|_{L^p(\Omega)} + \|D^\alpha g\|_{L^p(\Omega)}\right)^p$$

$$\leq \left[\left(\sum_{|\alpha|\leq m} \|D^\alpha f\|^p_{L^p(\Omega)}\right)^{1/p} + \left(\sum_{|\alpha|\leq m} \|D^\alpha g\|^p_{L^p(\Omega)}\right)^{1/p}\right]^p$$

$$\leq \left[\|f\|_{W^{m,p}(\Omega)} + \|g\|_{W^{m,p}(\Omega)}\right]^p$$

from which the third condition follows. It remains to prove the completeness of the normed space $W^{m,p}(\Omega)$. Let $(f_n)_{n\in\mathbb{N}}$ be a Cauchy sequence in $W^{m,p}(\Omega)$. Then, $(D^\alpha f_n)_{n\in\mathbb{N}}$ is a Cauchy sequence in $L^p(\Omega)$ for any multi-index α with $|\alpha| \leq m$. Since $L^p(\Omega)$ is a Banach space, there is $f^\alpha \in L^p(\Omega)$ with $D^\alpha f_n \to f^\alpha$ in $L^p(\Omega)$. It remains to show that $f^\alpha = D^\alpha f$ where $f = f^0$. We show this for $\alpha = (1, 0, \ldots, 0)$. For any $\varphi \in C_0^\infty(\Omega)$ we have

$$\left|\int_\Omega (f^\alpha \varphi + f D^\alpha \varphi)\mathrm{d}x\right| \leq \int_\Omega |f^\alpha - D^\alpha f_n|\,|\varphi|\mathrm{d}x + \left|\int_\Omega (D^\alpha f_n \varphi + f_n D^\alpha \varphi)\mathrm{d}x\right|$$

$$+ \int_\Omega |f - f_n|\,|D^\alpha \varphi|\mathrm{d}x.$$

The second term on the right hand side equals zero since $D^\alpha f_n$ is the weak derivative of f_n. The convergence of f_n to f and $D^\alpha f_n$ to f^α in $L^p(\Omega)$, respectively, imply the convergence in $L^1(\Omega)$, thus the first and third term on the right hand side tend to zero for $n \to \infty$. The left hand side

is non-negative and independent of n, consequently we conclude that it is zero, which means f^α is the weak derivative of f.
The properties for the inner product in $H^m(\Omega)$, see Definition 1.9, follow from its definition and the observation that $(v,v)_{H^m(\Omega)} = \|v\|^2_{W^{m,2}(\Omega)}$ for all $v \in W^{m,2}(\Omega)$. ■

Definition 1.16. We denote by $W_0^{m,p}(\Omega)$ the closure of $C_0^\infty(\Omega)$ in the norm $\|\cdot\|_{W^{m,p}(\Omega)}$. Analogously, $H_0^m(\Omega)$ is the closure of $C_0^\infty(\Omega)$ in $H^m(\Omega)$.

In the following, for $f \in W^{m,p}(\omega)$, $m \geq 0$, $1 \leq p \leq \infty$, the norm will be shortly denoted by $\|f\|_{m,p,\omega} := \|f\|_{W^{m,p}(\omega)}$ and a seminorm is introduced by

$$|f|_{m,p,\omega} := \left(\int_\omega \sum_{|\alpha|=m} |D^\alpha f(x)|^p \, \mathrm{d}x \right)^{1/p} \qquad \text{if } 1 \leq p < \infty$$

and

$$|f|_{m,p,\omega} := \max_{|\alpha|=m} \left(\operatorname{ess\,sup}_{x\in\Omega} |D^\alpha f(x)| \right) \qquad \text{if } p = \infty.$$

In case, $\omega = \Omega$ and/or $p = 2$ we omit Ω and p, respectively. Thus,

$$\|f\|_m = \|f\|_{m,2,\Omega}, \quad \|f\|_{m,p} = \|f\|_{m,p,\Omega}, \quad |f|_m = |f|_{m,2,\Omega}, \quad |f|_{m,p} = |f|_{m,p,\Omega}.$$

Lemma 1.7 (Poincaré inequality). *There is a positive constant $c_P \leq 2a$ such that*

$$\|v\|_{0,p} \leq c_P |v|_{1,p} \quad \textit{for all } v \in W_0^{1,p}(\Omega).$$

Here, 'a' denotes the diameter of Ω in an arbitrary but fixed direction.

Proof. Suppose the Poincaré inequality is true for functions from $C_0^\infty(\Omega)$. Then, for any $v \in W_0^{1,p}(\Omega)$ there is a sequence $v_n \in C_0^\infty(\Omega)$ with $v_n \to v$ in $W_0^{1,p}(\Omega)$. We have

$$\begin{aligned}\|v\|_{0,p} &\leq \|v - v_n\|_{0,p} + \|v_n\|_{0,p} \leq \|v - v_n\|_{0,p} + c_P |v_n|_{1,p} \\ &\leq (1 + c_P)\|v - v_n\|_{1,p} + c_P |v|_{1,p}.\end{aligned}$$

Since $\|v - v_n\|_{1,p}$ tends to zero for $n \to \infty$, the Poincaré inequality is true for functions from $W_0^{1,p}(\Omega)$.

In order to prove the Poincaré inequality for functions $v \in C_0^\infty(\Omega)$ we extend $v : \Omega \to \mathbb{R}$ by setting $v(x) = 0$ for all $x \in \mathbb{R}^d \backslash \Omega$. Let $\Omega \subset [-a, +a] \times \mathbb{R}^{d-1}$, then

$$v(x) = \int_{-a}^{x_1} \frac{\partial v}{\partial x_1}(s, x_2, \ldots, x_d) \, \mathrm{d}s$$

and Hölder's inequality with $1/p + 1/q = 1$ implies

$$|v(x)| \leq \left(\int_{-a}^{x_1} \left| \frac{\partial v}{\partial x_1}(s, x_2, \ldots, x_d) \right|^p \mathrm{d}s \right)^{1/p} \left(\int_{-a}^{x_1} 1^q \mathrm{d}s \right)^{1/q},$$

$$|v(x)|^p \leq (2a)^{p/q} \int_{-a}^{+a} \left| \frac{\partial v}{\partial x_1}(s, x_2, \ldots, x_d) \right|^p \mathrm{d}s,$$

$$\int_{-a}^{+a} |v(x)|^p \,\mathrm{d}x_1 \leq (2a)^{1+p/q} \int_{-a}^{+a} \left| \frac{\partial v}{\partial x_1}(s, x_2, \ldots, x_d) \right|^p \mathrm{d}s,$$

$$\int_{\mathbb{R}^d} |v(x)|^p \,\mathrm{d}x \leq (2a)^p \int_{\mathbb{R}^d} \left| \frac{\partial v}{\partial x_1}(x) \right|^p \mathrm{d}x.$$

Using $v(x) = 0$ for $x \in \mathbb{R}^d \backslash \Omega$ the inequality follows. ■

The Poincaré inequality allows to show that the seminorm $|\cdot|_{m,p}$ is equivalent to the norm $\|\cdot\|_{m,p}$ on $W_0^{m,p}(\Omega)$. Let $v \in W_0^{m,p}(\Omega)$, then $D^\alpha v \in W_0^{1,p}(\Omega)$ for $|\alpha| \leq m-1$. We apply successively the inequality

$$|D^\alpha v|_{0,p} \leq c_P |D^\alpha v|_{1,p} \quad \text{for } |\alpha| = 0, 1, \ldots, m-1$$

to estimate the lower derivatives by the highest derivative and obtain

$$|v|_{m,p} \leq \|v\|_{m,p} = \left(\sum_{|\alpha| \leq m} |D^\alpha v|_{0,p}^p \right)^{1/p} \leq C|v|_{m,p},$$

that is, the equivalence of the seminorm to the norm.

An important tool will be the Sobolev embedding theorems which state that functions from $W^{m,p}(\Omega)$ belong to W^{m-1,p^*} for some $p^* > p$. Moreover, it can be shown that for suitable numbers m, p functions from $W^{m,p}(\Omega)$ belong to classical spaces of continuous and continuously differentiable functions, respectively.

Definition 1.17. We introduce the space of Hölder continuous functions with Hölder exponent $\beta \in (0, 1]$

$$C^{0,\beta}(\overline{\Omega}) := \left\{ u \in C^0(\overline{\Omega}) : \sup_{x,y \in \Omega, x \neq y} \frac{|u(x) - u(y)|}{|x-y|^\beta} < \infty \right\}$$

equipped with the norm

$$\|v\|_{C^{0,\beta}(\overline{\Omega})} = \sup_{x \in \Omega} |u(x)| + \sup_{x,y \in \Omega, x \neq y} \frac{|u(x) - u(y)|}{|x-y|^\beta}.$$

Hölder continuous functions with a Hölder exponent $\beta = 1$ are also called Lipschitz continuous.

Exercise 1.1. Consider $f : [-1, +1] \to \mathbb{R}$ given by $f(x) = \sqrt{|x|}$. Prove that $f \in C^{0,1/2}([-1, +1])$ and that there is no $\beta \in (1/2, +1]$ with $f \in C^{0,\beta}([-1, +1])$.

Exercise 1.2. Let $f : \mathbb{R} \to \mathbb{R}$ be given by

$$f(x) = \begin{cases} x^2 \sin \dfrac{1}{x^2} & \text{for } x \neq 0, \\ 0 & \text{for } x = 0. \end{cases}$$

Show that f is differentiable at any point $x \in \mathbb{R}$ and determine the first derivative $f'(x)$. Is f Lipschitz continuous on $\mathbb{R}$?

Theorem 1.5 (Sobolev embedding theorem). *Let $\Omega \subset \mathbb{R}^d$ be a domain with Lipschitz continuous boundary. Then, we have the following continuous embeddings for $m \geq 0$, $1 \leq p \leq \infty$,*

$$W^{m,p}(\Omega) \hookrightarrow L^{p^*}(\Omega) \quad \textit{with } \frac{1}{p^*} = \frac{1}{p} - \frac{m}{d}, \quad \textit{if } m < \frac{d}{p},$$

$$W^{m,p}(\Omega) \hookrightarrow L^q(\Omega) \quad \textit{for all } q \in [1, \infty), \quad \textit{if } m = \frac{d}{p},$$

$$W^{m,p}(\Omega) \hookrightarrow C^{0,m-d/p}(\overline{\Omega}), \quad \textit{if } \frac{d}{p} < m < \frac{d}{p} + 1,$$

$$W^{m,p}(\Omega) \hookrightarrow C^{0,\alpha}(\overline{\Omega}) \quad \textit{for all } \alpha \in (0,1), \quad \textit{if } m = \frac{d}{p} + 1,$$

$$W^{m,p}(\Omega) \hookrightarrow C^{0,1}(\overline{\Omega}), \quad \textit{if } m > \frac{d}{p} + 1.$$

Proof. See, Adams (1975). ■

The last three embeddings have to be understood in the following sense. A function $v \in W^{m,p}(\Omega)$ can be modified on a set of measure zero to a continuous function $\tilde{v}$ and the 'embedding operator' is continuous, that is, in case of $W^{m,p}(\Omega) \hookrightarrow C^{0,\alpha}(\overline{\Omega})$ we have

$$\|\tilde{v}\|_{C^{0,\alpha}(\overline{\Omega})} \leq C \|v\|_{W^{m,p}(\Omega)} \quad \text{for all } v \in W^{m,p}(\Omega).$$

Often we do not distinguish between v and $\tilde{v}$, and write v instead of $\tilde{v}$.

Example 1.5. Theorem 1.5 shows that H^1 functions in 1d are continuous. This is not the case in the 2d case as shown in the following example. Let $\Omega = \{(x_1, x_2) \in \mathbb{R}^2 : x_1^2 + x_2^2 < 1\}$ and $f(r) = \log(\log(1/r))$ for $r \neq 0$. Using polar coordinates $(x_1, x_2) = (r \cos \varphi, r \sin \varphi)$ we compute

$$|f|_1^2 = \int_\Omega |\nabla f|^2 \, \mathrm{d}x = \int_0^{2\pi} \int_0^1 \left| \frac{\partial f}{\partial r} \right|^2 r \, \mathrm{d}r \, \mathrm{d}\varphi = 2\pi \int_0^1 \frac{\mathrm{d}r}{r \log^2(1/r)} < \infty.$$

Thus, $v \in H^1(\Omega)$ but has a singularity at $r = 0$.

Example 1.6. Consider the unit ball, $\Omega = \{x \in \mathbb{R}^d : x_1^2 + \cdots + x_d^2 < 1\}$ and the function $f : \Omega \backslash \{0\} \to \mathbb{R}$ with $f(x) = (\sum_{i=1}^d x_i^2)^{\alpha/2} = r^\alpha$. We want to know for which α the function

f belongs to $H^1(\Omega)$. In the case, $d = 2$ we use polar coordinates to get

$$|f|_1^2 = \int_\Omega |\nabla f|^2 \,\mathrm{d}x = \int_0^{2\pi} \int_0^1 \left|\frac{\partial f}{\partial r}\right|^2 r\,\mathrm{d}r\,\mathrm{d}\varphi = 2\pi\alpha^2 \int_0^1 \frac{\mathrm{d}r}{r^{1-2\alpha}} < \infty$$

provided that $\alpha > 0$. In the 3d case, $d = 3$ and spherical coordinates $(x_1, x_2, x_3) = (r\cos\varphi\sin\theta, r\sin\varphi\sin\theta, r\cos\theta)$ imply

$$\begin{aligned} |f|_1^2 &= \int_\Omega |\nabla f|^2 \,\mathrm{d}x = \int_0^{2\pi} \int_0^\pi \int_0^1 \left|\frac{\partial f}{\partial r}\right|^2 r^2 \sin\theta \,\mathrm{d}r\,\mathrm{d}\theta\,\mathrm{d}\varphi \\ &= 2\pi \cdot 2 \int_0^1 \frac{\mathrm{d}r}{r^{-2\alpha}} < \infty \end{aligned}$$

provided that $\alpha > -1/2$. This shows that functions $f : \Omega \subset \mathbb{R}^d \to \mathbb{R}$ belonging to $H^1(\Omega)$ can have a singularity of type r^α for $\alpha > -1/2$ in the 3d case, however, not in the 2d case where $\alpha > 0$ is needed. Note that the singularity considered in Example 1.5 is weaker than the singularity r^α with $\alpha < 0$.

In general, functions in Sobolev spaces are defined up to a set of measure zero. Since the d-dimensional measure of $\Gamma = \partial\Omega$ is zero, it makes no sense, on the first glance, to speak on boundary values of functions $v \in W^{m,p}(\Omega)$. However, one can show that there is a positive constant $C(\Omega)$ such that

$$\|v\|_{0,\Gamma} := \left(\int_\Gamma v^2 \,\mathrm{d}\gamma\right)^{1/2} \leq C(\Omega)\,\|v\|_{1,\Omega} \quad \text{for all } v \in C^\infty(\overline{\Omega}).$$

Since $C^\infty(\overline{\Omega})$ is dense in $H^1(\Omega)$, this inequality states that there is a continuous linear mapping $\mathrm{tr} : H^1(\Omega) \to L^2(\Gamma)$, called the trace operator. In this sense, we say that functions from $H^1(\Omega)$ have traces ('boundary values') in $L^2(\Gamma)$. This also gives another characterization of $H_0^1(\Omega)$ as

$$H_0^1(\Omega) = \{v \in H^1(\Omega) : v = 0 \text{ on } \Gamma\}.$$

Note that the image $\mathrm{tr}(H^1(\Omega))$ belongs to $L^2(\Gamma)$ but is a proper subspace in $L^2(\Gamma)$. For a precise characterization of the image $\mathrm{tr}(W^{m,p}(\Omega))$, fractional Sobolev spaces $W^{s,p}(\Omega)$, $0 < s$, $1 \leq p < \infty$ have to be introduced (see Adams (1975)).

Theorem 1.6. *Let $1 \leq p < \infty$ and $\Omega \subset \mathbb{R}^d$ be a bounded domain with Lipschitz continuous boundary Γ. Then the trace operator,* $\mathrm{tr} : W^{1,p}(\Omega) \to W^{1-1/p,p}(\Gamma)$ *is surjective and* $W_0^{1,p}(\Omega) = \{v \in W^{1,p}(\Omega) : \mathrm{tr}(v) = 0\}$.

Proof. See Theorem 7.53 in Adams (1975). ■

As a special case we set $H^{1/2}(\Gamma) := W^{1/2,2}(\Gamma)$. Then, the first statement of Theorem 1.6 means that every function in $H^{1/2}(\Gamma)$ is the trace of a function in $H^1(\Omega)$. This type of results is useful in case we seek solutions of partial differential equations in a domain Ω that satisfy given values on the boundary $\Gamma = \partial\Omega$. In general, we have

Corollary 1.1. *Let* $1 \leq p < \infty$ *and* $\Omega \subset \mathbb{R}^d$ *be a bounded domain with Lipschitz continuous boundary* Γ. *Then, there is a positive constant* C *such that for all* $g \in W^{1-1/p,p}(\Gamma)$ *there exists* $u_g \in W^{1,p}(\Omega)$ *satisfying*

$$\mathrm{tr}(u_g) = g \quad \textit{and} \quad \|u_g\|_{W^{1,p}(\Omega)} \leq C\|g\|_{W^{1-1/p,p}(\Gamma)}.$$

The function, u_g *is called extension or lifting of* g *in* $W^{1,p}(\Omega)$.

CHAPTER 2

ELLIPTIC SCALAR PROBLEMS

In this chapter, we introduce the concept of weak (variational) formulations and weak solutions of elliptic boundary value problems of second order. The fundamental results on existence, uniqueness and well-posedness of weak solutions are presented. As a basic tool for a finite element approximation, the standard Galerkin method for elliptic problems and their abstract error estimate are introduced.

2.1. A GENERAL ELLIPTIC PROBLEM OF SECOND ORDER

Let $\Omega \subset \mathbb{R}^d$, d is the dimension, be an open and bounded domain with Lipschitz continuous boundary $\Gamma = \partial\Omega$. We consider the second order elliptic equation

$$-\sum_{i,j=1}^{d} \frac{\partial}{\partial x_j}\left(a_{ij}(x)\frac{\partial u}{\partial x_i}\right) + \sum_{i=1}^{d} b_i(x)\frac{\partial u}{\partial x_i} + c(x)u = f(x) \qquad x \in \Omega, \tag{2.1}$$

where u is an unknown scalar function. Here, the diffusion, advection/convection and reaction coefficients, $a_{ij}(x)$, $b_i(x)$ and $c(x)$, respectively, and the source term $f(x)$ are assumed to be sufficiently smooth. For continuously differentiable functions $u : \Omega \to \mathbb{R}$, the mixed derivatives satisfy

$$\frac{\partial^2 u}{\partial x_j \partial x_i} = \frac{\partial^2 u}{\partial x_i \partial x_j}, \qquad i,j = 1,\ldots,d.$$

Thus, we can assume without losing generality that the matrix a_{ij} is symmetric, that is, $a_{ij} = a_{ji}$, $i,j = 1,\ldots,d$. Then, the differential equation (2.1) is called elliptic at $x \in \Omega$ if the matrix $a_{ij}(x)$ is positive definite. We assume that the differential equation is uniformly elliptic in Ω, that is, there is a positive constant a_0 such that

$$\sum_{i,j=1}^{d} a_{ij}(x)\xi_i\xi_j \geq \alpha_0 \sum_{i=1}^{d} \xi_i^2 \quad \text{for all } x \in \Omega, \xi \in \mathbb{R}^d.$$

This requirement is slightly stronger than assuming that the differential equation is elliptic for all $x \in \Omega$.

To complete Equation (2.1), boundary conditions need to be specified. The most commonly used boundary conditions to specify a solution of the elliptic Equation (2.1) are

- Dirichlet boundary condition

$$u = g_D \quad \text{on } \Gamma,$$

  where $g_D : \Gamma \to \mathbb{R}$ is a given function,
- Neumann boundary condition

$$\sum_{i,j=1}^{d} a_{ij}\frac{\partial u}{\partial x_i} n_j = g_N \quad \text{on } \Gamma,$$

  where $\mathbf{n} = (n_1, \ldots, n_d)^T$ is the outward normal vector on Γ and $g_N : \Gamma \to \mathbb{R}$ is a given flux,
- Robin boundary condition

$$\sum_{i,j=1}^{d} a_{ij}\frac{\partial u}{\partial x_i} n_j = \sigma(u_\infty - u) \quad \text{on } \Gamma,$$

where $\sigma > 0$ and u_∞ are given functions on Γ. In the simplest case, the solution of the elliptic Equation (2.1) subject to one of the above boundary conditions is sought. In applications, mixed boundary conditions may also appear, and in which different types of conditions are posed on different parts of the boundary. The concept of weak solutions of a boundary value problem will be discussed next.

2.2. Weak solution

We begin with the Poisson problem by imposing $a_{ij}(x) = \delta_{ij}$, $i, j = 1, \ldots, d$, $b_i(x) = 0$, $i = 1, \ldots, d$, $c(x) = 0$ in the elliptic Equation (2.1). Using homogeneous Dirichlet boundary condition, it becomes

$$-\Delta u = f \quad \text{in } \Omega, \qquad u = 0 \quad \text{on } \Gamma, \tag{2.2}$$

where the Laplace operator Δ is given by

$$\Delta u := \sum_{i=1}^{d} \frac{\partial^2 u}{\partial x_i^2}.$$

We know from the theory of partial differential equations, Equation (2.2) has a unique classical solution $u \in C^2(\Omega) \cap C(\overline{\Omega})$ provided that $f \in C(\Omega)$ and the boundary Γ is smooth enough. However, both the smoothness assumption on the source f and on the boundary are often not fulfilled in practical applications. For example, the following problem

$$-\Delta u = f \quad \text{in } \Omega = (-2,2)^d, \qquad u = 0 \quad \text{on } \Gamma$$

with $f = 1 + \operatorname{sgn}(1 - |x|)$ does not have a classical solution $u \in C^2(\Omega) \cap C(\overline{\Omega})$ since the source f is only piecewise continuous.

Partial differential equations with 'non-smooth' data appear in many practical applications. Since a classical solution of these equations does not exist in general, we are motivated to relax the differentiability requirements of the solution and the data.

Let $u \in C^2(\Omega) \cap C(\overline{\Omega})$ be a classical solution of Equation (2.2), and assume that $f \in C(\Omega)$. Upon multiplying Equation (2.2) by a function $v \in C_0^\infty(\Omega)$ and integrating over Ω, we obtain

$$-\int_\Omega \Delta u\, v\, \mathrm{d}x = \int_\Omega f\, v\, \mathrm{d}x.$$

Integrating by parts, the Laplace term becomes

$$\begin{aligned} -\int_\Omega \Delta u v \mathrm{d}x &= -\sum_{i=1}^d \int_\Omega \frac{\partial}{\partial x_i}\left(\frac{\partial u}{\partial x_i}\right) v \mathrm{d}x \\ &= \sum_{i=1}^d \left[\int_\Omega \frac{\partial u}{\partial x_i}\frac{\partial v}{\partial x_i}\mathrm{d}x - \int_\Gamma \frac{\partial u}{\partial x_i} n_i v \mathrm{d}s\right] = \int_\Omega \nabla u \cdot \nabla v \mathrm{d}x. \end{aligned}$$

Here, the boundary integral vanishes since the test function, v vanishes on the boundary Γ and we end up with

$$\int_\Omega \nabla u \cdot \nabla v \mathrm{d}x = \int_\Omega f\, v \mathrm{d}x \quad \text{for all } v \in C_0^\infty(\Omega).$$

Moreover, the classical solution u also satisfies

$$\int_\Omega \nabla u \cdot \nabla v \mathrm{d}x = \int_\Omega f\, v \mathrm{d}x \quad \text{for all } v \in H_0^1(\Omega), \tag{2.3}$$

since $C_0^\infty(\Omega)$ is dense in $H_0^1(\Omega)$. Now, note that the solution u need not be in $C^2(\Omega) \cap C(\overline{\Omega})$ for the integrals in Equation (2.3) to be meaningful, and it is enough to seek the solution u in the space $H_0^1(\Omega)$.

Definition 2.1. A function $u \in H_0^1(\Omega)$ satisfying Equation (2.3) is called a *weak solution* or a *generalized solution* of Equation (2.2). Note that the partial derivatives in the integrals of Equation (2.3) have to be understood as weak derivatives.

Hence, the weak formulation of the elliptic Problem (2.2) reads
Find $u \in V := H_0^1(\Omega)$ such that

$$a(u,v) := \int_\Omega \nabla u \cdot \nabla v \mathrm{d}x = \int_\Omega f\, v \mathrm{d}x =: F(v) \quad \text{for all } v \in V. \tag{2.4}$$

It is not difficult to check that the mappings $a : V \times V \longrightarrow \mathbb{R}$ and $F : V \longrightarrow \mathbb{R}$ are bilinear and linear, respectively. Further, it is clear from Equation (2.3) that if u is the classical solution of Problem (2.2) then it is also a weak solution of Problem (2.2). However, the converse is not true when the weak solution is not smooth enough. Roughly speaking, we increase the set of possible solutions (to the class of generalized solutions) without losing the uniqueness (as we will see below).

Next, we shall prove the existence and uniqueness of a weak solution of the elliptic Problem (2.2) using the Lax–Milgram theorem.

Definition 2.2. Let V be a Hilbert space. A bilinear form $a : V \times V \longrightarrow \mathbb{R}$ is said to be *continuous* (bounded) if there exist a constant $\beta > 0$ such that

$$|a(u,v)| \leq \beta \|u\|_V \|v\|_V \quad \text{for all } u,\ v \in V$$

and *coercive* or V*-elliptic*, if there exist a constant $\alpha > 0$ such that

$$a(u,u) \geq \alpha \|u\|_V^2 \quad \text{for all } u \in V.$$

Under the additional assumption that the bilinear form is symmetric, that is,

$$a(u,v) = a(v,u) \quad \text{for all } u, v \in V,$$

the mapping $(u,v) \mapsto a(u,v)$ defines an inner product on V and the associated norm $v \mapsto \sqrt{a(v,v)}$ is called energy norm.

Theorem 2.1 (Lax–Milgram). *Let V be a Hilbert space, $a(\cdot,\cdot)$ be a continuous, coercive bilinear form, and $F(\cdot)$ be a continuous linear functional. Then, there exists a unique $u \in V$ such that*

$$a(u,v) = F(v) \quad \textit{for all } v \in V. \tag{2.5}$$

Proof. For each $u \in V$ the mapping $v \mapsto a(u,v)$ is linear and continuous on V, thus there exists a unique functional $Au \in V^*$ (V^* is the dual space of V) such that

$$a(u,v) = \langle Au, v\rangle \quad \text{for all } v \in V. \tag{2.6}$$

Further, we have

$$\|Au\|_{V^*} := \sup_{0 \neq v \in V} \frac{|\langle Au, v\rangle|}{\|v\|_V} \leq \beta \|u\|_V.$$

Consequently, the linear mapping $A : V \longrightarrow V^*$ is continuous with the operator norm

$$\|A\|_{\mathcal{L}(V,V^*)} \leq \beta.$$

Now, from Theorem 1.2 (Riesz representation theorem), there is a unique linear mapping $R : V^* \longrightarrow V$ such that for each $f \in V^*$

$$(Rf, v) = \langle f, v\rangle, \qquad \|Rf\|_V = \|f\|_{V^*} \quad \text{for all } v \in V. \tag{2.7}$$

Comparing (2.6) and (2.7), the solution of the weak form (2.5) is equivalent to solve the equation, $RAu = Rf$. In other words, we must prove that the equation, $RAu = Rf$ has a unique solution $u \in V$. We prove this by showing that, for an appropriate value of $\rho > 0$, the mapping $T : V \to V$ with

$$T(v) := v - \rho(RAv - Rf) \quad \text{for all } v \in V$$

is contractive. If T is a contraction, then there exists a unique $u \in V$ such that $T(u) = u - \rho(RAu - Rf) = u$, that is, $RAu = Rf$. To show that T is a contraction, let $v = v_1 - v_2$, for any $v_1, v_2 \in V$. Then,

$$\begin{aligned}
\|Tv_1 - Tv_2\|_V^2 &= \|v_1 - v_2 - \rho(RAv_1 - RAv_2)\|_V^2 \\
&= \|v - \rho(RAv)\|_V^2 \\
&= \|v\|_V^2 - 2\rho(RAv, v) + \rho^2\|RAv\|_V^2 \\
&= \|v\|_V^2 - 2\rho\langle Av, v\rangle + \rho^2\langle Av, RAv\rangle \\
&= \|v\|_V^2 - 2\rho a(v, v) + \rho^2 a(v, RAv) \\
&\le \|v\|_V^2 - 2\rho\alpha\|v\|_V^2 + \rho^2\beta\|v\|_V\,\|RAv\|_V \\
&\le (1 - 2\rho\alpha + \rho^2\beta^2)\|v\|_V^2 \\
&= (1 - 2\rho\alpha + \rho^2\beta^2)\|v_1 - v_2\|_V^2.
\end{aligned}$$

Therefore, the mapping T is contractive provided $(1 - 2\rho\alpha + \rho^2\beta^2) < 1$, that is, $\rho \in (0, 2\alpha/\beta^2)$, and this completes the proof. ∎

Remark 2.1. It follows from the coercivity of the bilinear form and (2.5) that

$$\|u\|_V \le \frac{1}{\alpha}\|f\|_{V^*},$$

where α is the coercivity constant. Therefore, the weak form (2.5) is well-posed in the sense that it has a unique solution, which depends continuously on the data f.

Theorem 2.2 (Existence theorem). *There exists a unique weak solution $u \in V$ for the homogeneous Dirichlet Problem* (2.2) *which satisfies Problem* (2.4).

Proof. To prove the existence of the weak solution, we apply Theorem 2.1 (Lax–Milgram). Therefore, it is enough to show that $a(\cdot,\cdot)$ and $F(\,\cdot\,)$ defined in Problem (2.4) satisfy the assumptions of the Lax–Milgram theorem.

For a fixed $u \in V$, the mapping $v \mapsto a(u, v)$ is linear, and similarly, for a fixed $v \in V$ the mapping $u \mapsto a(u, v)$ is linear. Therefore, $a(\cdot,\cdot)$ is bilinear. Applying Schwarz inequality, first for integrals and then for sums, we have

$$\begin{aligned}
|a(u, v)| = \left|\int_\Omega \nabla u \cdot \nabla v \,\mathrm{d}x\right| &\le \sum_{i=1}^{d} \int_\Omega \left|\frac{\partial u}{\partial x_i}\frac{\partial v}{\partial x_i}\right| \mathrm{d}x \\
&\le \sum_{i=1}^{d} \left(\int_\Omega \left|\frac{\partial u}{\partial x_i}\right|^2 \mathrm{d}x\right)^{1/2} \left(\int_\Omega \left|\frac{\partial v}{\partial x_i}\right|^2 \mathrm{d}x\right)^{1/2}
\end{aligned}$$

$$\leq \left(\sum_{i=1}^{d}\int_{\Omega}\left|\frac{\partial u}{\partial x_i}\right|^2 \mathrm{d}x\right)^{1/2}\left(\sum_{i=1}^{d}\int_{\Omega}\left|\frac{\partial v}{\partial x_i}\right|^2 \mathrm{d}x\right)^{1/2}$$
$$= |u|_1|v|_1 \leq \|u\|_V\|v\|_V.$$

Thus, the bilinear form $a(\cdot,\cdot)$ is continuous with $\beta = 1$. Next, the coercivity of $a(\cdot,\cdot)$ follows from

$$a(v,v) = \sum_{i=1}^{d}\int_{\Omega}\left|\frac{\partial v}{\partial x_i}\right|^2 \mathrm{d}x = |v|_1^2 \geq \frac{1}{1+c_{\mathrm{P}}^2}\|v\|_V^2,$$

where in the last step, the norm equivalence of the seminorm to the norm in $H_0^1(\Omega)$ has been used. Now, coercivity follows by setting $\alpha = 1/(1+c_{\mathrm{P}}^2)$. It remains to prove that $F(\cdot)$ is linear and bounded. Linearity of F is immediate, the boundedness of F follows for $f \in L^2(\Omega)$ by means of Schwarz inequality

$$|F(v)| = \left|\int_{\Omega} f\, v\mathrm{d}x\right| \leq \|f\|_0\|v\|_0 \leq \|f\|_0\|v\|_V.$$

Hence, the bilinear form $a(\cdot,\cdot)$ is continuous and coercive, the linear form $F(\cdot)$ is continuous. The existence of a unique solution for Problem (2.2) which satisfies Problem (2.4) follows from Theorem 2.1 (Lax–Milgram). ∎

Example 2.1. We now consider a more general elliptic equation. Let $\Omega \subset \mathbb{R}^d$ be a bounded domain with a boundary Γ split into mutually disjoint parts of Dirichlet, Neumann and Robin type. Consider the elliptic equation of second order with mixed boundary conditions

$$\begin{aligned} -\Delta u + \mathbf{b}\cdot\nabla u + cu = f \quad \text{in } \Omega, \qquad & u = 0 && \text{on } \Gamma_{\mathrm{D}}, \\ \frac{\partial u}{\partial \mathbf{n}} = g \quad \text{on } \Gamma_{\mathrm{N}}, \quad & \frac{\partial u}{\partial \mathbf{n}} = \sigma(u_\infty - u) && \text{on } \Gamma_{\mathrm{R}}. \end{aligned} \tag{2.8}$$

The data, $\mathbf{b} = (b_1,\ldots,b_d)$, c, σ are assumed to be sufficiently smooth, $f \in L^2(\Omega)$, $g \in L^2(\Gamma_{\mathrm{N}})$, and $u_\infty \in L^2(\Gamma_{\mathrm{R}})$. Furthermore, we suppose that the $(d-1)$-dimensional measure of Γ_{D} is positive.

We now define

$$V := H^1_{\Gamma_{\mathrm{D}}} = \left\{v \in H^1(\Omega) \,:\, v = 0 \quad \text{on } \Gamma_{\mathrm{D}}\right\}.$$

Note that on the subspace $V \subset H^1(\Omega)$, the seminorm $|\cdot|_1$ is equivalent to the norm $\|\cdot\|_1$. As in the case of the Poisson equation, we multiply the differential equation by a test function $v \in V$ which vanishes on Γ_{D}, integrate over Ω, integrate by parts the Laplace term, and finally incorporate the Neumann and Robin type boundary conditions

$$\begin{aligned} -\int_{\Omega}\Delta u v\mathrm{d}x &= \int_{\Omega}\nabla u\cdot\nabla v\mathrm{d}x - \int_{\Gamma_{\mathrm{N}}\cup\Gamma_{\mathrm{R}}}\frac{\partial u}{\partial \mathbf{n}} v\mathrm{d}s \\ &= \int_{\Omega}\nabla u\cdot\nabla v\mathrm{d}x + \int_{\Gamma_{\mathrm{R}}}\sigma\, u v\mathrm{d}s - \int_{\Gamma_{\mathrm{R}}}\sigma\, u_\infty v\mathrm{d}s - \int_{\Gamma_{\mathrm{N}}} g v\mathrm{d}s. \end{aligned}$$

Collecting all terms we obtain a weak form of Problem (2.8)

Find $u \in V$, such that

$$a(u,v) = F(v) \quad \text{for all } v \in V, \tag{2.9}$$

where

$$a(u,v) = \int_\Omega (\nabla u \cdot \nabla v + (\mathbf{b} \cdot \nabla u + cu)v)\, \mathrm{d}x + \int_{\Gamma_\mathrm{R}} \sigma\, u\, v\, \mathrm{d}s,$$

$$F(v) = \int_\Omega f\, v\, \mathrm{d}x + \int_{\Gamma_\mathrm{N}} g\, v\, \mathrm{d}s + \int_{\Gamma_\mathrm{R}} \sigma\, u_\infty\, v\, \mathrm{d}s.$$

Remark 2.2. Note that the boundary data g and $\sigma\, u_\infty$ need not to be in $L^2(\Gamma_\mathrm{N})$ and $L^2(\Gamma_\mathrm{R})$, respectively. It is sufficient to be in $H^{-1/2}(\Gamma_\mathrm{N})$ and $H^{-1/2}(\Gamma_\mathrm{R})$, respectively. In this case, we may write the boundary integrals in $F(v)$ as $\langle g, v\rangle_{\Gamma_\mathrm{N}}$ and $\langle \sigma\, u_\infty, v\rangle_{\Gamma_\mathrm{R}}$, where $\langle \cdot,\cdot\rangle_\Gamma$ denotes the duality paring between $H^{-1/2}(\Gamma)$ and $H^{1/2}(\Gamma)$.

We next verify the assumptions of Theorem 2.1 (Lax–Milgram) to prove the existence of a unique solution of Problem (2.9). We observe that

$$|a(u,v)| \le \sum_{i=1}^d \int_\Omega \left| \frac{\partial u}{\partial x_i} \frac{\partial v}{\partial x_i} \right| \mathrm{d}x + \sum_{i=1}^d \sup_{x\in\Omega} |b_i(x)| \int_\Omega \left| \frac{\partial u}{\partial x_i} v \right| \mathrm{d}x$$
$$+ \sup_{x\in\Omega} |c(x)| \int_\Omega |u\, v|\, \mathrm{d}x + \sup_{x\in\Gamma_\mathrm{R}} |\sigma(x)| \int_{\Gamma_\mathrm{R}} |u\, v|\, \mathrm{d}x$$

and apply Cauchy–Schwarz inequality (first for integrals then for sums) to get

$$|a(u,v)| \le |u|_1\, |v|_1 + \sum_{i=1}^d \|b_i\|_\infty \left(\int_\Omega \left| \frac{\partial u}{\partial x_i} \right|^2 \mathrm{d}x \right)^{1/2} \|v\|_0$$
$$+ \|c\|_\infty \|u\|_0 \|v\|_0 + \|\sigma\|_{\infty,\Gamma_\mathrm{R}} \|u\|_{0,\Gamma_\mathrm{R}} \|v\|_{0,\Gamma_\mathrm{R}}$$
$$\le |u|_1\, |v|_1 + \left(\sum_{i=1}^d \|b_i\|_\infty^2 \right)^{1/2} |u|_1 \|v\|_0 + \|c\|_\infty \|u\|_0 \|v\|_0 + C_t^2 \|\sigma\|_{\infty,\Gamma_\mathrm{R}} \|u\|_1 \|v\|_1$$
$$\le \beta \|u\|_1 \|v\|_1, \quad \text{where} \quad \beta = 1 + \left(\sum_{i=1}^d \|b_i\|_\infty^2 \right)^{1/2} + \|c\|_\infty + C_t^2 \|\sigma\|_{\infty,\Gamma_\mathrm{R}}.$$

Here, the continuity constant of the trace operator $v \in H^1(\Omega) \mapsto v|_{\Gamma_\mathrm{R}} \in L^2(\Gamma_\mathrm{R})$ has been denoted by C_t, that is,

$$\|v\|_{0,\Gamma_\mathrm{R}} \le C_t \|v\|_1 \quad \text{for all } v \in H^1(\Omega).$$

Next, we seek sufficient conditions on the data that guarantee the coercivity of the bilinear form $a(\cdot,\cdot)$. We start with the identity

$$a(v,v) = \sum_{i=1}^{d} \int_{\Omega} \left| \frac{\partial v}{\partial x_i} \right|^2 \mathrm{d}x + \sum_{i=1}^{d} \int_{\Omega} b_i(x) \frac{\partial v}{\partial x_i} v \,\mathrm{d}x + \int_{\Omega} c v^2 \,\mathrm{d}x + \int_{\Gamma_{\mathrm{R}}} \sigma v^2 \,\mathrm{d}s.$$

Applying integration by parts to the second term, we get

$$\begin{aligned}
\sum_{i=1}^{d} \int_{\Omega} b_i(x) \frac{\partial v}{\partial x_i} v \,\mathrm{d}x &= \frac{1}{2} \sum_{i=1}^{d} \int_{\Omega} b_i(x) \frac{\partial v^2}{\partial x_i} \,\mathrm{d}x \\
&= \frac{1}{2} \int_{\Gamma} \sum_{i=1}^{d} b_i(x) n_i(x) v^2 \,\mathrm{d}s - \frac{1}{2} \int_{\Omega} \sum_{i=1}^{d} \frac{\partial b_i(x)}{\partial x_i} v^2 \,\mathrm{d}x \\
&= \frac{1}{2} \int_{\Gamma_{\mathrm{N}} \cup \Gamma_{\mathrm{R}}} \mathbf{b} \cdot \mathbf{n} v^2 \,\mathrm{d}s - \frac{1}{2} \int_{\Omega} \operatorname{div} \mathbf{b}\, v^2 \,\mathrm{d}x
\end{aligned}$$

which lead us to

$$a(v,v) = |v|_1^2 + \int_{\Omega} \left(c - \frac{1}{2} \operatorname{div} \mathbf{b} \right) v^2 \,\mathrm{d}x + \int_{\Gamma_{\mathrm{N}} \cup \Gamma_{\mathrm{R}}} \mathbf{b} \cdot \mathbf{n} v^2 + \int_{\Gamma_{\mathrm{R}}} \sigma v^2 \,\mathrm{d}s.$$

Suppose that $\mathbf{b}$ and c satisfy the inequalities

$$\begin{aligned}
c - \frac{1}{2} \operatorname{div} \mathbf{b} \geq 0 \quad & \text{for all } x \in \Omega, \\
\mathbf{b} \cdot \mathbf{n} \geq 0 \quad & \text{for all } x \in \Gamma_{\mathrm{N}} \cup \Gamma_{\mathrm{R}},
\end{aligned} \tag{2.10}$$

then with $\sigma(x) > 0$, we obtain

$$a(v,v) \geq |v|_1^2. \tag{2.11}$$

Using the Poincare inequality, we obtain

$$\|v\|_1^2 = \|v\|_0^2 + |v|_1^2 \leq (1 + c_{\mathrm{P}}^2)|v|_1^2$$

and conclude the coercivity of the bilinear form

$$a(v,v) \geq \frac{1}{1 + c_{\mathrm{P}}^2} \|v\|_1^2 \quad \text{for all } v \in V. \tag{2.12}$$

Finally, the application of Schwarz inequality and the continuity of the trace operator yield the continuity of $F(\cdot)$

$$\begin{aligned}
|F(v)| &\leq \int_{\Omega} |f\, v| \,\mathrm{d}x + \int_{\Gamma_{\mathrm{N}}} |g\, v| \,\mathrm{d}s + \int_{\Gamma_{\mathrm{R}}} |\sigma u_\infty\, v| \,\mathrm{d}s \\
&\leq \|f\|_0 \|v\|_0 + \|g\|_{0,\Gamma_{\mathrm{N}}} \|v\|_{0,\Gamma_{\mathrm{N}}} + \|\sigma u_\infty\|_{0,\Gamma_{\mathrm{R}}} \|v\|_{0,\Gamma_{\mathrm{R}}} \\
&\leq (\|f\|_0 + C_t \|g\|_{0,\Gamma_{\mathrm{N}}} + C_t \|\sigma u_\infty\|_{0,\Gamma_{\mathrm{R}}}) \|v\|_1 \leq C \|v\|_1.
\end{aligned}$$

Since the bilinear form $a(\cdot,\cdot)$ is continuous and coercive, and the linear form $F(v)$ is continuous, the existence of a unique weak solution for Problem (2.8) follows from Theorem 2.1 (Lax–Milgram).

Remark 2.3. Note that the Dirichlet boundary condition has to be imposed on the function space, whereas the Neumann and Robin type boundary conditions have been incorporated in the weak formulation naturally. Thus, the Dirichlet boundary condition is called *essential boundary condition*, whereas the Neumann and the Robin type boundary conditions are called *natural boundary conditions.*

2.3. Standard Galerkin method

In this section we consider approximations of the solution of an abstract problem class which covers the examples of the previous section. The solution of the abstract problem belongs to some infinite dimensional Hilbert space whereas its approximation lives in a finite dimensional subspace. The construction of finite dimensional spaces by means of finite elements will be discussed in Chapter 3.

Let V be a real Hilbert space, the bilinear form $a : V \times V \to \mathbb{R}$ be continuous and coercive (but not necessarily symmetric), and the linear form $F : V \to \mathbb{R}$ be continuous. We consider the problem

$$\text{Find } u \in V, \text{ such that } a(u,v) = F(v) \quad \text{for all } v \in V. \tag{2.13}$$

According to Theorem 2.1 (Lax–Milgram) there exists a unique solution $u \in V$.

Let V_h be a finite dimensional subspace of V. Here, h is a discretization parameter with the notion that the discrete solution will converge to the continuous solution as $h \to 0$. The standard Galerkin method consists of restricting Equation (2.13) to the finite dimensional space V_h. Thus, the discrete problem reads

$$\text{Find } u_h \in V_h, \text{ such that } a(u_h, v_h) = F(v_h) \quad \text{for all } v_h \in V_h. \tag{2.14}$$

The bilinear form $a(\cdot,\cdot)$ and the linear form $F(\,\cdot\,)$ are continuous on V_h since $V_h \subset V$. The coercivity of $a(\cdot,\cdot)$ on V implies the coercivity of $a(\cdot,\cdot)$ on V_h. As a consequence, Theorem 2.1 (Lax–Milgram) can be applied to Problem (2.14) and there is a unique solution $u_h \in V_h$.

Let us show that Problem (2.14) is equivalent to a linear algebraic system of equations. We denote by $\{\phi_i\}$, $i = 1, \ldots, N$ a basis of V_h. Setting $v_h = \phi_i$ in (2.14), we get

$$a(u_h, \phi_i) = F(\phi_i), \quad i = 1, \ldots, N. \tag{2.15}$$

We now write the discrete solution u_h and its gradient in terms of the basis of V_h as

$$u_h(x) = \sum_{j=1}^{N} U_j\, \phi_j(x), \qquad \nabla u_h(x) = \sum_{j=1}^{N} U_j\, \nabla \phi_j(x), \tag{2.16}$$

where U_j, $j = 1, \ldots, N$, are the unknown coefficients to be determined. Substituting (2.16) in (2.15) leads to

$$\sum_{j=1}^{N} U_j a(\phi_j, \phi_i) = F(\phi_i), \quad i = 1, \ldots, N.$$

Setting $a_{ij} := a(\phi_j, \phi_i)$ and $f_i := F(\phi_i)$, we get the system of algebraic equations

$$AU = b,$$

where the stiffness matrix, the solution vector and the load vector, respectively, are given by

$$A = \begin{bmatrix} a_{11} & a_{12} & \cdots & a_{1N} \\ a_{21} & a_{22} & \cdots & a_{2N} \\ \vdots & \vdots & \ddots & \vdots \\ a_{N1} & a_{N2} & \cdots & a_{NN} \end{bmatrix}, \qquad U = \begin{bmatrix} U_1 \\ U_2 \\ \vdots \\ U_N \end{bmatrix}, \qquad b = \begin{bmatrix} f_1 \\ f_2 \\ \vdots \\ f_N \end{bmatrix}.$$

Solving the system of algebraic equations, we obtain the discrete solution u_h. The system of algebraic equations is equivalent to (2.15), but on the first glance not to (2.14). We show that it is also equivalent to (2.14). Let $v_h \in V_h$ be arbitrary. Then, there is a representation

$$v_h = \sum_{i=1}^{N} V_i \phi_i$$

and, we have

$$\begin{aligned} a(u_h, v_h) &= a\left(u_h, \sum_{i=1}^{N} V_i \phi_i\right) = \sum_{i=1}^{N} V_i a(u_h, \phi_i) = \sum_{i=1}^{N} V_i F(\phi_i) \\ &= F\left(\sum_{i=1}^{N} V_i \phi_i\right) = F(v_h). \end{aligned}$$

Remark 2.4. The matrix A will be symmetric when the bilinear form $a(\cdot,\cdot)$ is symmetric, that is, $a(u,v) = a(v,u)$ for all $u, v \in V$.

Remark 2.5. The matrix A will be positive definite when the bilinear form $a(\cdot,\cdot)$ is coercive. Indeed, for $\xi \in \mathbb{R}^N$, we have

$$\begin{aligned} \xi^T A \xi &= \sum_{i=1}^{N} \sum_{j=1}^{N} \xi_j a_{ij} \xi_i = \sum_{i=1}^{N} \sum_{j=1}^{N} \xi_j a(\phi_j, \phi_i) \xi_i \\ &= a\left(\sum_{j=1}^{N} \xi_j \phi_j, \sum_{i=1}^{N} \xi_i \phi_i\right) \\ &\geq \alpha \left\| \sum_{i=1}^{N} \xi_i \phi_i \right\|_V^2 \geq 0. \end{aligned}$$

Further, $\xi^T A \xi$ will be equal to zero only for $\sum_{i=1}^{N} \xi_i \phi_i = 0$, which, due to the linear independency of ϕ_i, is equal to $\xi = (\xi_1, \ldots, \xi_N) = 0$.

Remark 2.6. The stability $\|u_h\|_V \leq \alpha^{-1}\|F\|_{V^*}$ of Equation (2.15) follows by means of the coercivity of $a(\cdot,\cdot)$, that is,

$$\alpha\|u_h\|_V^2 \leq a(u_h,u_h) = F(u_h) \leq \|F\|_{V^*}\|u_h\|_V.$$

2.4. Abstract error estimate

Let u be the continuous solution of the elliptic Problem (2.13) and u_h be the solution of the discrete Problem (2.14). Then we estimate the error $\|u-u_h\|_V$ by means of the following lemma.

Lemma 2.1 (Cea's lemma). *Let V be a Hilbert space. Suppose the bilinear form $a(\cdot,\cdot)$ is continuous and coercive, and the linear form $F(\cdot)$ is continuous. Then, for the unique solutions u and u_h of* (2.13) *and* (2.14), *respectively, we have*

$$\|u-u_h\|_V \leq \frac{\beta}{\alpha}\inf_{v_h\in V_h}\|u-v_h\|_V.$$

Here, β is the continuity constant and α is the coercivity constant of $a(\cdot,\cdot)$.

Proof. Since $a(u,v)=F(v)$ for all $v\in V$, $a(u_h,v)=F(v)$ for all $v\in V_h$ and $V_h\subset V$, we have (by subtracting) the Galerkin orthogonality

$$a(u-u_h,v)=0 \quad \text{for all } v\in V_h. \tag{2.17}$$

Due to the coercivity, we have

$$\begin{aligned}\alpha\|u-u_h\|_V^2 &\leq a(u-u_h,u-u_h)\\ &= a(u-u_h,u-v_h)+a(u-u_h,v_h-u_h) \quad \text{for all } v_h\in V_h.\end{aligned}$$

Since $u_h, v_h\in V_h$, we have $v_h-u_h\in V_h$. Now by using Galerkin orthogonality and the continuity of $a(\cdot,\cdot)$, we get

$$\begin{aligned}\alpha\|u-u_h\|_V^2 &\leq \beta\|u-u_h\|_V\|u-v_h\|_V\\ \|u-u_h\|_V &\leq \frac{\beta}{\alpha}\|u-v_h\|_V \quad \text{for all } v_h\in V_h.\end{aligned}$$

Therefore,

$$\|u-u_h\|_V \leq \frac{\beta}{\alpha}\inf_{v_h\in V_h}\|u-v_h\|_V.$$

■

Remark 2.7. Lemma 2.1 shows that u_h is quasioptimal in the sense that the error $\|u-u_h\|_V$ is proportional to the error of the best approximation.

CHAPTER 3

FINITE ELEMENT SPACES

In applications, the discrete spaces $V_h \subset V$ have been designed by decomposing the computational domain Ω into a finite set of subdomains K (triangles, quadrilaterals, tetrahedrons, hexahedrons, etc.) and considering a function space $\mathcal{P}_K$ (often polynomials of a certain degree) defined on K. In this chapter we discuss how nodal functionals uniquely determine the functions from $\mathcal{P}_K$ locally and how they impose the continuity and differentiability properties globally. In the simplest case, we end up with spaces of continuous, piecewise linear functions.

3.1. Simplices and barycentric coordinates

First we introduce some useful notations.

Definition 3.1. A d-simplex is the convex hull in $\mathbb{R}^d$ of $d+1$ points $\{a_j\}_{j=0}^d$ (let us assume that they are given as column vectors), such that the matrix

$$A = \begin{bmatrix} a_0 & a_1 & \cdots & a_d \\ 1 & 1 & \cdots & 1 \end{bmatrix} \in \mathbb{R}^{(d+1)\times(d+1)}$$

is nonsingular, that is, $\det A \neq 0$.

For example, an interval is a 1d simplex, a triangle is a 2d simplex, and a tetrahedron is a 3d simplex. Since $\det A = \det[(a_1 - a_0) \quad \dots \quad (a_d - a_0)]$, the determinant of the matrix A is equal to the volume of the parallelepiped spanned by the vectors $(a_1 - a_0)$, $(a_2 - a_0)$, $\dots$, $(a_d - a_0)$ and we have $\det A = d!|K|$, where $|K|$ is the Lebesgue measure of the d-simplex K in $\mathbb{R}^d$, the interval length in one dimension, the area in two dimensions and the volume in three dimensions.

Definition 3.2. Let $\{a_j\}_{j=0}^d$ be the $(d+1)$ generating points of a d-simplex in $\mathbb{R}^d$. The barycentric coordinates of any $x \in \mathbb{R}^d$ with respect to these points are the numbers $\{\lambda_j\}_{j=0}^d$ such that

$$\sum_{j=0}^{d} \lambda_j a_j = x, \qquad \sum_{j=0}^{d} \lambda_j = 1.$$

In other words, the barycentric coordinates λ_0, λ_1, ..., λ_d are the solution of the linear system of equations

$$\begin{bmatrix} a_0 & a_1 & \cdots & a_d \\ 1 & 1 & \cdots & 1 \end{bmatrix} [\lambda] = \begin{bmatrix} x \\ 1 \end{bmatrix}. \tag{3.1}$$

There is a nice geometrical interpretation of the barycentric coordinates as indicated in Figure 3.1. Let x be a point of the d-simplex K and $K_i(x)$ be the d-simplex with vertices a_j, $j \neq i$, and x. Then, solving the algebraic system (3.1) by Cramer's rule, we obtain

$$\lambda_i(x) = \frac{|K_i(x)|}{|K|}.$$

From this formula we conclude that $\lambda_i(x)$ is an affine function of x, vanishes on the face of the d-simplex K opposite to a_i and equals 1 at the vertex a_i. Any d-simplex $K \subset \mathbb{R}^d$ can be considered as the image of an affine map $F_K : \widehat{K} \to \mathbb{R}^d$ of a reference simplex $\widehat{K}$ spanned by the vertices $\hat{a}_0, \ldots, \hat{a}_d$. It is convenient to work with $\hat{a}_i = e_i$, $i = 0, \ldots, d$, where $e_0 = (0,0,\ldots,0)^T$, $e_1 = (1,0,\ldots,0)^T$, $e_2 = (0,1,\ldots,0)^T$, ..., $e_d = (0,\ldots,0,1)^T$ as shown in Figure 3.2. Taking into consideration that $a_i = F_K(e_i)$, $i = 0, \ldots, d$, we find the representation for F_K

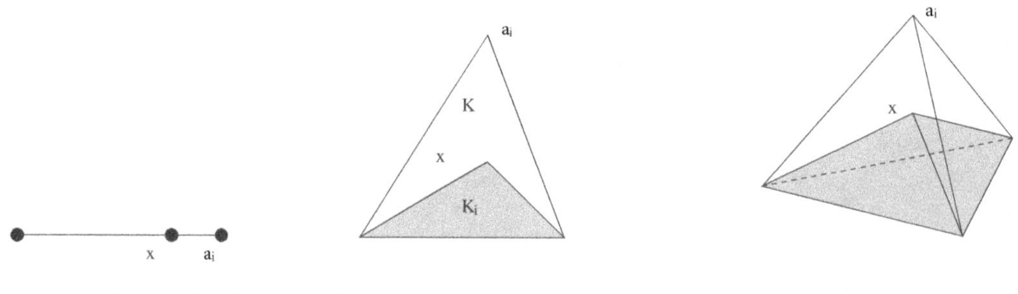

FIGURE 3.1 Geometrical interpretation of the barycentric coordinates.

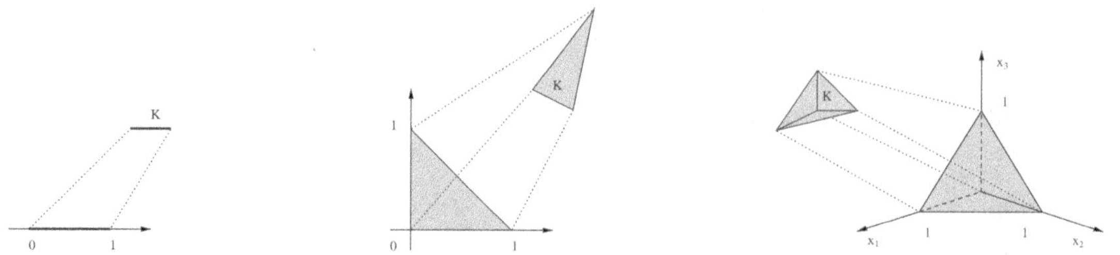

FIGURE 3.2 Reference simplices $\widehat{K}$ in one, two and three dimensions.

$$F_K(\hat{x}) = B_K\hat{x} + b_K, \qquad B_K = \begin{bmatrix} a_1 - a_0 & \cdots & a_d - a_0 \end{bmatrix}, \quad b_K = a_0.$$

Lemma 3.1. *Let $\lambda_0, \ldots, \lambda_d$ be the barycentric coordinates of the d-simplex $K \subset \mathbb{R}^d$ and $F_K : \widehat{K} \to K$ be the affine mapping from the reference simplex $\widehat{K}$ to the d-simplex K. Then, the functions $\lambda_i \circ F_K : \widehat{K} \to \mathbb{R}^d$, $i = 0, \ldots, d$, are the barycentric coordinates of the reference simplex $\widehat{K}$.*

Proof. Starting with the definition of barycentric coordinates $\lambda_0, \ldots, \lambda_d$, we get using $x = F_K(\hat{x}) = B_K\hat{x} + b_K$

$$\sum_{j=0}^{d} \lambda_j a_j = \sum_{j=0}^{d} \lambda_j (B_K\hat{a}_j + b_K) = \sum_{j=0}^{d} \lambda_j B_K\hat{a}_j + b_K = B_K\hat{x} + b_K$$

thus, the functions $\hat{\lambda}_j = \lambda_j \circ F_K$ are the unique solution of the system

$$\sum_{j=0}^{d} \hat{\lambda}_j \hat{a}_j = \hat{x}, \qquad \sum_{j=0}^{d} \hat{\lambda}_j = 1,$$

that is, the barycentric coordinates on the reference cell. ■

3.2. Simplicial finite elements and local spaces

For simplicity, let $\Omega \subset \mathbb{R}^d$ be a polyhedral domain. We start with a decomposition $\mathcal{T}_h$ of Ω into open cells K with $\overline{\Omega} = \bigcup_{K \in \mathcal{T}_h} \overline{K}$. In this section, cells will be triangles if $d = 2$, tetrahedra if $d = 3$ and, in general, d-simplices. The set of all polynomials of degree less than or equal to k will be denoted by

$$P_k(K) := \left\{ p : K \to \mathbb{R} \mid p(x) = \sum_{|\alpha| \le k} c_\alpha x^\alpha, \quad x \in K \right\},$$

where $c_\alpha \in \mathbb{R}$ are constant coefficients.

Definition 3.3. A triple $(K, \mathcal{P}_K, \Sigma_K)$ is called finite element, where K is an open cell in $\mathcal{T}_h$, $\mathcal{P}_K$ is a local function space of finite dimension with $\dim \mathcal{P}_K = n_{\text{loc}} + 1$ and Σ_K is a set of degrees of freedom (dof) given as linear nodal functionals $N_i(v)$, $i = 0, \ldots, n_{\text{loc}}$, $v \in \mathcal{P}_K$. Σ_K is assumed to be $\mathcal{P}_K$ unisolvent, that is, for arbitrary given values d_i, $i = 0, \ldots, n_{\text{loc}}$, there is a unique $v \in \mathcal{P}_K$, such that $N_i(v) = d_i$, $i = 0, \ldots, n_{\text{loc}}$.

Definition 3.4. A finite element $(K, \mathcal{P}_K, \Sigma_K)$ is called a Lagrange finite element if there is a set of points $\{a_0, a_1, \ldots, a_{n_{\text{loc}}}\}$ that define the set of dof Σ_K by the nodal functionals $N_i(v) = v(a_i)$, $i = 0, \ldots, n_{\text{loc}}$. The points a_i, $i = 0, \ldots, n_{\text{loc}}$ are called nodes of the finite element.

Let the cell K be a triangle with vertices a_0, a_1 and a_2. In case of the Courant element, we set $\mathcal{P}_K := P_1(K)$ with $\dim(P_1(K)) = 3$. In order to fix the function on each triangle locally in $P_1(K)$ we define the set of dof by means of the linear nodal functionals

$$N_0(v) := v(a_0), \qquad N_1(v) := v(a_1), \qquad N_2(v) := v(a_2). \tag{3.2}$$

Since the three vertices of K are not collinear, the problem of looking for a function $v \in P_1(K)$ satisfying (3.2) for arbitrary given values $v(a_i)$, $i = 0,1,2$, has a unique solution. Consequently, the nodal functionals (3.2) are $P_1(K)$ unisolvent. Taking into consideration that $P_1(K) = \text{span}\{\lambda_0, \lambda_1, \lambda_2\}$ and that the barycentric coordinates satisfy $\lambda_i(a_j) = \delta_{ij}$, $i,j = 0,1,2$, we can explicitly express

$$v = \sum_{i=0}^{2} N_i(v)\,\lambda_i \quad \text{for all } v \in P_1(K).$$

The extension of the Courant triangle to d-dimensions is straightforward. Let K be a d-simplex with vertices $a_0, \ldots, a_d$, let the local function space be $\mathcal{P}_K := P_1(K) = \text{span}\{\lambda_0, \ldots, \lambda_d\}$ with $\dim(\mathcal{P}_K) = d+1$, and let the set of dof, Σ_K be defined by the nodal functionals $N_i(v) := v(a_i)$, $i = 0, \ldots, d$. Any function from $\mathcal{P}_K$ can be represented by a linear combination (with coefficients γ_i) of the basis. Then, the $\mathcal{P}_K$ unisolvence of Σ_K is equivalent to the unique solvability of the linear system of equations

$$\begin{aligned}
\gamma_0\lambda_0(a_0) + \cdots + \gamma_d\lambda_d(a_0) &= v(a_0) = N_0(v)\\
&\;\;\vdots\\
\gamma_0\lambda_0(a_d) + \cdots + \gamma_d\lambda_d(a_d) &= v(a_d) = N_d(v).
\end{aligned}$$

From the property $\lambda_i(a_j) = \delta_{ij}$ of the barycentric coordinates, we immediately conclude the unique solution $\gamma_i = N_i(v)$, $i = 0, \ldots, d$.

In the following, let $(K, \mathcal{P}_K, \Sigma_K)$ be an arbitrary finite element as mentioned in Definition 3.3. The $\mathcal{P}_K$ unisolvence of the set of dof $N_i(v)$, $i = 0, \ldots, n_{\text{loc}}$, guarantees that there are unique functions $\varphi_j \in \mathcal{P}_K$, $j = 0, \ldots, n_{\text{loc}}$, satisfying $N_i(\varphi_j) = \delta_{ij}$ for $i,j = 0, \ldots, n_{\text{loc}}$. We call these functions $\varphi_0, \ldots, \varphi_{n_{\text{loc}}}$ as local canonical basis of $\mathcal{P}_K$. As we saw above the local canonical basis of $\mathcal{P}_K = P_1(K)$ consists just of the barycentric coordinates. Having a canonical basis of the local space $\mathcal{P}_K = \text{span}\{\varphi_0, \ldots, \varphi_{n_{\text{loc}}}\}$ we see that for arbitrary given values d_i, $i = 0, \ldots, n_{\text{loc}}$

$$v = \sum_{j=0}^{n} d_j\,\varphi_j$$

satisfies

$$N_i(v) = N_i\left(\sum_{j=0}^{n} d_j\varphi_j\right) = \sum_{j=0}^{n} d_j N_i(\varphi_j) = \sum_{j=0}^{n} d_j\delta_{ij} = d_i, \qquad i = 0, \ldots, n_{\text{loc}}.$$

Therefore, the set of dof is $\mathcal{P}_K$ unisolvent.

A quadratic element on a d-simplex with vertices $a_0,\ldots,a_d$ will be considered next. For $0 \le i < j \le d$, let the midpoints of edges be denoted by $a_{ij} = (a_i + a_j)/2$. The finite element is given by

- K is a d-simplex,
- $\mathcal{P}_K = P_2(K)$, $\dim P_2(K) = (d+1)(d+2)/2$,
- $\Sigma_K = \{v(a_i),\ 0 \le i \le d;\ v(a_{ij}),\ 0 \le i < j \le d\}$.

The canonical basis functions consist of functions

$$\varphi_i := \lambda_i(2\lambda_i - 1), \quad i = 0,\ldots,d$$

associated with the vertex a_i such that $\varphi_i(a_j) = \delta_{ij}$ and $\varphi_i(a_{kl}) = 0$. The remaining functions

$$\varphi_{ij} := 4\lambda_i\lambda_j, \quad 0 \le i < j \le d$$

are associated with the midpoints of edges and satisfy $\varphi_{ij}(a_{kl}) = \delta_{ik}\delta_{jl}$ and $\varphi_{ij}(a_k) = 0$. Therefore, the set of dof, Σ_K is $P_2(K)$ unisolvent.

The continuous, piecewise linear and the continuous, piecewise quadratic element are examples of Lagrange finite elements (see Definition 3.4).

For finite elements and local spaces of higher polynomial degree, we refer to Section 3.9 and Ciarlet (2002); Braess (2007); Brenner and Scott (2008) and Boffi et al. (2013).

3.3. Construction of finite element spaces

Let $\mathcal{T}_h$ be a decomposition of a polyhedral domain Ω into nonoverlapping open cells $K \in \mathcal{T}_h$. We need some assumptions guaranteeing, roughly speaking, that the decomposition is geometrically conform.

Definition 3.5. A decomposition of the domain $\Omega \subset R^d$, $d = 2$ and $d = 3$, into cells $K \in \mathcal{T}_h$ is admissible if

(i) $\overline{\Omega} = \bigcup_{K \in \mathcal{T}_h} \overline{K}$,
(ii) any nonempty intersection of two cells $\overline{K} \cap \overline{K'}$ is either a vertex, or an edge, or a face of both cells $K, K' \in \mathcal{T}_h$.

See Figure 3.3 for a forbidden and admissible decomposition in 2d.

Starting with a finite element $(K, \mathcal{P}_K, \Sigma_K)$, the finite element space V_h is designed in such a way that the restriction $v_h|_K$ of each $v_h \in V_h$ belongs to the finite dimensional space $\mathcal{P}_K$. The $\mathcal{P}_K$ unisolvence of the set of dof Σ_K defines $v_h|_K$ uniquely on K. We still have to set the global properties of V_h to guarantee the inclusion $V_h \subset H_0^1(\Omega)$ or $V_h \subset H^1(\Omega)$. For this, the following theorem is very helpful.

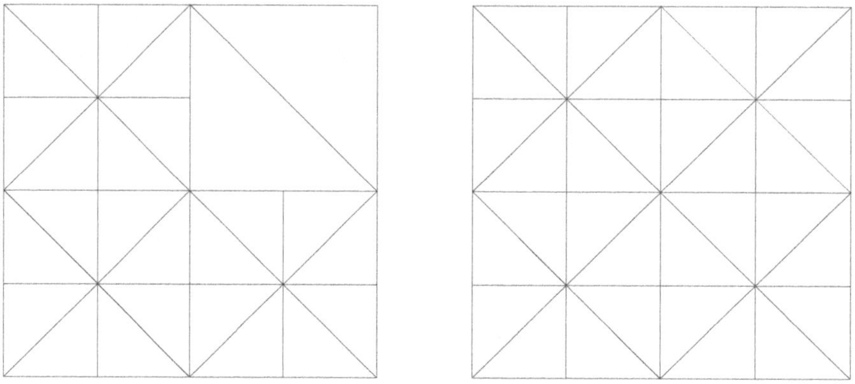

FIGURE 3.3 Forbidden (left) and admissible (right) decomposition $\mathcal{T}_h$.

Theorem 3.1. *If for every $K \in \mathcal{T}_h$, $\mathcal{P}_K \subset H^1(K)$, and $V_h \subset C(\overline{\Omega})$, then $V_h \subset H^1(\Omega)$. If in addition $v = 0$ on $\partial\Omega$ for all $v \in V_h$, then $V_h \subset H_0^1(\Omega)$.*

Proof. Let $v \in V_h$ be given. We have to show that v has weak derivatives $D^\alpha v \in L^2(\Omega)$ for $|\alpha| = 1$. A possible candidate would be the piecewise defined function $w := D^\alpha(v|_K) \in L^2(K)$. Now for any $\varphi \in C_0^\infty(\Omega)$ we have

$$\int_\Omega (w\varphi + vD^\alpha\varphi)\,\mathrm{d}x = \sum_{K\in\mathcal{T}_h}\int_K (D^\alpha(v|_K)\varphi + vD^\alpha\varphi)\,\mathrm{d}x = \sum_{K\in\mathcal{T}_h}\int_{\partial K} v|_K\varphi\, n_K^\alpha\,\mathrm{d}\gamma,$$

where n_K^α is the i-th component of the outer unit normal $\mathbf{n}_K$ such that $\alpha_i = 1$ and $|\alpha| = 1$. The integral over each boundary face $E \subset \partial\Omega$ vanishes since $\varphi \in C_0^\infty(\Omega)$. For each inner face $E = \partial K \cap \partial K'$, the integral in the sum appears twice where $v|_K\varphi = v|_{K'}\varphi$ on E and $n_K^\alpha = -n_{K'}^\alpha$ on E. As a consequence

$$\int_\Omega (w\varphi + vD^\alpha\varphi)\,\mathrm{d}x = 0 \quad \text{for all } \varphi \in C_0^\infty(\Omega),$$

which means that $w = D^\alpha v$ is the weak derivative of v. ■

Now we come back to the definition of a finite element space generated by a set of finite elements $\{(K, \mathcal{P}_K, \Sigma_K) : K \in \mathcal{T}_h\}$ where $\mathcal{T}_h$ is an admissible decomposition of the polyhedral domain $\Omega \subset \mathbb{R}^d$. According to Theorem 3.1, the continuity across the faces of neighbouring cells is sufficient for $V_h \subset H^1(\Omega)$. This can be realized by using enough dof associated with the faces of an element. For example, in case of the linear element on d-simplices, discussed in Section 3.2, the face $E \subset \partial K$ is a $(d-1)$ simplex and the d number of face dof determine a function from $P_1(E)$ uniquely. The generated finite element space of piecewise linear functions is defined as

$$V_h := \{v : \Omega \to \mathbb{R} : v|_K \in P_1(K) \text{ for all } K \in \mathcal{T}_h,\ v \text{ continuous at inner vertices}\}.$$

If we need a finite element space satisfying homogeneous boundary conditions of Dirichlet type we define

$$V_{0h} := \{v : \Omega \to \mathbb{R} : v|_K \in P_1(K) \text{ for all } K \in \mathcal{T}_h,$$
$$v \text{ continuous at inner vertices, } v = 0 \text{ at boundary vertices}\}.$$

The functions from V_{0h} vanish not only at the boundary nodes but identically on the whole boundary. Theorem 3.1 states $V_{0h} \subset H_0^1(\Omega)$.

A similar argumentation holds for the quadratic element. The restriction of a function $v \in P_2(K)$ on a face $E \subset \partial K$ is a quadratic function in $\mathbb{R}^{d-1}$ and uniquely defined by its $d(d+1)/2$ values at the vertices and midpoints of edges of E. This guarantees the continuity across faces and the fulfilment of homogeneous boundary conditions along the boundary. The discrete counterparts of $H^1(\Omega)$ and $H_0^1(\Omega)$ become

$$V_h := \{v : \Omega \to \mathbb{R} : v|_K \in P_2(K) \text{ for all } K \in \mathcal{T}_h, v \text{ continuous at inner nodes}\}.$$
$$V_{0h} := \{v : \Omega \to \mathbb{R} : v|_K \in P_2(K) \text{ for all } K \in \mathcal{T}_h,$$
$$v \text{ continuous at inner nodes, } v = 0 \text{ at boundary nodes}\}.$$

Here, a node is a vertex or a midpoint of an edge of the triangulation $\mathcal{T}_h$.

For the solution of fourth order equations in Chapter 5 by conforming finite elements we have to construct finite element spaces $V_h \subset H^2(\Omega)$. For this, the following theorem is helpful.

Theorem 3.2. *If for every $K \in \mathcal{T}_h$, $\mathcal{P}_K \subset H^2(K)$, and $V_h \subset C^1(\overline{\Omega})$, then $V_h \subset H^2(\Omega)$. If in addition $v = \partial_{\mathbf{n}} v = 0$ on $\partial\Omega$ for all $v \in V_h$, then $V_h \subset H_0^2(\Omega)$.*

Proof. The arguments are similar to those given to prove Theorem 3.1. ∎

3.4. The concept of mapped finite elements: Affine mappings

In Section 3.1 we saw that any d-simplex $K \in \mathcal{T}_h$ can be considered as the image of a reference d-simplex $\widehat{K}$ under an affine mapping $F_K : \widehat{K} \to K$ with $F_K(\hat{x}) = B_K\hat{x} + b_K$, $B_K \in \mathbb{R}^{d\times d}$ and $b_K \in \mathbb{R}^d$. In the case of the Courant element or the quadratic element discussed in Section 3.2, the local approximation space and the set of dof can also be considered as 'mapped' from the reference objects

$$K = F_K(\widehat{K}), \qquad \mathcal{P}_K = \{p = \hat{p} \circ F_K^{-1} : \hat{p} \in \widehat{P}\}, \qquad \Sigma_K = \{v(F_K(\hat{a}_i))\}. \tag{3.3}$$

Since the mapping $F_K^{-1} : K \to \widehat{K}$ is affine, we have for $k \in \mathbb{N}_0$

$$\hat{p} \in P_k(\widehat{K}) \quad \Rightarrow \quad p = \hat{p} \circ F_K^{-1} \in P_k(K).$$

Moreover, $F_K : \widehat{K} \to K$ maps the nodes $\hat{a}_i$, $i = 0,\dots,d$, of the reference d-simplex $\widehat{K}$ onto the nodes a_i, $i = 0,\dots,d$, of the d-simplex, consequently the dof satisfy

$$v(a_i) = v(F_K(\hat{a}_i)) = \hat{v}(\hat{a}_i) \quad i = 0,\dots,d.$$

In case of a quadratic element, the affine mapping $F_K : \widehat{K} \to K$ maps a midpoint $\hat{a}_{ij}$ of an edge of the reference simplex onto the midpoint a_{ij} of the associated edge of K, since

$$a_{ij} = \frac{a_i + a_j}{2} = \frac{F_K(\hat{a}_i) + F_K(\hat{a}_j)}{2} = F_K\left(\frac{\hat{a}_i + \hat{a}_j}{2}\right) = F_K(\hat{a}_{ij}).$$

Therefore, the dof connected with the midpoint of edges can also be considered as 'mapped' from the reference object.

The concept of mapped finite elements simplifies the assembling process of the matrix A of the linear system of algebraic equations associated with the discrete Problem (2.14) (see Section 2.3). We now explain this aspect in more detail for the Poisson Problem (2.2). Let $\{\Phi_i\}$, $i = 1,\dots,n_{\text{glob}}$, be the basis of the finite element space V_h. The discrete problem corresponds a linear system of algebraic equations with the coefficient matrix

$$a_{ij} = a(\Phi_j, \Phi_i) = \int_\Omega \nabla\Phi_j \cdot \nabla\Phi_i \, \mathrm{d}x, \qquad i,j = 1,\dots,n_{\text{glob}}.$$

We split the integral into a sum over the local contribution resulting from a cell $K \in \mathcal{T}_h$

$$\int_\Omega \nabla\Phi_j \cdot \nabla\Phi_i \, \mathrm{d}x = \sum_{K \in \mathcal{T}_h} \int_K \nabla\Phi_j \cdot \nabla\Phi_i \, \mathrm{d}x$$

where it is enough to compute the local contributions (local element matrix)

$$\int_K \nabla\varphi_l \cdot \nabla\varphi_k \, \mathrm{d}x \qquad k,l = 0,\dots,n_{\text{loc}}$$

for the local basis functions in $\mathcal{P}_K$. Here, $\dim \mathcal{P}_K = n_{\text{loc}} + 1$ and the notations $\varphi_k := \Phi_i|_K$ and $\varphi_l := \Phi_j|_K$ are used. Let $F_K : \widehat{K} \to K$ with $F_K(\hat{x}) = B_K\hat{x} + b_K$ be the affine mapping from the reference d-simplex $\widehat{K}$ onto K and $\hat{\varphi}_l = \varphi_l \circ F_K$, $l = 0,\dots,n_{\text{loc}}$. Then the chain rule gives

$$(\hat{\nabla}\hat{\varphi}_l)(\hat{x}) = B_K^T \left((\nabla\varphi_l)(F_K(\hat{x}))\right)$$

such that the transformation of the integral leads to

$$\int_K \nabla\varphi_l \cdot \nabla\varphi_k \, \mathrm{d}x = \int_{\widehat{K}} (\hat{\nabla}\hat{\varphi}_l)^T (B_K^T B_K)^{-1} \hat{\nabla}\hat{\varphi}_k \det B_K \, \mathrm{d}\hat{x}.$$

Applying a quadrature rule for the integral over $\widehat{K}$ shows that the derivatives of the local basis functions $\hat{\varphi}_l$ have to be evaluated at the quadrature points only once for all $K \in \mathcal{T}_h$. More details on quadrature rules can be find in Section 9.2.

3.5. Finite elements on rectangular and brick meshes

In this section, we use the concept of mapped finite elements (3.3) to derive finite elements on rectangular and brick meshes. We first introduce local finite element spaces on the reference cell $\widehat{K} = (-1,+1)^d$. Note that a rectangle or a brick with faces parallel to the hyperplanes $x_i = 0$, $i = 1,\ldots,d$, is the image of the reference cell under an affine mapping $F_K(\hat{x}) = B_K\hat{x} + b_K$, where $B_K \in \mathbb{R}^{d\times d}$ is an invertible diagonal matrix and $b_K \in \mathbb{R}^d$. The set of all polynomials of degree less than or equal to k, $k \in \mathbb{N}$, in each variable will be denoted by

$$Q_k(\widehat{K}) := \left\{ p : \widehat{K} \to \mathbb{R} \mid p(\hat{x}) = \sum_{|\alpha|_\infty \le k} c_\alpha \hat{x}^\alpha \right\},$$

where $|\alpha|_\infty = \max_{1\le i\le d} |\alpha_i|$. The dimension is $\dim Q_k(\widehat{K}) = (k+1)^d$ and the inclusion $P_k(\widehat{K}) \subset Q_k(\widehat{K}) \subset P_{kd}(\widehat{K})$ holds. Consider the nodal set

$$\widehat{M}_k := \left\{ b = (-1,\ldots,-1) + 2\left(\frac{i_1}{k},\ldots,\frac{i_d}{k}\right) : i_j \in \{0,1,\ldots,k\}, 1 \le j \le d \right\}.$$

Lemma 3.2. *A polynomial $p \in Q_k(\widehat{K})$ is uniquely determined by its values on the nodal set $\widehat{M}_k$. The set of dof $\widehat{\Sigma} = \{v(b) : b \in \widehat{M}_k\}$ is $Q_k(\widehat{K})$ unisolvent and $(\widehat{K},\widehat{P},\widehat{\Sigma})$ with $\widehat{P} = Q_k(\widehat{K})$ is a finite element.*

Proof. In 1d, the statement follows from the Lagrange interpolation of $k+1$ values by a polynomial of degree k. For higher dimension $d \ge 2$ apply tensor product arguments. ∎

Lemma 3.3. *Let $\widehat{K} = (-1,+1)^d$, $F_K : \widehat{K} \to K$ be an affine bijective mapping, $K = F_K(\widehat{K})$ and $Q_k(K) := \{p = \hat{p} \circ F_K^{-1} : \hat{p} \in Q_k(\widehat{K})\}$ the space of mapped Q_k functions. Then,*

$$Q_k(K) = \left\{ p : K \to \mathbb{R} \mid p(x) = \sum_{|\alpha|_\infty \le k} c_\alpha x^\alpha \right\}.$$

Proof. An affine mapping transforms polynomials of degree less than or equal to k in each variable into a polynomial of degree less than or equal to k in each variable. Thus, the class of functions stays unchanged. ∎

In the following we restrict ourselves to two- and three-dimensional cases, that is, for $d = 2$ and $d = 3$. Let the vertices of the reference square $\widehat{K}$ be denoted by $\hat{a}_0 = (-1,-1)^T$, $\hat{a}_1 = (+1,-1)^T$, $\hat{a}_2 = (+1,+1)^T$ and $\hat{a}_3 = (-1,+1)^T$. The canonical basis functions $\varphi_i \in Q_1(\widehat{K})$, $i = 0,\ldots,3$, on $\widehat{K}$ are given by

$$\varphi_0(\hat{x}) = \frac{(1-\hat{x}_1)(1-\hat{x}_2)}{4}, \qquad \varphi_1(\hat{x}) = \frac{(1+\hat{x}_1)(1-\hat{x}_2)}{4},$$

$$\varphi_2(\hat{x}) = \frac{(1+\hat{x}_1)(1+\hat{x}_2)}{4}, \qquad \varphi_3(\hat{x}) = \frac{(1-\hat{x}_1)(1+\hat{x}_2)}{4}.$$

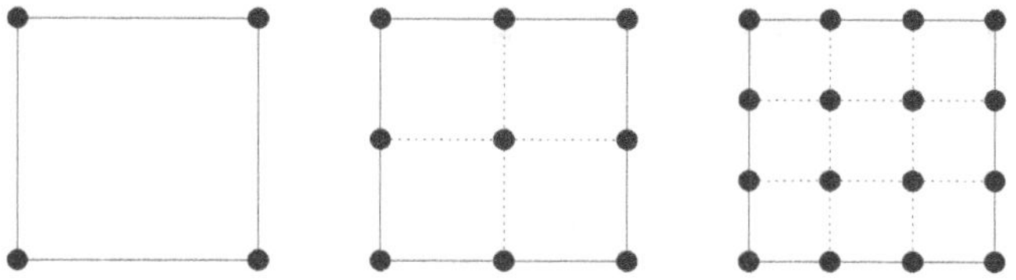

FIGURE 3.4 Nodal sets $\widehat{M}_1$, $\widehat{M}_2$, $\widehat{M}_3$ in two space dimensions.

These functions satisfy $\varphi_i(\hat{a}_j) = \delta_{ij}$, $i,j = 0,\dots,3$. Thus, for arbitrary points a_i, $i = 0,\dots,3$, the function $F_K : \widehat{K} \to \mathbb{R}^2$ given by

$$x = F_K(\hat{x}) = \sum_{i=0}^{3} a_i\varphi_i(\hat{x}) = \frac{a_0 + a_1 + a_2 + a_3}{4} + \frac{-a_0 + a_1 + a_2 - a_3}{4}\hat{x}_1 + \frac{-a_0 - a_1 + a_2 + a_3}{4}\hat{x}_2 + \frac{a_0 - a_1 + a_2 - a_3}{4}\hat{x}_1\hat{x}_2$$

maps the vertices $\hat{a}_i$ onto a_i, $i = 0,\dots,3$. As one can see from the representation, the mapping F_K is affine if and only if

$$a_2 - a_1 = a_3 - a_0,$$

that is, the quadrilateral with the vertices $a_0,\dots,a_3$ has equal opposite edges and the four points build a parallelogram. For the parallelogram case, the affine mapping $F_K : \widehat{K} \to K$ becomes

$$F_K(\hat{x}) = \frac{1}{2}\begin{bmatrix} a_1 - a_0 & a_3 - a_0 \end{bmatrix}\begin{bmatrix} \hat{x}_1 \\ \hat{x}_2 \end{bmatrix} + \frac{a_0 + a_1 + a_2 + a_3}{4}.$$

It is invertible if the area of the parallelogram $|K| = \det\begin{bmatrix} a_1 - a_0 & a_3 - a_0 \end{bmatrix}$ is positive (note that the vertices $a_0,\dots,a_3$ have been numbered in a counter-clockwise manner). In case of an axi-parallel rectangular of size $h_x \times h_y$, we have $a_2 - a_1 = a_3 - a_0 = (0, h_y)^T$ and $a_2 - a_3 = a_1 - a_0 = (h_x, 0)^T$ and the mapping specifies to

$$F_K(\hat{x}) = \frac{1}{2}\begin{bmatrix} h_x & 0 \\ 0 & h_y \end{bmatrix}\begin{bmatrix} \hat{x}_1 \\ \hat{x}_2 \end{bmatrix} + \frac{a_0 + a_1 + a_2 + a_3}{4}.$$

Next we discuss the three-dimensional case. Let the vertices of the reference cube $\widehat{K}$ be denoted by $\hat{a}_0 = (-1,-1,-1)^T$, $\hat{a}_1 = (+1,-1,-1)^T$, $\hat{a}_2 = (+1,+1,-1)^T$, $\hat{a}_3 = (-1,+1,-1)^T$, $\hat{a}_4 = (-1,-1,+1)^T$, $\hat{a}_5 = (+1,-1,+1)^T$, $\hat{a}_6 = (+1,+1,+1)^T$ and $\hat{a}_7 = (-1,+1,+1)^T$. Then the canonical basis functions $\varphi_i \in Q_1(\widehat{K})$, $i = 0,\dots,7$, on $\widehat{K}$ are given by

$$\varphi_0(\hat{x}) = \frac{(1-\hat{x}_1)(1-\hat{x}_2)(1-\hat{x}_3)}{8}, \qquad \varphi_1(\hat{x}) = \frac{(1+\hat{x}_1)(1-\hat{x}_2)(1-\hat{x}_3)}{8},$$

$$\varphi_2(\hat{x}) = \frac{(1+\hat{x}_1)(1+\hat{x}_2)(1-\hat{x}_3)}{8}, \qquad \varphi_3(\hat{x}) = \frac{(1-\hat{x}_1)(1+\hat{x}_2)(1-\hat{x}_3)}{8},$$

$$\varphi_4(\hat{x}) = \frac{(1-\hat{x}_1)(1-\hat{x}_2)(1+\hat{x}_3)}{8}, \qquad \varphi_5(\hat{x}) = \frac{(1+\hat{x}_1)(1-\hat{x}_2)(1+\hat{x}_3)}{8},$$
$$\varphi_6(\hat{x}) = \frac{(1+\hat{x}_1)(1+\hat{x}_2)(1+\hat{x}_3)}{8}, \qquad \varphi_7(\hat{x}) = \frac{(1-\hat{x}_1)(1+\hat{x}_2)(1+\hat{x}_3)}{8}.$$

These functions satisfy $\varphi_i(\hat{a}_j) = \delta_{ij}$, $i,j = 0,\dots,7$. Thus, for arbitrary points a_i, $i = 0,\dots,7$, the function $F_K : \widehat{K} \to \mathbb{R}^3$ given by

$$\begin{aligned}
x = F_K(\hat{x}) &= \sum_{i=0}^{7} a_i \varphi_i(\hat{x}) \\
&= \frac{a_0 + a_1 + a_2 + a_3 + a_4 + a_5 + a_6 + a_7}{8} \\
&+ \frac{-a_0 + a_1 + a_2 - a_3 - a_4 + a_5 + a_6 - a_7}{8}\hat{x}_1 \\
&+ \frac{-a_0 - a_1 + a_2 + a_3 - a_4 - a_5 + a_6 + a_7}{8}\hat{x}_2 \\
&+ \frac{-a_0 - a_1 - a_2 - a_3 + a_4 + a_5 + a_6 + a_7}{8}\hat{x}_3 \\
&+ \frac{a_0 - a_1 + a_2 - a_3 + a_4 - a_5 + a_6 - a_7}{8}\hat{x}_1\hat{x}_2 \\
&+ \frac{a_0 - a_1 - a_2 + a_3 - a_4 + a_5 + a_6 - a_7}{8}\hat{x}_1\hat{x}_3 \\
&+ \frac{a_0 + a_1 - a_2 - a_3 - a_4 - a_5 + a_6 + a_7}{8}\hat{x}_2\hat{x}_3 \\
&+ \frac{-a_0 - a_1 - a_2 + a_3 + a_4 - a_5 + a_6 - a_7}{8}\hat{x}_1\hat{x}_2\hat{x}_3
\end{aligned}$$

maps the vertices $\hat{a}_i$ onto a_i, $i = 0,\dots,7$. A short calculation shows that the mapping F_K is affine if and only if

$$a_3 = a_2 - a_1 + a_0,\ a_6 = a_2 - a_1 + a_5,\ a_4 = a_0 - a_1 + a_5,\ a_7 = a_4 - a_5 + a_6,$$

that is, the points $a_0,\dots,a_7$ are the vertices of a parallelepiped. In this case, the affine mapping $F_K(\hat{x})$ specifies to

$$F_K(\hat{x}) = \frac{1}{2}\begin{bmatrix} a_1 - a_0 & a_3 - a_0 & a_4 - a_0 \end{bmatrix}\begin{bmatrix} \hat{x}_1 \\ \hat{x}_2 \\ \hat{x}_3 \end{bmatrix} + \frac{1}{8}\sum_{i=0}^{7} a_i.$$

It is invertible if the volume of the parallelepiped does not vanish, that is,

$$|K| = \det\begin{bmatrix} a_1 - a_0 & a_3 - a_0 & a_4 - a_0 \end{bmatrix} \neq 0.$$

In case of an axi-parallel brick of size $h_x \times h_y \times h_z$, we have

$$a_5 - a_1 = a_6 - a_2 = a_7 - a_3 = a_4 - a_0 = (0,0,h_z)^T,$$
$$a_2 - a_1 = a_3 - a_0 = a_6 - a_5 = a_7 - a_4 = (0,h_y,0)^T,$$
$$a_2 - a_3 = a_1 - a_0 = a_5 - a_4 = a_6 - a_7 = (h_x,0,0)^T$$

and the mapping becomes

$$F_K(\hat{x}) = \frac{1}{2}\begin{bmatrix} h_x & 0 & 0 \\ 0 & h_y & 0 \\ 0 & 0 & h_z \end{bmatrix}\begin{bmatrix} \hat{x}_1 \\ \hat{x}_2 \\ \hat{x}_3 \end{bmatrix} + \frac{1}{8}\sum_{i=0}^{7} a_i.$$

3.6. Mapped finite elements: General bijective mappings

In the last two sections we considered mapped finite elements where the bijective mapping was affine. In this section, we extend the concept of mapped finite elements to allow general bijective mappings.

Lemma 3.4. *Let $(\widehat{K},\widehat{P},\widehat{\Sigma})$ be a Lagrangian finite element with $\widehat{\Sigma} = \{\hat{p}(\hat{a}_i) : i = 0,\ldots,n_{\mathrm{loc}}\}$ and $F_K : \widehat{K} \to K$ be a bijective mapping. We set*

$$K = F_K(\widehat{K}),$$
$$\mathcal{P}_K = \{p : K \to \mathbb{R} : p = \hat{p} \circ F_K^{-1},\ \hat{p} \in \widehat{P}\},$$
$$\Sigma_K = \{p(F_K(\hat{a}_i)) : i = 0,\ldots,n_{\mathrm{loc}}\}.$$

Then, the set of dof Σ_K is $\mathcal{P}_K$ unisolvent.

Proof. We consider the bijections

$$\hat{x} \in \widehat{K} \mapsto x = F_K(\hat{x}) \in K, \qquad \hat{p} \in \widehat{P} \mapsto p = \hat{p} \circ F_K^{-1} \in \mathcal{P}_K.$$

Let $\hat{\varphi}_i$, $i = 0,\ldots,n_{\mathrm{loc}}$, be a basis in $\widehat{P}$. Then, for all $p \in \mathcal{P}_K$ and all $x \in K$ it holds

$$p(x) = \hat{p}(\hat{x}) = \sum_{i=0}^{n_{\mathrm{loc}}} \hat{p}(\hat{a}_i)\hat{\varphi}_i(\hat{x}) = \sum_{i=0}^{n_{\mathrm{loc}}} p(a_i)\varphi_i(x),$$

therefore

$$p = \sum_{i=0}^{n_{\mathrm{loc}}} p(a_i)\varphi_i \quad \text{for all } p \in \mathcal{P}_K.$$

The functions $\varphi_i := \hat{\varphi}_i \circ F_K^{-1}$, $i = 0,\ldots,n_{\text{loc}}$, are linearly independent since

$$\sum_{i=0}^{n_{\text{loc}}} \alpha_i \varphi_i = 0 \quad \Rightarrow \quad \sum_{i=0}^{n_{\text{loc}}} \alpha_i \hat{\varphi}_i = 0 \quad \Rightarrow \quad \alpha_i = 0,\ i = 0,\ldots,n_{\text{loc}},$$

and build a basis in $\mathcal{P}_K$. Consequently, the set of dof Σ_K is $\mathcal{P}_K$ unisolvent. ∎

As we already discussed in Section 3.5, the vertices of the reference square $(-1,+1) \times (-1,+1)$ can be mapped onto vertices a_i, $i = 0,\ldots,3$, of an arbitrary quadrilateral by the mapping $F_K : \widehat{K} \to \mathbb{R}^2$ given by

$$F_K(\hat{x}) = \sum_{i=0}^{3} a_i \varphi_i(\hat{x}), \tag{3.4}$$

where $\varphi_i \in Q_1(\widehat{K})$, $i = 0,\ldots,3$, denote the canonical basis functions on $\widehat{K}$. If the vertices form a parallelogram (a parallelepiped in the d-dimensional case) the mapping becomes affine and one-to-one. It can be shown that, in general, the mapping $F_K : \widehat{K} \to F_K(\widehat{K})$ is bijective if and only if the vertices a_i, $i = 0,\ldots,3$, form a convex quadrilateral, see for example, Matthies and Tobiska (2002). We show an example in which the mapping F_K defined in Equation (3.4) is not injective and $K = F_K(\widehat{K})$ is different from the quadrilateral formed by a_i, $i = 0,\ldots,3$.

Example 3.1. Let $a_0 = (1/2,1/2)^T$, $a_1 = (+1,-1)^T$, $a_2 = (+1,+1)^T$ and $a_3 = (-1,+1)^T$ be the vertices of the nonconvex quadrilateral Q. Consider the associated mapping $F_K : \widehat{K} \to \mathbb{R}^2$ given by Equation (3.4). The vertices a_i, $i = 0,\ldots,3$, are the images of the vertices $\hat{a}_i$, $i = 0,\ldots,3$. In particular, we have $(-1,-1) \mapsto (1/2,1/2)$, however, it holds also $(1/3,1/3) \mapsto (1/2,1/2)$, thus F_K is not injective on $\widehat{K} = (-1,+1) \times (-1,+1)$. Moreover, the inner point $(0,0) \in \widehat{K}$ is mapped to $F_K((0,0)^T) = (3/8,3/8)^T \notin Q$ which shows that $K = F_K(\widehat{K}) \neq Q$.

3.7. Mapped Q_k finite element

Let, as in Lemma 3.2, $\widehat{K} = (-1,+1)^d$, $\widehat{P} = Q_k(\widehat{K})$ and $\widehat{\Sigma}_K = \{v(b) : b \in \widehat{M}_k\}$. Assume that a_i, $i = 0,\ldots,2^d - 1$, are the images of the nodal set $\widehat{M}_1$ via a bijection $F_K : \widehat{K} \to K$ with $F_K \in Q_1(\widehat{K})^d$. Since the restriction of the mapping F_K onto a face in 2d is affine, the points a_i, $i = 0,\ldots,3$, generate a convex quadrilateral K. In 3d the edges of the hexahedra K are still straight lines, however, the faces are, in general, not plane. We refer to Bernardi (1989), Matthies and Tobiska (2002) for sufficient conditions guaranteeing the bijectivity of $F_K : \widehat{K} \to K$.

Definition 3.6. Suppose that $F_K \in Q_1(\widehat{K})^d$ is a bijection. The mapped Q_k finite element consists of

$$K = F_K(\widehat{K}), \quad \mathcal{P}_K = \{p = \hat{p} \circ F_K^{-1} : \hat{p} \in Q_k(\widehat{K})\}, \quad \Sigma_K = \{v(F_K(b)) : b \in \widehat{M}_k\}.$$

See Figure 3.5 for the case $d = 2$ and $k = 3$.

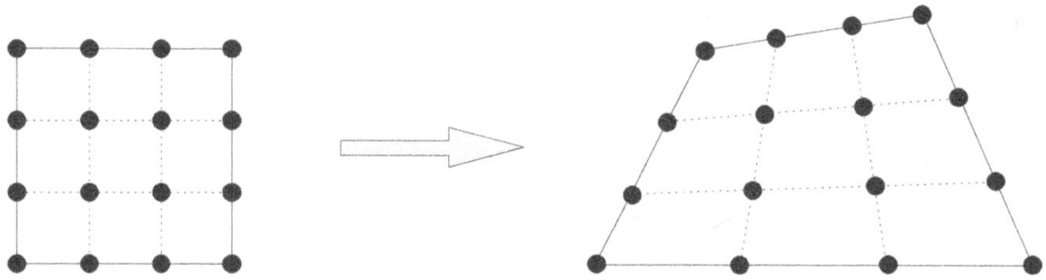

FIGURE 3.5 Mapped Q_3 finite element in two space dimensions.

Remark 3.1. Note that for mapped finite elements, the functions $p \in \mathcal{P}_K$ are not necessary polynomials even in the case if $\widehat{P}$ is a set of polynomials. Indeed the bijection $F_K : \widehat{K} \to K$ with

$$F_K(\hat{x}) = \begin{bmatrix} \frac{1}{4}(1+\hat{x}_1)(3-\hat{x}_2) \\ \frac{1}{2}(1+\hat{x}_2) \end{bmatrix}$$

maps the reference square $\widehat{K} = (-1,+1) \times (-1,+1)$ onto the trapezoid with vertices $a_0 = (0,0)^T$, $a_1 = (2,0)^T$, $a_2 = (1,1)^T$ and $a_3 = (0,1)^T$. The polynomial $\hat{p} = \hat{x}_1$ is mapped onto $p(x) = (\hat{p} \circ F_K^{-1})(x) = -1 + 2x_1/(2-x_2)$ which is neither a polynomial nor a polynomial in each variable, that is, $p \notin P_k(K)$ and $p \notin Q_k(K)$ for all $k \in \mathbb{N}_0$. Although for the mapped Q_k finite element not all functions in $\mathcal{P}_K$ are polynomials, we have at least $P_k(K) \subset \mathcal{P}_K$. This follows from

$$p \in P_k(K) \quad \Rightarrow \quad \hat{p} = p \circ F_K \in Q_k(\widehat{K}).$$

3.8. Isoparametric finite elements

The advantage of isoparametric elements consists in allowing triangles and tetrahedra with curved boundaries for a better domain approximation and using arbitrary convex quadrilaterals and general hexahedra.

Definition 3.7. A family of finite elements $(K, \mathcal{P}_K, \Sigma_K)$ is called an isoparametric family if for each $K \in \mathcal{T}_h$ an isoparametric mapping $F_K : \widehat{K} \to \mathbb{R}^d$ exists with $F_K \in \widehat{P}^d$ and F_K injective, such that

$$\begin{aligned} K &= F_K(\widehat{K}), \\ \mathcal{P}_K &= \{p : K \to \mathbb{R} : p = \hat{p} \circ F_K^{-1},\ \hat{p} \in \widehat{P}\}, \\ \Sigma_K &= \{p(F_K(\hat{a}_i)) : i = 0, \ldots, n_{\text{loc}}\}. \end{aligned}$$

The simplest example of an isoparametric family of elements is given by the Courant element where $\widehat{K}$ is a d-simplex, $\widehat{P} = P_1(\widehat{K})$ and $F_K \in P_1(\widehat{K})^d$. Another example is the mapped Q_1 element, where $\widehat{K} = (-1,+1)^d$ is the reference cube, $\widehat{P} = Q_1(\widehat{K})$ and $F_K \in Q_1(\widehat{K})^d$.

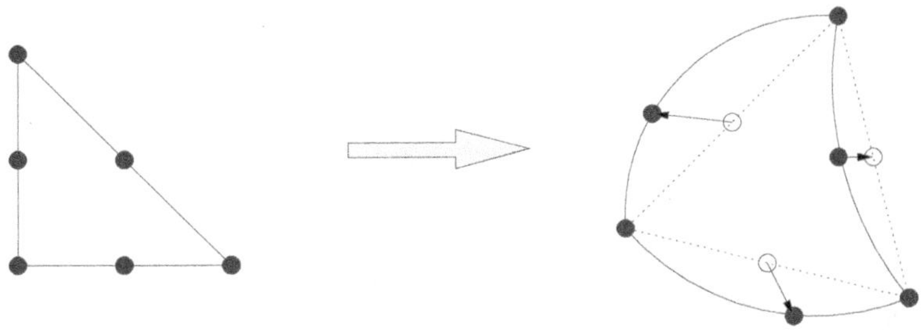

FIGURE 3.6 Curved isoparametric P_2 triangle.

The construction of these families is based on the space of linear and multilinear functions, respectively.

Finally, we consider the case of quadratic functions and for simplicity only in two dimensions. The isoparametric P_2 element is a mapped $P_2(\widehat{K})$ element for which $F_K \in P_2(\widehat{K})^2$. We follow Section 3.2 and use the canonical basis functions on the reference triangle $\hat{\lambda}_i(2\hat{\lambda}_i - 1)$, $i = 0, 1, 2$, associated with the vertices, and $4\hat{\lambda}_i\hat{\lambda}_j$, $0 \leq i < j \leq 2$, associated with the midpoints of edges. The mapping $F_K : \widehat{K} \to K$ with

$$F_K(\hat{x}) = \sum_{i=0}^{2} a_i\hat{\lambda}_i(\hat{x})(2\hat{\lambda}_i(\hat{x}) - 1) + \sum_{0\leq i<j\leq 2} a_{ij}4\hat{\lambda}_i(\hat{x})\hat{\lambda}_j(\hat{x})$$

maps the reference triangle onto a curved triangle K provided that the points a_{ij}, $0 \leq i < j \leq 2$ on the curved edges are sufficiently close to the midpoints of the straight triangle, see Figure 3.6. In case that $a_{ij} = (a_i + a_j)/2$, $0 \leq i < j \leq 2$, the mapping F_K becomes affine and $K = F_K(\widehat{K})$ the straight triangle. As shown in Ciarlet (2002), the requirement

$$\left| a_{ij} - \frac{a_i + a_j}{2} \right| = \mathcal{O}(h_K^2), \qquad 0 \leq i < j \leq 2$$

is sufficient for the bijectivity of F_K. For more details including the approximation order of isoparametric elements we refer to Bernardi (1989), Ciarlet (2002), Ern and Guermond (2004).

3.9. FURTHER EXAMPLES OF FINITE ELEMENT SPACES IN C^0 AND C^1

Here, we consider further examples of finite elements useful in applications.

Serendipity element

Let K be a rectangle with edges parallel to the coordinate axes. The vertices are denoted by a_i, the midpoint of edges by b_i, $i = 0, \ldots, 3$ and the centroid by $c = (a_0 + a_1 + a_2 + a_3)/4$.

The Serendipity element is a reduced Q_2 rectangular element, in which $p(c)$ is replaced by a weighted sum of the other dof

$$\mathcal{P}_K := \left\{ p \in Q_2(K) : 4p(c) + \sum_{i=0}^{3} p(a_i) - 2\sum_{i=0}^{3} p(b_i) = 0 \right\}.$$

It can be shown that $P_2(K) \subset \mathcal{P}_K \subset Q_2(K)$, for details see Ciarlet (2002). The set of dof is given by the nodal functionals

$$N_i(v) = v(a_i), \qquad N_{i+4}(v) = v(b_i), \quad i = 0,\ldots,3.$$

Lemma 3.5. *The set of dof Σ_K is $\mathcal{P}_K$ unisolvent. The generated finite element space V_h belongs to $C^0(\overline{\Omega}) \cap H^1(\Omega)$.*

Proof. Since $\dim \mathcal{P}_K = 8$ it is sufficient to show that $v \in \mathcal{P}_K$ with $N_i(v) = 0$, $i = 0,\ldots,7$ implies $v \equiv 0$. The restriction of v on an edge is a quadratic function in one variable vanishing at the two vertices and at the midpoint. Thus, v is equal to zero along this edge. This argument holds for any edge and v vanishes along the boundary ∂K. We get $v = C\varphi_8$, where C is a constant and φ_8 is the function in the canonical basis of $Q_2(K)$ satisfying $\varphi_8(c) = 1$. However, $v \in \mathcal{P}_K$, that is, $v(c)$ a linear combination of $N_i(v) = 0$, $i = 0,\ldots,7$, resulting in $v \equiv 0$. The continuity of the generated finite element space over inner edges follows from the unique quadratic Lagrange interpolation in 1d. ■

Remark 3.2. One can think of two possible extensions of the Serendipity element on a rectangle to a convex quadrilateral K with the vertices a_i, $i = 0,\ldots,3$, which differ in the definition of the local space $\mathcal{P}_K$. Let $F_K : \widehat{K} \to K$ be the bijective (in general, not affine) mapping from the reference cell $\widehat{K} = (-1,+1)^2$ onto K. Then, the vertices a_i, the midpoint of edges b_i, and the centroid, c are the images of the associated points $\hat{a}_i$, $\hat{b}_i$ and $\hat{c}$ of the reference element for $i = 0,\ldots,3$. For the unmapped version, a subset of polynomials from $Q_2(K)$ is used whereas in the mapped version the functions in $\mathcal{P}_K$ are not neccessarily polynomials (see Remark 3.1). More precisely, we have for the unmapped case

$$\mathcal{P}_K^{\text{unmap}} := \left\{ p \in Q_2(K) : 4p(c) + \sum_{i=0}^{3} p(a_i) - 2\sum_{i=0}^{3} p(b_i) = 0 \right\}$$

and for the mapped case

$$\mathcal{P}_K := \left\{ p = \hat{p} \circ F_K^{-1} : \hat{p} \in Q_2(\widehat{K}), 4\hat{p}(\hat{c}) + \sum_{i=0}^{3} \hat{p}(\hat{a}_i) - 2\sum_{i=0}^{3} \hat{p}(\hat{b}_i) = 0 \right\}.$$

Note that in the unmapped case, the local basis functions depend on the convex quadrilateral K and have to be recomputed for each cell K which can be quite expensive. In the mapped case, the local basis function on the reference cell are fixed and no recomputation is needed. However, the mapping $F_K : \widehat{K} \to K$ is not affine, which can lead to reduced orders of convergence for the mapped version on general meshes. It is shown in Matthies (2001) that the convergence order for the mapped version can be retained on sequences of uniformly refined meshes. For more details, see Arnold et al. (2001, 2002) and Matthies (2001).

Argyris triangle

Let $\Omega \subset \mathbb{R}^2$ be a polygon and be decomposed into triangles K. We denote the vertices of $K \in \mathcal{T}_h$ by a_i, $i = 0, 1, 2$, the midpoints of edges opposite to a_i by b_i, $i = 0, 1, 2$, as indicated in Figure 3.7. As the local approximation space we choose the space of polynomials of degree less than or equal to five, that is, $\mathcal{P}_K = P_5(K)$, where $\dim \mathcal{P}_K = 21$. We define the set of dof Σ_K by the 21 nodal functionals

$$N_i^\alpha(v) = D^\alpha v(a_i), \quad i = 0,1,2, \; |\alpha| \le 2 \qquad \text{(18 dof)},$$

$$N_i(v) = \frac{\partial v}{\partial n}(b_i), \quad i = 0,1,2 \qquad \text{(3 dof)}.$$

Lemma 3.6. *The set of dof Σ_K in the Argyris element is $\mathcal{P}_K$ unisolvent. The generated finite element space V_h belongs to $C^1(\overline{\Omega}) \cap H^2(\Omega)$.*

Proof. We have to show that $v \in \mathcal{P}_K$ with $N_i^\alpha(v) = 0$, $i = 0,1,2$, $|\alpha| \le 2$, $N_i(v) = 0$, $i = 0,1,2$, implies $v = 0$. The restriction of $v \in \mathcal{P}_K$ onto an edge $E \subset \partial K$ is a polynomial of degree less than or equal to five in one variable. The function and its tangential derivatives up to second order vanishes at the two endpoints of E. The associated 1d Hermite interpolation is unique and equal to zero. Thus, $v|_E = 0$. Repeating the argument for the other two edges we obtain $v = 0$ on ∂K. Now the restriction of $w := \partial_n v$ onto an edge $E \subset \partial K$ is a polynomial of degree less than or equal to four in one variable. The function $w \in P_4(E)$ and its tangential derivatives vanishes at the two endpoints, in addition w vanishes at the midpoint of E. Therefore, $w|_E = 0$. The argument holds again for any $E \subset \partial K$ such that $w = 0$ on ∂K. Now the polynomial v can be represented as $v = q(\lambda_0\lambda_1\lambda_2)^2$ with a polynomial q. Since $v \in P_5(K)$ we conclude that $q \equiv 0$.

For the $C^1(\overline{\Omega})$ continuity of V_h we consider a common edge $E = \partial K \cap \partial K'$ of K and its adjacent triangle K'. The two 1d Hermite interpolations discussed above produce on both sides of E the same polynomials of degree five and four, respectively. Thus, $V_h \subset C^1(\overline{\Omega})$ and Theorem 3.1 implies $C^1(\overline{\Omega}) \cap H^2(\Omega)$. ■

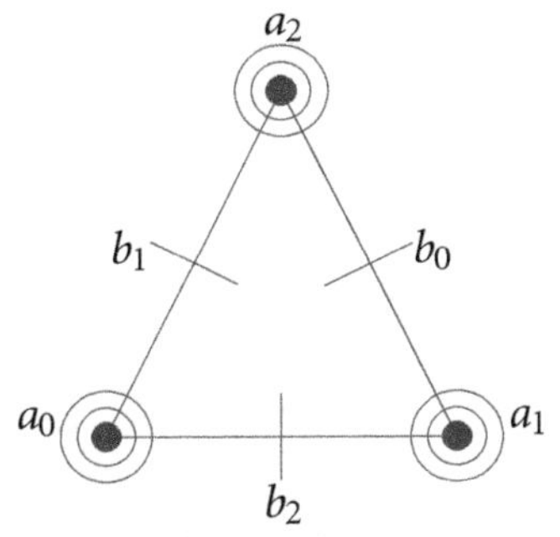

FIGURE 3.7 Degrees of freedom of the Argyris triangle. Derivatives up to second order at the vertices and normal derivatives at the midpoints of edges.

In order to construct a finite element space $V_{0h} \subset H_0^2(\Omega)$, we can use the result of Theorem 3.2. We define

$$V_{0h} := \left\{ v \in V_h : v = \frac{\partial v}{\partial n} = 0 \text{ on } \partial\Omega \right\}.$$

Let E be a boundary edge connecting the vertices a_i and a_j. We know that $v \in V_h$ restricted to the edge E belongs to $P_5(E)$. Then, setting the following dof to zero

$$v(a_i),\ Dv(a_i)(a_j - a_i),\ D^2v(a_i)(a_j - a_i, a_j - a_i),$$
$$v(a_j),\ Dv(a_j)(a_i - a_j),\ D^2v(a_j)(a_i - a_j, a_i - a_j),$$

guarantees that $v \in V_h$ equals zero on E. The normal derivative of $v \in V_h$ restricted to the edge E belongs to $P_4(E)$ and becomes zero along E if the dof

$$Dv(a_i)((a_j - a_i)^{\perp}),\ D^2v(a_i)((a_j - a_i)^{\perp}, a_j - a_i),\ \ Dv(b_k)((a_j - a_i)^{\perp})$$
$$Dv(a_j)((a_i - a_j)^{\perp}),\ D^2v(a_j)((a_i - a_j)^{\perp}, a_i - a_j)$$

are set to zero. Here, we used $b_k = (a_i + a_j)/2$ and the fact that $(a_j - a_i)$ is tangential and $(a_j - a_i)^{\perp}$ is normal to E. Note that, at the vertex a_i in which two boundary edges meet under an angle different from π all six dof are set to zero. In case, two boundary edges meet at a_i under an angle of π the associated outer normals are equal and $D^2v(a_i)((a_j - a_i)^{\perp}, (a_j - a_i)^{\perp})$ is one of the 'free' boundary dof.

Bell's triangle

Let $\Omega \subset \mathbb{R}^2$ be a polygon and be decomposed into triangles K. We denote the vertices of K by a_i, $i = 0, 1, 2$, as indicated in Figure 3.8. Bell's triangle is a reduced Argyris element in the sense that $\mathcal{P}_K^{\text{Bell}} \subset \mathcal{P}_K^{\text{Argyris}}$ and $\Sigma_K^{\text{Bell}} \subset \Sigma_K^{\text{Argyris}}$. We define the local approximation space as

$$\mathcal{P}_K := \{p \in P_5(K) : \partial_n p|_E \in P_3(E) \text{ for all edges } E \subset \partial K\}$$

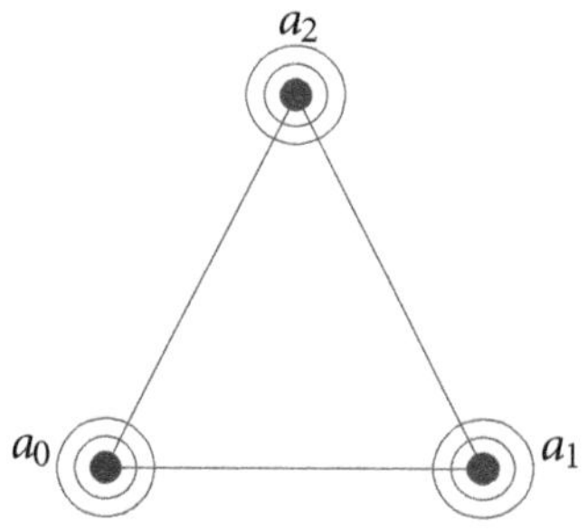

Figure 3.8 Degrees of freedom for the Bell triangle. Function values and derivatives up to second order at the vertices.

and the set of dof Σ_K are given by the 18 nodal functionals

$$N_i^\alpha(v) = D^\alpha v(a_i), \quad i = 0, 1, 2, \ |\alpha| \leq 2.$$

We have $P_4(K) \subset \mathcal{P}_K \subset P_5(K)$. Before proving that the set of dof Σ_K in the Bell element is $\mathcal{P}_K$ unisolvent, we give an equivalent characterization of the local approximation space $\mathcal{P}_K$.

Lemma 3.7. *Let E be an edge with endpoints a_i, a_j and midpoint b_k. Furthermore, we suppose $p \in P_4(E)$. Then, $p \in P_3(E)$ if and only if $N_E(p) = 0$ where*

$$N_E(p) := 4(p(a_i) + p(a_j)) - 8p(b_k) + Dp(a_i)(a_j - a_i) + Dp(a_j)(a_i - a_j).$$

Proof. Map E onto $(-1, +1)$ by $x = x(t) = (a_i + a_j)/2 + t(a_j - a_i)/2$ and show the equivalence for $\hat{p}(t) = p(x(t))$. For details, see Ciarlet (2002). ∎

Lemma 3.7 allows us to characterize the local approximation space of the Bell triangle as

$$\mathcal{P}_K = \{p \in P_5(K) : N_E(\partial_n p) = 0 \text{ for any } E \subset \partial K\}.$$

One can think that the Bell element is constructed from the Argyris element by elimination of the normal derivative dof using the three equations, $N_E(\partial_n p) = 0$.

Lemma 3.8. *The set of dof Σ_K in the Bell element is $\mathcal{P}_K$ unisolvent. The generated finite element space V_h belongs to $C^1(\overline{\Omega}) \cap H^2(\Omega)$.*

Proof. Similar arguments as in the proof of Lemma 3.7 are used. ∎

CHAPTER 4

INTERPOLATION AND DISCRETIZATION ERROR

Cea's lemma provides an estimate for the discretization error of the finite element method in terms of the best approximation of $u \in V$ in the finite element space V_h. Further, the error of best approximation can be bounded by the interpolation error, that is,

$$\|u - u_h\|_V \leq C \inf_{v_h \in V_h} \|u - v_h\|_V \leq C\|u - \Pi_h u\|_V$$

for any interpolation $\Pi_h u \in V_h$ depending on the finite element space V_h and the regularity of $u \in V$. This chapter studies the interpolation error in Sobolev spaces in different norms. We follow the concept of affine equivalent finite elements that allows to derive the desirable estimates on a reference cell first and then to transform it to a general cell by affine mappings. Using duality arguments we show in addition for a second order elliptic model problem that the discretization error in the L^2 norm behaves one order better than in the H^1 norm.

4.1. Transformation formulas

Let K be a cell in an admissible decomposition of the domain Ω, see Definition 3.5, and $\hat{K}$ be its reference cell. Consider a bijective affine mapping $F_K : \widehat{K} \to K$ with $F_K(\hat{x}) = B_K\hat{x} + b_K$, where $B_K \in \mathbb{R}^{d\times d}$ and $b_K \in \mathbb{R}^d$. We define the correspondence between functions $\hat{v}$ and v defined over $\widehat{K}$ and K, respectively, by

$$\hat{v} : \widehat{K} \to \mathbb{R} \quad \Longrightarrow \quad v = \hat{v} \circ F_K^{-1} : K \to \mathbb{R},$$
$$v : K \to \mathbb{R} \quad \Longrightarrow \quad \hat{v} = v \circ F_K : \widehat{K} \to \mathbb{R}.$$

It turns out that the seminorms of v and $\hat{v}$ can be bounded by the seminorms of $\hat{v}$ and v, respectively, in the associated Sobolev spaces $W^{m,p}$.

Theorem 4.1. *Let $v \in W^{m,p}(K)$ for some integer $m \geq 0$ and $1 \leq p \leq \infty$. Then, $\hat{v} \in W^{m,p}(\widehat{K})$ and it holds*

$$|\hat{v}|_{m,p,\widehat{K}} \leq C\|B_K\|^m |\det B_K|^{-1/p} |v|_{m,p,K}.$$

Contrary, if $\hat{v} \in W^{m,p}(\widehat{K})$ then $v \in W^{m,p}(K)$ and it holds

$$|v|_{m,p,K} \leq C\|B_K^{-1}\|^m |\det B_K|^{1/p} |\hat{v}|_{m,p,\widehat{K}}.$$

Here, $\|\cdot\|$ denotes the matrix norm induced by the Euclidian vector norm.

Proof. Consider the case $1 \leq p < \infty$. Modifications for $p = \infty$ are straightforward. Let $\hat{v}$ and v be smooth enough. We recall that the m-th derivative of a function $\hat{v} : \widehat{K} \subset \mathbb{R}^d \to \mathbb{R}$ is a multilinear mapping $D^m\hat{v}(\hat{x}) \in \mathcal{L}(\underbrace{\mathbb{R}^d, \ldots, \mathbb{R}^d}_{m\text{-times}}; \mathbb{R})$. For any multiindex α with $|\alpha| = m$ and with some basis vectors e_{α_i}, $i = 1, \ldots, m$, of $\mathbb{R}^d$ it holds

$$D^\alpha \hat{v}(\hat{x}) = D^m\hat{v}(\hat{x})(e_{\alpha_1}, \ldots, e_{\alpha_m}).$$

Further, we get the bound

$$|D^\alpha \hat{v}(\hat{x})| \leq \|D^m\hat{v}(\hat{x})\| = \sup_{\|\xi_i\| \leq 1, 1 \leq i \leq m} |D^m\hat{v}(\hat{x})(\xi_1, \ldots, \xi_m)|.$$

Applying the chain rule on $\hat{v}(\hat{x}) = v(F_K(\hat{x}))$ we get that for any vectors $\xi_i \in \mathbb{R}^d$, $1 \leq i \leq m$,

$$D^m\hat{v}(\hat{x})(\xi_1, \ldots, \xi_m) = D^m v(x)(B_K\xi_1, \ldots, B_K\xi_m)$$

holds true from which

$$\|D^m\hat{v}(\hat{x})\| \leq \|D^m v(x)\| \, \|B_K\|^m$$

follows. Summarizing and using the formula for transformation of multiple integrals, we obtain for $1 \leq p < \infty$ with constants C_1, C_2, C

$$\begin{aligned}
|\hat{v}|^p_{m,p,\widehat{K}} &= \int_{\widehat{K}} \sum_{|\alpha|=m} |D^\alpha \hat{v}(\hat{x})|^p \, d\hat{x} \leq C_1 \int_{\widehat{K}} \|D^m\hat{v}(\hat{x})\|^p \, d\hat{x} \\
&\leq C_1 \int_{\widehat{K}} \|D^m v(x)\|^p \, \|B_K\|^{mp} \, d\hat{x} = C_1 |\det B_K|^{-1} \, \|B_K\|^{mp} \int_K \|D^m v(x)\|^p \, dx \\
&\leq C_1 C_2 |\det B_K|^{-1} \, \|B_K\|^{mp} \int_K \sum_{|\alpha|=m} |D^\alpha v(x)|^p \, dx \\
&\leq C |\det B_K|^{-1} \, \|B_K\|^{mp} |v|^p_{m,p,K}.
\end{aligned}$$

Taking the inequality to the power of $1/p$ and using density arguments to extend the validity to functions from Sobolev spaces, the first statement of the theorem is proved. The second statement follows just by changing the notation $K \to \widehat{K}$, $\widehat{K} \to K$, $B_K \to B_K^{-1}$, $|\det B_K^{-1}| = |\det B_K|^{-1}$. ■

Our next step is to estimate the quantities $\det B_K$, $\|B_K\|$ and $\|B_K^{-1}\|$ in terms of geometric quantities of K. We set $h_K = \operatorname{diam}(K)$, $\hat{h} = \operatorname{diam}(\widehat{K})$,

$$\rho_K = \sup\{\operatorname{diam}(S) : S \text{ ball contained in } K\},$$
$$\hat{\rho} = \sup\{\operatorname{diam}(S) : S \text{ ball contained in } \widehat{K}\}.$$

Lemma 4.1. *Let $F_K : \widehat{K} \to K$ with $F_K(\hat{x}) = B_K\hat{x} + b_K$ be a bijective affine mapping. Then, the following geometric relations hold true*

$$\det B_K = \frac{|K|}{|\widehat{K}|}, \qquad \|B_K\| \le \frac{h_K}{\hat{\rho}}, \qquad \|B_K^{-1}\| \le \frac{\hat{h}}{\rho_K}.$$

Proof. We use the formula for the transformation of multiple integrals and obtain

$$|K| = \int_K 1 \,\mathrm{d}x = \int_{\widehat{K}} 1 \det B_K \,\mathrm{d}\hat{x} = |\widehat{K}| \det B_K.$$

Next we observe that

$$\|B_K\| := \sup_{0 \neq \xi} \frac{\|B_K\xi\|}{\|\xi\|} = \frac{1}{\hat{\rho}} \sup_{\|\xi\| = \hat{\rho}} \|B_K\xi\|.$$

Since $\hat{\rho}$ is the diameter of the largest ball in $\widehat{K}$, we find for any given vector $\xi \in \mathbb{R}^d$ with $\|\xi\| = \hat{\rho}$ two points $\hat{\xi}_1, \hat{\xi}_2 \in \widehat{K}$ such that $\xi = \hat{\xi}_1 - \hat{\xi}_2$. Now

$$\|B_K\xi\| = \|B_K(\hat{\xi}_1 - \hat{\xi}_2)\| = \|B_K\hat{\xi}_1 - B_K\hat{\xi}_2\| \le h_K.$$

The third statement follows by changing the role of $\widehat{K}$ and K. ■

4.2. Affine equivalent finite elements

Let us assume that we have a decomposition of $\Omega \subset \mathbb{R}^d$ into simplices $K \in \mathcal{T}_h$ such that $\overline{\Omega} = \bigcup_{K \in \mathcal{T}_h} \overline{K}$.

Definition 4.1. A family of finite elements $(K, \mathcal{P}_K, \Sigma_K)_{K \in \mathcal{T}_h}$ is called affine equivalent if a fixed reference element $(\widehat{K}, \widehat{P}, \widehat{\Sigma})$ exists such that for any $K \in \mathcal{T}_h$ there is a bijective affine mapping $F_K : \widehat{K} \to K$ with $F_K(\hat{x}) = B_K\hat{x} + b_K$ satisfying the following properties

$$K = F_K(\widehat{K}),$$
$$\mathcal{P}_K = \left\{ p : K \to \mathbb{R} : p = \hat{p} \circ F_K^{-1}, \hat{p} \in \widehat{P} \right\},$$
$$\Sigma_K = \left\{ N(p) : N(p) = N(\hat{p} \circ F_K^{-1}) = \widehat{N}(\hat{p}),\ \widehat{N}(\hat{p}) \in \widehat{\Sigma} \right\}.$$

Example 4.1. Consider the family of finite elements $(K, \mathcal{P}_K, \Sigma_K)_{K \in \mathcal{T}_h}$ where $K \subset \mathbb{R}^2$ is a triangle with vertices a_0, a_1, a_2, $\mathcal{P}_K = \operatorname{span}(\lambda_0, \lambda_1, \lambda_2, \lambda_0\lambda_1\lambda_2)$, and the dof Σ_K given by

$$N_i(v) = v(a_i), \quad i = 0,1,2, \qquad N_3(v) = \frac{1}{|K|} \int_K v(x) \,\mathrm{d}x.$$

Here, λ_i, $i = 0,1,2$, denote the barycentric coordinates of K. The affine mapping from the reference cell $\widehat{K}$ with the vertices $\hat{a}_0$, $\hat{a}_1$, $\hat{a}_2$ is given by

$$F_K = \sum_{i=0}^{2} a_i \hat{\lambda}_i, \qquad \hat{\lambda}_i(\hat{a}_j) = \delta_{ij}, \quad i,j = 0,1,2,$$

where $\hat{\lambda}_i$, $i = 0,1,2$, are the barycentric coordinates of the reference cell. Since $\lambda_i = \hat{\lambda}_i \circ F_K^{-1}$, the property

$$\mathcal{P}_K = \left\{p : K \to \mathbb{R} : p = \hat{p} \circ F_K^{-1}, \hat{p} \in \widehat{\mathcal{P}}\right\}$$

of an affine equivalent family of finite elements is satisfied. Further, the dof are invariant with respect to affine mappings. We have

$$N_i(v) = v(a_i) = v(F_K(\hat{a}_i)) = \hat{v}(\hat{a}_i) = \widehat{N}_i(\hat{v}), \quad i = 0,1,2$$

and

$$N_3(v) = \frac{1}{|K|}\int_K v(x)\,\mathrm{d}x = \frac{1}{|K|}\int_{\widehat{K}} v(F_K(\hat{x}))\det B_K\,\mathrm{d}\hat{x} = \frac{1}{|\widehat{K}|}\int_{\widehat{K}} \hat{v}(\hat{x})\,\mathrm{d}\hat{x} = \widehat{N}_3(\hat{v}).$$

As a consequence, the described family of finite elements is affine equivalent.

Example 4.2. Consider the family of Morley triangles which will be studied in detail in Section 5.4. We have $K \subset \mathbb{R}^2$ as a simplex with vertices a_0, a_1, a_2, midpionts of edges b_0, b_1, b_2, $\mathcal{P}_K = P_2(K)$, and the dof given by

$$N_i(v) = v(a_i), \qquad N_{i+3}(v) = \frac{\partial v}{\partial n}(b_i), \quad i = 0,1,2.$$

The first three dof are invariant with respect to affine mappings

$$N_i(v) = v(a_i) = v(F_K(\hat{a}_i)) = \hat{v}(\hat{a}_i) = \widehat{N}_i(\hat{v}), \quad i = 0,1,2.$$

However, $N_{i+3}(v)$, $i = 0,1,2$, do not satisfy the property of an affine family since for $\lambda_i = \hat{\lambda}_i \circ F_K^{-1}$

$$N_{i+3}(\lambda_i) = \frac{\partial \lambda_i}{\partial n}(b_i) = -\frac{1}{d_i}, \qquad \widehat{N}_{i+3}(\hat{\lambda}_i) = \frac{\partial \hat{\lambda}_i}{\partial \hat{n}}(\hat{b}_i) = -\frac{1}{\hat{d}_i},$$

where d_i ($\hat{d}_i$) denotes the distance of a_i ($\hat{a}_i$) from the straight line through a_{i+1} ($\hat{a}_{i+1}$) and a_{i+2} ($\hat{a}_{i+2}$) counted modulo 3. In general, $d_i \neq \hat{d}_i$.

4.3. Canonical interpolation

Let $(K, \mathcal{P}_K, \Sigma_K)$ be a finite element with $\Sigma_K = \{N_i : i = 0,\ldots,n_{\text{loc}}\}$ and $\dim \mathcal{P}_K = n_{\text{loc}} + 1$. Since the set of dof is $\mathcal{P}_K$ unisolvent there are local basis functions $\varphi_j \in \mathcal{P}_K$ with $N_i(\varphi_j) = \delta_{ij}$, for $i,j = 0,\dots,n_{\text{loc}}$. Then, the canonical interpolation $\Pi_K v \in \mathcal{P}_K$ on K is uniquely defined by the requirements

$$N_i(\Pi_K v) = N_i(v), \quad i = 0,\dots,n_{\text{loc}}$$

and can be represented as

$$\Pi_K v = \sum_{i=0}^{n_{\text{loc}}} N_i(v)\,\varphi_i.$$

The domain of definition of the interpolation operator Π_K is the intersection of all domains of definition of the nodal functionals. Using the affine mapping $F_K : \widehat{K} \to K$ from the reference cell $\widehat{K}$ to the cell K we define an interpolation operator $\widehat{\Pi}_K$ on the reference domain by

$$\hat{v} \mapsto \widehat{\Pi}_K \hat{v} := \left(\Pi_K\left(\hat{v} \circ F_K^{-1}\right)\right) \circ F_K = \widehat{\Pi_K v}. \tag{4.1}$$

Note that, the operator $\widehat{\Pi}_K$ depends by construction on K as we can see from the diagram and as we indicate by a subscript.

$$\begin{array}{ccc} \hat{v} & \longrightarrow & \widehat{\Pi}_K \hat{v} \\ \downarrow F_K^{-1} & & \uparrow F_K \\ v & \longrightarrow & \Pi_K v \end{array}$$

Next, we show that for a family of affine equivalent finite elements this interpolation operator coincides with the canonical interpolation on the reference cell.

Lemma 4.2. *Let $(K, \mathcal{P}_K, \Sigma_K)_{K\in\mathcal{T}_h}$ be an affine equivalent family of finite elements. Then, $\widehat{\Pi}_K$ defined in (4.1) is equal to the canonical interpolation operator $\widehat{\Pi} := \Pi_{\widehat{K}}$ on the reference cell. As a consequence, the local basis functions $\hat{\varphi}_j = \varphi_j \circ F_K$ on the reference cell do not depend on K.*

Proof. For an affine equivalent family of finite elements it holds $N_i(v) = \widehat{N}_i(\hat{v})$ and in particular

$$N_i(\varphi_j) = \widehat{N}_i(\hat{\varphi}_j) = \delta_{ij}, \quad i,j = 0,\dots,n_{\text{loc}}.$$

Using the representation for $\Pi_K v$ we obtain

$$\widehat{\Pi}_K \hat{v} = \widehat{\Pi_K v} = \sum_{i=0}^{n_{\text{loc}}} N_i(v)\,(\varphi_i \circ F_K) = \sum_{i=0}^{n_{\text{loc}}} \widehat{N}_i(\hat{v})\,\hat{\varphi}_i.$$

The interpolation operator $\widehat{\Pi}_K$ satisfies

$$\widehat{N}_j(\widehat{\Pi}_K \hat{v}) = \sum_{i=0}^{n_{\text{loc}}} \widehat{N}_i(\hat{v})\widehat{N}_j(\hat{\varphi}_i) = \widehat{N}_j(\hat{v}), \quad j = 0,\dots,n_{\text{loc}},$$

that is, the same equations as the canonical interpolation $\widehat{\Pi}$ on the reference cell. Uniqueness of the canonical interpolation implies $\widehat{\Pi}_K = \widehat{\Pi}$. As solution of the linear algebraic equations

$$\widehat{N}_i(\hat{\varphi}_j) = \delta_{ij}, \quad i,j = 0,\dots,n_{\text{loc}},$$

the local basis functions $\hat{\varphi}_j$, $j = 0,\dots,n_{\text{loc}}$, do not carry information on K. ∎

4.4. Local and global interpolation error

In this section we will derive error estimates for the canonical interpolation of a family of affine equivalent finite elements. First, we recall a result from functional analysis.

Theorem 4.2 (Bramble–Hilbert lemma). *Let $\Omega \subset \mathbb{R}^d$ be open with Lipschitz continuous boundary. For some integer $k \geq 0$ and some number $1 \leq p \leq \infty$ let L be a continuous linear operator from $W^{k+1,p}(\Omega)$ into a normed space Y with*

$$L(p) = 0 \quad \text{for all } p \in P_k(\Omega).$$

Then, there is a constant $C(\Omega)$ such that

$$\|Lv\|_Y \leq C(\Omega)\|L\|\,|v|_{k+1,p,\Omega}, \tag{4.2}$$

where $\|L\| = \sup\{\|Lv\|_Y : \|v\|_{k+1,p,\Omega} = 1,\, v \in W^{k+1,p}(\Omega)\}$.

Proof. See Braess (2007) or Theorem 4.1.3 of Ciarlet (2002). ■

Remark 4.1. Since $|p|_{k+1,p,\Omega} = 0$ for any polynomial $p \in P_k(\Omega)$ we see that $L(p) = 0$ is necessary for (4.2). The Bramble–Hilbert lemma even states that $L(p) = 0$ is sufficient for (4.2).

Theorem 4.3. *Let $(\widehat{K}, \widehat{P}, \widehat{\Sigma})$ be the reference element of an affine equivalent family of finite elements. Assume that for some integers $m \geq 0$ and $k \geq 0$ and for some numbers $1 \leq p, q \leq \infty$ the following hypotheses are fulfilled:*

- *the nodal functionals $\widehat{N}_i : W^{k+1,p}(\widehat{K}) \to \mathbb{R}$, $i = 0, \ldots, n_{\text{loc}}$, are linear continuous,*
- *$W^{k+1,p}(\widehat{K})$ is continuously embedded into $W^{m,q}(\widehat{K})$,*
- *$P_k(\widehat{K}) \subset \widehat{P} \subset W^{m,q}(\widehat{K})$.*

There is a positive constant $C = C(\widehat{K}, \widehat{P}, \widehat{\Sigma})$ such that for the family of affine equivalent finite elements $(K, \mathcal{P}_K, \Sigma_K)$ and all $v \in W^{k+1,p}(K)$

$$|v - \Pi_K v|_{m,q,K} \leq C|K|^{1/q-1/p}\frac{h_K^{k+1}}{\rho_K^m}|v|_{k+1,p,K}, \tag{4.3}$$

where $\Pi_K v$ is the canonical $\mathcal{P}_K$ interpolation of v.

Proof. Since the nodal functionals N_i, $i = 0, \ldots, n_{\text{loc}}$ are continuous linear forms on $v \in W^{k+1,p}(K)$ and the local basis functions φ_i, $i = 0, \ldots, n_{\text{loc}}$, belong to $W^{m,q}(K)$ the canonical $\mathcal{P}_K$ interpolation $\Pi_K : W^{k+1,p}(K) \to W^{m,q}(K)$

$$\Pi_K v = \sum_{i=0}^{n_{\text{loc}}} N_i(v)\,\varphi_i$$

is welldefined. We use the transformation from K to the reference cell $\widehat{K}$ and the property $\widehat{\Pi_K v} = \widehat{\Pi}\hat{v}$ (see (4.1) and Lemma 4.2) to get

$$|v - \Pi_K v|_{m,q,K} \le C|\det B_K|^{1/q} \|B_K^{-1}\|^m |\hat{v} - \widehat{\Pi_K v}|_{m,q,\widehat{K}}$$
$$\le C|\det B_K|^{1/q} \|B_K^{-1}\|^m |\hat{v} - \widehat{\Pi}\hat{v}|_{m,q,\widehat{K}}.$$

Next, we study the linear mapping $L : W^{k+1,p}(\widehat{K}) \to Y := W^{m,q}(\widehat{K})$ given by $L(\hat{v}) = \hat{v} - \widehat{\Pi}\hat{v}$. The mapping L is continuous due to the continuous embedding $W^{k+1,p}(\widehat{K}) \hookrightarrow W^{m,q}(\widehat{K})$, the continuity of the nodal functionals and the independency of the local basis functions $\hat{\varphi}_i$ from K. Moreover,

$$\|L(\hat{v})\|_{m,q,\widehat{K}} \le \|\hat{v}\|_{m,q,\widehat{K}} + \left\| \sum_{i=0}^{n_{\mathrm{loc}}} \widehat{N}_i(\hat{v})\hat{\varphi}_i \right\|_{m,q,\widehat{K}}$$
$$\le C\|\hat{v}\|_{k+1,p,\widehat{K}} + \sum_{i=0}^{n_{\mathrm{loc}}} |\widehat{N}_i(\hat{v})| \; \|\hat{\varphi}_i\|_{m,q,\widehat{K}}$$
$$\le C\|\hat{v}\|_{k+1,p,\widehat{K}}.$$

As a result of $P_k(\widehat{K}) \subset \widehat{P}$, the interpolation preserves polynomials of degree k, that is, $\widehat{\Pi}\hat{p} = \hat{p}$ for all $\hat{p} \in P_k(\widehat{K})$, and therefore $L(\hat{p}) = 0$ for all $\hat{p} \in P_k(\widehat{K})$. Applying Theorem 4.2 (Bramble–Hilbert lemma) we obtain

$$|\hat{v} - \widehat{\Pi}\hat{v}|_{m,q,\widehat{K}} \le \|L(\hat{v})\|_{m,q,\widehat{K}} \le C|\hat{v}|_{k+1,p,\widehat{K}}.$$

It remains to transform back to the original cell K

$$|v - \Pi_K v|_{m,q,K} \le C|\det B_K|^{1/q} \|B_K^{-1}\|^m |\hat{v}|_{k+1,p,\widehat{K}},$$
$$\le C|\det B_K|^{1/q-1/p} \|B_K^{-1}\|^m \|B_K\|^{k+1} |v|_{k+1,p,K},$$
$$|v - \Pi_K v|_{m,q,K} \le C|K|^{1/q-1/p} \frac{h_K^{k+1}}{\rho_K^m} |v|_{k+1,p,K},$$

where in the last step we used the estimates of Lemma 4.1. ■

The local $\mathcal{P}_K$ interpolation defined for functions on $K \in \mathcal{T}_h$ can be extended to a global interpolation for functions defined on Ω in the finite element space V_h. The global interpolation $\Pi_h : W^{k+1,p}(\Omega) \to V_h$ is then given as $(\Pi_h v)|_K = \Pi_K v$ for $K \in \mathcal{T}_h$.

Definition 4.2. A family of finite elements $\{\mathcal{T}_h\}$ is called shape regular if there is a positive constant C such that

$$\frac{h_K}{\rho_K} \le C \quad \text{for all } K \in \mathcal{T}_h, \text{ for all } \mathcal{T}_h \in \{\mathcal{T}_h\}.$$

The family of finite elements $\{\mathcal{T}_h\}$ is called quasiuniform if it is shape-regular and there is a positive constant C such that

$$Ch \le h_K \quad \text{for all } K \in \mathcal{T}_h,$$

where $h = \max_{K \in \mathcal{T}_h} h_K$.

Remark 4.2. Quasiuniform families of finite elements do not allow adaptive refinements as shown in Figure 4.1.

Remark 4.3. For shape-regular families there are positive constants C_1, C_2 such that

$$C_1 \rho_K^d \leq |K| \leq C_2 h_K^d.$$

Theorem 4.4. *Let $(\widehat{K}, \widehat{P}, \widehat{\Sigma})$ be the reference element of a shape-regular, affine equivalent family of finite elements with $\overline{\Omega} = \bigcup_{K \in \mathcal{T}_h} \overline{K}$. Assume that for some integers m and k with $0 \leq m \leq k+1$ the following two hypotheses are fulfilled:*

- *nodal functionals $\widehat{N}_i : H^{k+1}(\widehat{K}) \to \mathbb{R}$, $i = 0, \ldots, n_{\mathrm{loc}}$, are linear continuous,*
- *$P_k(\widehat{K}) \subset \widehat{P} \subset H^m(\widehat{K})$.*

Then, there is a positive constant C such that for all $v \in H^{k+1}(\Omega)$

$$\|v - \Pi_h v\|_{m,\Omega} \leq C h^{k+1-m} |v|_{k+1,\Omega}, \quad m = 0, 1, \tag{4.4}$$

$$\left(\sum_{K \in \mathcal{T}_h} \|v - \Pi_h v\|_{m,K}^2 \right)^{1/2} \leq C h^{k+1-m} |v|_{k+1,\Omega}, \quad m = 2, \ldots, k+1, \tag{4.5}$$

where $\Pi_h v$ is the global canonical interpolation in $V_h \subset C(\overline{\Omega})$.

Proof. For $p = q = 2$ and $k+1 \geq m$, the space $W^{k+1,2}(\widehat{K}) = H^{k+1}(\widehat{K})$ is continuously embedded into $W^{m,2}(\widehat{K}) = H^m(\widehat{K})$. We apply Theorem 4.3. The shape-regularity allows to estimate ρ_K from below in terms of h_K. Further, we use $h_K \leq h$ to obtain

$$|v - \Pi_K v|_{m,K} \leq C h^{k+1-m} |v|_{k+1,K}$$

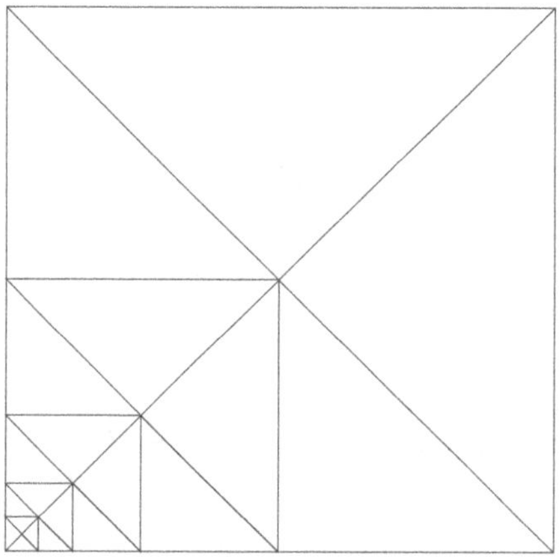

FIGURE 4.1 Illustration of an adaptive shape regular family of finite elements being not quasiuniform.

from which

$$\|v - \Pi_K v\|_{m,K} = \left(\sum_{j=0}^{m} |v - \Pi_K v|_{j,K}^2 \right)^{1/2} \leq C \left(\sum_{j=0}^{m} h^{2(k+1-j)} \right)^{1/2} |v|_{k+1,K}$$
$$\leq C h^{k+1-m} |v|_{k+1,K}$$

follows. Taking into consideration that $\Pi_h v \in H^1(\Omega)$ we obtain for $m = 0, 1$, the global interpolation estimate (4.4)

$$\|v - \Pi_h v\|_{m,\Omega} = \left(\sum_{K \in \mathcal{T}_h} |v - \Pi_K v|_{m,K}^2 \right)^{1/2} \leq C h^{k+1-m} |v|_{k+1,\Omega}$$

and in a similar way (4.5). ■

Example 4.3. We consider the solution $u \in H_0^1(\Omega)$ of the weak formulation of the Poisson problem

$$-\Delta u = f \quad \text{in } \Omega, \qquad u = 0 \quad \text{on } \Gamma = \partial\Omega$$

for data $f \in L^2(\Omega)$. Let $u_h \in V_h$ be the solution of the discrete problem using continuous, piecewise polynomials of degree $k \geq 1$. If $u \in H_0^1(\Omega) \cap H^{k+1}(\Omega)$ we have the unique solvability of the continuous and discrete problems (Lax–Milgram theorem) and the error bound (Cea's lemma/interpolation error)

$$\|u - u_h\|_1 \leq C \inf_{v_h \in V_h} \|u - v_h\|_1 \leq C \|u - \Pi_h u\|_1 \leq C h^k |u|_{k+1}.$$

Note that for the convergence order k, the regularity $u \in H^{k+1}(\Omega)$ of the weak solution has been used. For convex polygons Ω and data $f \in L^2(\Omega)$ it can be shown that the weak solution belongs to $H^2(\Omega)$, that is, piecewise linear finite elements converge of first order. This regularity result, $u \in H^2(\Omega)$, is not true for an L-shaped domain or in case of changing boundary conditions (mixed Dirichlet–Neumann type) indicated in Figure 4.2. Common approaches to handle singularities are techniques to refine *a priori* or *a posteriori* the underlying mesh. For details, we refer to Shaĭdurov (1995); Ainsworth and Oden (2000) and Verfürth (2013).

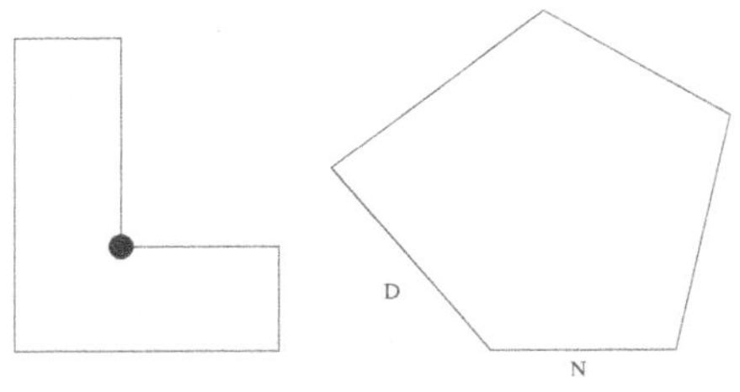

FIGURE 4.2 Cases for which $u \notin H^2(\Omega)$. Nonconvex domain (left) and mixed Dirichlet–Neumann type of boundary conditions (right).

4.5. Improved L^2 error estimates by duality

For the Poisson problem with a weak solution in $H_0^1(\Omega) \cap H^{k+1}(\Omega)$ we have seen in Example 4.3 that the convergence order of continuous, piecewise polynomials of degree k in the H^1 norm is k. Because of the continuous embedding $H^1(\Omega) \hookrightarrow L^2(\Omega)$ the convergence order with respect to the L^2 norm is at least k. Comparing this with the interpolation error estimate (4.4) we expect the order $k+1$ with respect to the L^2 norm. One way to show that the L^2 error of the finite element method converges one order better than the H^1 error would be to prove first the quasioptimality of the Galerkin method with respect to the L^2 norm. Then, in a second step, we could bound the approximation error by the interpolation error and gain the desired additional power of h. It turns out, however, that quasioptimality of the Galerkin method for the Poisson problem with respect to the L^2 norm does not hold. Thus we need another concept.

In order not to overload the representation we explain the new ideas for the simplest case of a second order elliptic problem. The continuous and discrete problem of the Poisson problem read as

$$\text{Find } u \in V = H_0^1(\Omega) \text{ such that } (\nabla u, \nabla v) = (f, v) \text{ for all } v \in V, \tag{4.6}$$

$$\text{Find } u_h \in V_h \subset V \text{ such that } (\nabla u_h, \nabla v_h) = (f, v_h) \text{ for all } v_h \in V_h. \tag{4.7}$$

The unique solvability of Problems (4.6) and (4.7) follow from Theorem 2.1. We conclude the Galerkin orthogonality from setting $v = v_h = z_h \in V_h$

$$(\nabla(u - u_h), \nabla z_h) = 0 \qquad \text{for all } z_h \in V_h.$$

In addition, let us consider the family of adjoint problems for given $g \in L^2(\Omega)$

$$\text{Find } z_g \in V \text{ such that } (\nabla v, \nabla z_g) = (g, v) \text{ for all } v \in V. \tag{4.8}$$

Definition 4.3. We call the family of adjoint problems as H^2 regular if for any given $g \in L^2(\Omega)$ there is a unique solution $z_g \in H_0^1(\Omega) \cap H^2(\Omega)$ satisfying

$$|z_g|_2 \leq C\|g\|_0.$$

Theorem 4.5 (Aubin–Nitsche). *Let the solution u of Problem (4.6) belong to $H_0^1(\Omega) \cap H^{k+1}(\Omega)$, let the finite element space $V_h \subset H_0^1(\Omega)$ consists of piecewise polynomials of degree k and let the family of adjoint problems (4.8) be H^2 regular. Then, the L^2 error estimate holds:*

$$\|u - u_h\|_0 \leq Ch^{k+1}|u|_{k+1}.$$

Proof. In the adjoint problem (4.8) we set the test function $v = u - u_h \in V$ and use the Galerkin orthogonality for an arbitrary test function $z_h \in V_h$

$$(g, u - u_h) = (\nabla(u - u_h), \nabla z_g) = (\nabla(u - u_h), \nabla(z_g - z_h)).$$

Let $z_h = \Pi_h^1 z_g \in V_h$ be the piecewise linear interpolation of $z_g \in H^2(\Omega)$. Then, we have the estimate

$$|(g, u - u_h)| \leq |u - u_h|_1 \, |z_g - \Pi_h^1 z_g|_1 \leq |u - u_h|_1 \, h|z_g|_2 \leq Ch|u - u_h|_1 \|g\|_0$$

from which

$$\|u-u_h\|_0 = \sup_{g\in L^2(\Omega)} \frac{|(g,u-u_h)|}{\|g\|_0} \leq Ch|u-u_h|_1 \leq Ch^{k+1}|u|_{k+1}$$

follows. ■

Remark 4.4. Note that we used the H^2 regularity of the adjoint problem which holds true, for example, for the homogeneous Dirichlet problem of a second order elliptic equation with smooth coefficients in convex polyhedrons. More detailed information about the regularity can be found in Grisvard (1992).

4.6. Interpolation of less smooth functions

The standard nodal interpolation is based on the evaluation of function values, thus it is welldefined for smooth enough functions. Theorem 1.5 (Sobolev embedding) tells us that $W^{m,p}(\Omega) \hookrightarrow C(\overline{\Omega})$ for $m > d/p$. As a result, the nodal interpolation is defined for functions from $W^{m,p}(\Omega)$ provided $m > d/p$. In the one-dimensional case, functions in $H^1(\Omega) = W^{1,2}(\Omega)$ are continuous. Examples 1.5 and 1.6 show that this is not the case in higher dimensions.

In the following we introduce an interpolation proposed by Scott and Zhang (1990) for less smooth functions. For keeping the representation simple, we restrict ourselves to the 2d case and to continuous, piecewise linear elements. Extensions to higher polynomial degrees and dimensions can be found in Scott and Zhang (1990) and Ern and Guermond (2004).

Consider a shape regular decomposition $\mathcal{T}_h$ of the domain Ω with vertices a_i, $i = 1,\dots,n_{\text{glob}}$, and global shape functions Φ_i, $i = 1,\dots,n_{\text{glob}}$, satisfying $\Phi_j(a_i) = \delta_{ij}$. Any node a_i is associated with an edge e_i touching a_i. For a given a_i there are several choices for e_i, we only require that $e_i \subset \partial\Omega$ if $a_i \in \partial\Omega$. Let a_{k_i} be the second node of the segment $e_i = [a_i\, a_{k_i}]$. Then, the restrictions of Φ_i and Φ_{k_i} onto e_i span the local space $P_1(e_i)$. We denote by $\psi_i \in P_1(e_i)$ the unique function satisfying

$$\int_{e_i} \Phi_i\, \psi_i\, d\gamma = 1, \qquad \int_{e_i} \Phi_{k_i}\, \psi_i\, d\gamma = 0.$$

Now the Scott–Zhang interpolation operator into the space of piecewise linears can be defined by

$$v \mapsto S_h v := \sum_{i=1}^{n_{\text{glob}}} \Phi_i \int_{e_i} v\, \psi_i\, d\gamma \in V_h.$$

Note that the interpolation can be constructed as long as $v|_{e_i} \in L^1(e_i)$. Due to the properties of the trace operator this is the case for functions $v \in W^{m,p}(\Omega)$ if $m > 1/p$, $1 \leq p < \infty$ or $m \geq 1$, $p = 1$, see Adams (1975). This technique of averaging and interpolation can be extended to higher order polynomials and any space dimension. The statements in the following are true for the general case of an interpolation into the space of continuous, piecewise polynomial functions of degree k, although the given arguments are related to the example above ($k = 1$ and $d = 2$).

The Scott–Zhang interpolation operator preserves homogeneous boundary conditions. The choice of edges e_i for boundary nodes a_i guarantees $e_i \subset \partial\Omega$. From this we conclude that if v vanishes on $\partial\Omega$ then $S_h v$ vanishes on $\partial\Omega$ too.

Further, the Scott–Zhang interpolation operator preserves finite element functions, that is, $S_h v_h = v_h$ for all $v_h \in V_h$. For the above example, we have

$$v = \sum_{j=1}^{n_{\text{glob}}} v(a_j)\, \Phi_j$$

and obtain, consequently,

$$\begin{aligned} S_h v &= \sum_{i=1}^{n_{\text{glob}}} \Phi_i \int_{e_i} v\, \psi_i \, \mathrm{d}\gamma = \sum_{i,j=1}^{n_{\text{glob}}} v(a_j)\, \Phi_i \int_{e_i} \Phi_j\, \psi_i \, \mathrm{d}\gamma \\ &= \sum_{i=1}^{n_{\text{glob}}} \left[v(a_i)\, \Phi_i \int_{e_i} \Phi_i\, \psi_i \, \mathrm{d}\gamma + v(a_{k_i}) \int_{e_i} \Phi_{k_i}\, \psi_i \, \mathrm{d}\gamma \right] \\ &= \sum_{i=1}^{n_{\text{glob}}} v(a_i)\, \Phi_i = v. \end{aligned}$$

Note that the Scott–Zhang interpolation operator is less local than the standard nodal interpolation since the value in one node a_i of a cell K can, by construction, depend on averaged values along an edge e_i which is not a part of ∂K. Therefore, we introduce the set Δ_K of elements in $\mathcal{T}_h$, sharing at least one vertex with K. The stability and interpolation properties are stated in the following theorem.

Theorem 4.6. *Let p and l satisfy $1 \leq p < +\infty$ and $l \geq 1$ if $p = 1$, and $l > 1/p$ otherwise. Then, there is a positive constant C independent of h such that the Scott–Zhang operator satisfies*

$$\begin{aligned} \|S_h v\|_{m,p,\Omega} &\leq C \|v\|_{l,p,\Omega} \quad \text{for all } v \in W^{l,p}(\Omega),\ 0 \leq m \leq \min(1,l), \\ \|v - S_h v\|_{m,p,K} &\leq C h_K^{k+1-m} |v|_{k+1,p,\Delta_K} \quad \text{for all } v \in W^{k+1,p}(\Delta_K),\ 0 \leq m \leq k+1, \end{aligned}$$

for all $K \in \mathcal{T}_h$.

Proof. See Scott and Zhang (1990). ■

CHAPTER 5

BIHARMONIC EQUATION

We study the biharmonic equation as an example of higher order elliptic problems. It turns out that conforming finite element spaces V_h need to belong to C^1 which is more difficult to satisfy than $V_h \subset C^0$. In order to satisfy this smoothness requirement, higher order polynomials are needed. Indeed, in the two-dimensional case, at least fifth degree polynomials (Argyris triangle) are needed on a triangular mesh, see Ženíšek (1974b). In three dimensions, even higher degree polynomials are needed. A three-dimensional conforming finite element with 220 dof per cell based on ninth degree polynomials has been proposed in Ženíšek (1974a). Such higher order degree polynomials lead to high dimensional finite element spaces and high computational costs. Indeed, to compute the matrix entries we have to integrate the product of second order derivatives of polynomials of degree 9, thus polynomials of degree 14. An exact quadrature over tetrahedrons for such polynomials would need the evaluation of the integrand at 236 quadrature points.

One way to construct finite element spaces $V_h \subset C^1$ with a reduced number of dof per cell K is to use local spaces $\mathcal{P}_K$ of piecewise polynomials (or even a class of more general functions) instead of polynomials. This leads to the concept of composite finite elements. Another way is to relax the strong requirement $V_h \subset C^1$ by considering nonconforming finite elements. Note that even nonconforming finite element spaces $V_h \not\subset C^0$ can be used to approximate the solution of fourth order equations. Alternatively, we could also use mixed finite element methods that are based on the reformulation of fourth order problems as systems of two equations of second order. For details of this approach we refer to Chapter 10 of Boffi et al. (2013).

5.1. Deflection of a thin clamped plate

The domain in the thin clamped plate example is a flat elastic object subjected to a load in the transversal direction. The resulting transversal deflection can be modeled using the Kirchhoff plate theory. As the thickness of the plate is assumed to be very thin compared to the other two dimensions, it is sufficient to describe the deformation of the plate under the load f by

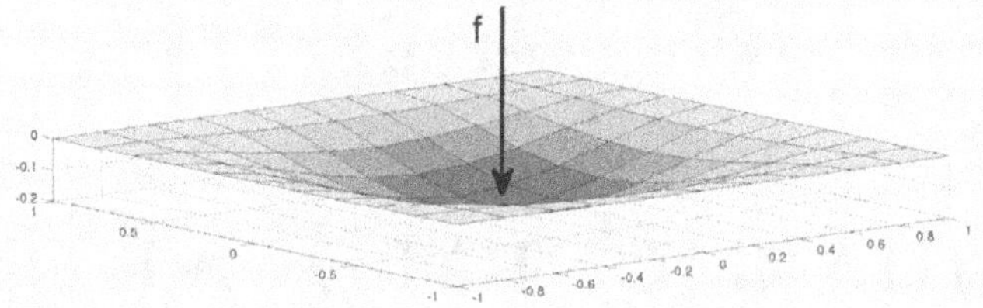

Figure 5.1 Deflection of a clamped plate under the load f.

its vertical deflection u (Figure 5.1) leading to the biharmonic equation with homogeneous boundary conditions

$$\Delta\Delta u = f \quad \text{in } \Omega, \qquad u = \frac{\partial u}{\partial n} = 0 \quad \text{on } \Gamma = \partial\Omega. \tag{5.1}$$

Then, a classical solution of the biharmonic equation with data $f \in C(\Omega)$ is a function $u \in C^4(\Omega) \cap C^1(\overline{\Omega})$ satisfying (5.1) in the sense of continuous functions.

5.2. Weak formulation of the biharmonic equation

We multiply the biharmonic equation in (5.1) by a test function $v \in C_0^\infty(\Omega)$, integrate over Ω and apply the integration by parts formula twice to get

$$\begin{aligned}\int_\Omega (\Delta\Delta u) v \, dx &= \int_\Gamma \frac{\partial \Delta u}{\partial n} v \, d\gamma - \int_\Omega \nabla(\Delta u) \nabla v \, dx \\ &= -\int_\Gamma \Delta u \frac{\partial v}{\partial n} d\gamma + \int_\Omega \Delta u \Delta v \, dx = \int_\Omega \Delta u \Delta v \, dx.\end{aligned}$$

Here, we used the test function $v \in C_0^\infty(\Omega)$ and its derivatives vanish in a neighbourhood of the boundary Γ. As a result, we see that a classical solution of the biharmonic equation satisfies

$$\int_\Omega \Delta u \Delta v \, dx = \int_\Omega f v \, dx \quad \text{for all } v \in C_0^\infty(\Omega).$$

Since $C_0^\infty(\Omega)$ is dense in $H_0^2(\Omega)$ and the mappings

$$v \mapsto \int_\Omega \Delta u \Delta v \, dx, \qquad v \mapsto \int_\Omega f v \, dx$$

are continuous on $H_0^2(\Omega)$, a classical solution satisfies also

$$\int_\Omega \Delta u \Delta v \, dx = \int_\Omega f v \, dx \quad \text{for all } v \in H_0^2(\Omega).$$

Therefore, the weak formulation of the biharmonic equation reads

Find $u \in V := H_0^2(\Omega)$ such that

$$a(u,v) := \int_\Omega \Delta u \Delta v \, dx = \int_\Omega fv \, dx =: F(v) \quad \text{for all } v \in H_0^2(\Omega). \tag{5.2}$$

Contrary, let $f \in C(\Omega)$ and let the weak solution $u \in H_0^2(\Omega)$ of (5.2) belong to $C^4(\Omega) \cap C^1(\overline{\Omega})$. Then, considering (5.2) for all $v \in C_0^\infty(\Omega) \subset H_0^2(\Omega)$ and applying integration by parts we obtain

$$\int_\Omega \Delta \Delta uv \, dx = \int_\Omega fv \, dx \quad \text{for all } v \in C_0^\infty(\Omega) \tag{5.3}$$

from which $\Delta \Delta u = f$ in Ω follows by a density argument. Further, the regularity assumption on $u \in H_0^2(\Omega)$ enforces the boundary conditions. Roughly speaking, a classical solution of Problem (5.1) is always a weak solution and a weak solution being sufficiently regular is a classical solution.

In order to study the existence and uniqueness of weak solutions we check the assumptions of the Lax–Milgram theorem. It is not difficult to see that the forms $a : V \times V \to \mathbb{R}$ and $F : V \to \mathbb{R}$ are bilinear and linear, respectively. Further, the continuity follows by applying Cauchy–Schwarz inequality

$$|a(u,v)| \le \|\Delta u\|_0 \, \|\Delta v\|_0 \le C|u|_2 \, |v|_2 \le C\|u\|_2 \, \|v\|_2 \quad \text{for all } u, v \in V,$$
$$|F(v)| \le \|f\|_0 \, \|v\|_0 \le C\|v\|_2 \quad \text{for all } v \in V.$$

In order to show the coercivity of $a : V \times V \to \mathbb{R}$ we need the following lemma.

Lemma 5.1. *For all $u, v \in H_0^2(\Omega)$ we have*

$$\int_\Omega \frac{\partial^2 u}{\partial x_i^2} \frac{\partial^2 v}{\partial x_j^2} dx = \int_\Omega \frac{\partial^2 u}{\partial x_i \partial x_j} \frac{\partial^2 v}{\partial x_i \partial x_j} dx \quad i, j = 1, \ldots, d.$$

Proof. Since $C_0^\infty(\Omega)$ is dense in $H_0^2(\Omega)$, it is enough to prove the statement for functions $u, v \in C_0^\infty(\Omega)$. Taking into consideration that u, v vanish in the neighbourhood of Γ, we get by integration by parts

$$\int_\Omega \frac{\partial^2 u}{\partial x_i^2} \frac{\partial^2 v}{\partial x_j^2} dx = -\int_\Omega \frac{\partial^3 u}{\partial x_i^2 \partial x_j} \frac{\partial v}{\partial x_j} dx = \int_\Omega \frac{\partial^2 u}{\partial x_i \partial x_j} \frac{\partial^2 v}{\partial x_i \partial x_j} dx \quad \text{for } i, j = 1, \ldots, d.$$

■

Lemma 5.2. *The bilinear form $a : V \times V \to \mathbb{R}$ is coercive, that is, there is a positive constant $m > 0$ such that*

$$a(v,v) \ge m\|v\|_2^2 \quad \textit{for all } v \in V.$$

Proof. Applying Lemma 5.1, we get

$$a(v,v) = \int_\Omega (\Delta v)^2\,\mathrm{d}x = \int_\Omega \left(\sum_{i=1}^d \frac{\partial^2 v}{\partial x_i^2}\right)\left(\sum_{j=1}^d \frac{\partial^2 v}{\partial x_j^2}\,\mathrm{d}x\right) = \int_\Omega \sum_{i,j=1}^d \frac{\partial^2 v}{\partial x_i^2}\frac{\partial^2 v}{\partial x_j^2}\,\mathrm{d}x$$
$$= \int_\Omega \sum_{i,j=1}^d \frac{\partial^2 v}{\partial x_i \partial x_j}\frac{\partial^2 v}{\partial x_i \partial x_j}\,\mathrm{d}x = |v|_2^2 \geq m\|v\|_2^2,$$

where the equivalence of the seminorm $|\cdot|_2$ to the norm $\|\cdot\|_2$ in V was used. ∎

Theorem 5.1. *Let $f \in L^2(\Omega)$. Then, there is a unique solution of* (5.2).

Proof. Application of Theorem 2.1 (Lax–Milgram). ∎

Remark 5.1. The necessary condition for minimizing the energy functional

$$\mathbb{E}(u) = \frac{1}{2}a(u,u) - F(u) = \int_\Omega \left[\frac{1}{2}(\Delta u)^2 - f\,u\right]\mathrm{d}x$$

over $V = H_0^2(\Omega)$ is just the weak formulation of the biharmonic equation. Due to the convexity of the energy functional, the minimization problem

$$\text{Find } u \in V := H_0^2(\Omega) \text{ with } \mathbb{E}(u) = \inf_{v\in V} \mathbb{E}(v)$$

is equivalent to the weak formulation of the biharmonic equation.

5.3. Conforming finite element methods

Let $V_h \subset V = H_0^2(\Omega)$ be a finite element space. Piecewise smooth functions belong to $H_0^2(\Omega)$ iff they are in $C^1(\overline{\Omega})$. For examples of C^1 elements, see Chapter 3. The discrete problem reads

$$\text{Find } u_h \in V_h \text{ such that } a(u_h, v_h) = F(v_h) \quad \text{for all } v_h \in V_h. \tag{5.4}$$

Theorem 5.2. *There is a unique solution $u_h \in V_h$ of the discrete problem* (5.4) *satisfying*

$$\|u - u_h\|_2 \leq C \inf_{v_h \in V_h} \|u - v_h\|_2.$$

Proof. Since $V_h \subset V$ all assumptions of Theorem 2.1 (Lax–Milgram) are satisfied and the error estimate follows from Lemma 2.1 (Cea's lemma). ∎

As for second order problem, the discrete problem (5.4) is equivalent to a linear algebraic system of equations. Let Φ_i, $i = 1, \ldots, n_{\text{glob}}$, be a basis of V_h. Then, $u_h \in V_h$ can be represented as

$$u_h = \sum_{j=1}^{n_{\text{glob}}} U_j \Phi_j.$$

Setting $v_h = \Phi_i$, $i = 1,\dots,n_{\text{glob}}$, we get the linear algebraic system of equations

$$\sum_{j=1}^{n_{\text{glob}}} a_{ij} U_j = f_i$$

for the unknown coefficients U_j, $j = 1,\dots,n_{\text{glob}}$. Here, the entries of the matrix $A = (a_{ij})$ and the entries of the right hand side vector $b = (f_i)$ are given by $a_{ij} = a(\Phi_j, \Phi_i)$ and $f_i = F(\Phi_i)$, $i, j = 1,\dots,n_{\text{glob}}$, respectively.

In the following, we present three examples of finite element spaces $V_h \subset C^1(\Omega)$ and discuss how homogeneous Dirichlet boundary condition can be realized by setting a selected set of dof equal to zero. We start with a simple rectangular element with 16 dof and polynomial local basis functions. Unfortunately, on general decompositions of Ω into triangles or tetrahedra the number of dof (and the polynomial degree of functions in $\mathcal{P}_K$) needed to guarantee C^1 continuity increases. One way to circumvent the use of these computationally expensive elements, for example, the 2d Argyris element (21 dof) or the 3d Ženíšek tetrahedral element (220 dof), is to use composite finite elements. The basic idea of composite finite elements is to relax the requirement that the local approximation space $\mathcal{P}_K$ consists of polynomials and getting more flexibility for constructing global C^1 finite element spaces. Here, we present examples of such a strategy in two and three space dimensions.

The Bogner–Fox–Schmit rectangle

Our first example of a conforming finite element method for solving the biharmonic equation is based on the Bogner–Fox–Schmit rectangle. Let $\Omega \subset \mathbb{R}^2$ be decomposed into axiparallel rectangles K. We denote the vertices of a single rectangle K by a_i, $i = 0,\dots,3$, as indicated in Figure 5.2.

As the local approximation space we choose the space of polynomials of degree less than or equal to three in each variable, more precisely $\mathcal{P}_K = Q_3(K)$ where $\dim \mathcal{P}_K = 16$. We define the

FIGURE 5.2 Degrees of freedom of the Bogner–Fox–Schmit rectangle. Values of the function, its first derivatives and its second mixed derivatives at the vertices.

set of dof Σ_K by 16 nodal functionals

$$N_i(v) := v(a_i), \qquad N_{4+i}(v) := \frac{\partial v}{\partial x_1}(a_i), \qquad N_{8+i}(v) := \frac{\partial v}{\partial x_2}(a_i),$$

$$N_{12+i}(v) := \frac{\partial^2 v}{\partial x_1\, \partial x_2}(a_i) \quad \text{for } i = 0,\dots,3.$$

Lemma 5.3. *The set of dof Σ_K in the Bogner–Fox–Schmit element is $\mathcal{P}_K$ unisolvent. The generated finite element space V_h belongs to $C^1(\overline{\Omega})$. We get $V_h \subset H_0^2(\Omega)$ by setting all dof on the boundary $\partial\Omega$ equal to zero.*

Proof. We first show that $v \in \mathcal{P}_K$ with $N_i(v) = 0$, $i = 0,\dots,15$, implies $v = 0$. The restriction of $v \in \mathcal{P}_K$ onto an edge of $K = (\alpha,\beta) \times (\gamma,\delta)$ is a polynomial of degree less than or equal three in one variable. The function and its tangential derivative vanishes at the two endpoints. The associated 1d Hermite interpolation is unique and equal to zero. This observation guarantees the continuity over the edge. Since the argument holds for any edge, $v \in \mathcal{P}_K$ with $N_i(v) = 0$, $i = 0,\dots,15$, vanishes along ∂K and can be represented as

$$v(x_1,x_2) = (x_1 - \alpha)(\beta - x_1)(x_2 - \gamma)(\delta - x_2)\, q(x_1,x_2), \qquad q \in Q_1(K).$$

The function

$$p(x_1,x_2) = (x_1 - \alpha)(\beta - x_1)(x_2 - \gamma)(\delta - x_2)$$

and its first derivatives vanishes at the four vertices a_i, $i = 0,\dots,3$. The mixed second derivative of p does not vanish at the vertices, thus

$$0 = \frac{\partial^2 v}{\partial x_1\, \partial x_2}(a_i) = \frac{\partial^2 p}{\partial x_1\, \partial x_2}(a_i)\, q(a_i) \quad \Rightarrow \quad q(a_i) = 0, \quad i = 0,\dots,3.$$

Having in mind that the nodal functionals $N_i(q)$, $i = 0,\dots,3$, define a set of dof which is $Q_1(K)$ unisolvent, we end up with $q = 0$ and consequently $v = 0$, that is, the set of dof is $\mathcal{P}_K$ unisolvent.

Along the edge $x_2 = \delta$, $x_1 \in (\alpha,\beta)$, the normal derivative $\partial v/\partial x_2$ belongs to $P_3(\alpha,\beta)$ and vanishes together with its tangential derivatives at the endpoints of the edge. Again, the associated unique Hermite interpolation becomes zero. Applying this argument for each edge, we see that the normal derivatives are continuous over the edges and $V_h \subset C^1(\overline{\Omega})$.

Let E be a boundary edge connecting the vertices a_1 and a_2. Then, we require eight dof to be zero, that is,

$$u_h(a_i) = \frac{\partial u_h}{\partial x_1}(a_i) = \frac{\partial u_h}{\partial x_2}(a_i) = \frac{\partial^2 u_h}{\partial x_1\, \partial x_2}(a_i) = 0, \quad i = 1,2.$$

Since $u_h|_E \in P_3(E)$ and $\frac{\partial u_h}{\partial n}|_E \in P_3(E)$, we conclude $u_h = \frac{\partial u_h}{\partial n} = 0$. ∎

The nodal functionals for the Bogner–Fox–Schmit rectangle can be replaced by the following ones (index for a_i is counted modulo 4)

$$\widetilde{N}_i(v) := v(a_i), \qquad \widetilde{N}_{4+i}(v) := Dv(a_i)(a_{i-1} - a_i),$$

$$\widetilde{N}_{8+i}(v) := Dv(a_i)(a_{i+1} - a_i), \qquad \widetilde{N}_{12+i}(v) := D^2 v(a_i)(a_{i-1} - a_i, a_{i+1} - a_i) \quad \text{with } 0 \le i \le 3.$$

The advantage of this new set is its affine equivalence to the reference cell. Indeed, let $F_K : \widehat{K} \to K$ with $F_K(\hat{x}) = B_K\hat{x} + b_K$, be the affine mapping from the reference cell to the general cell where $a_i = F_K(\hat{a}_i)$, $i = 0, \ldots, 3$. Then, applying the chain rule, we obtain for $\hat{v}(\hat{x}) = v(F_K(\hat{x}))$, for example,

$$Dv(a_i)(a_{i+1} - a_i) = \widehat{D}\hat{v}(\hat{a}_i)B_K^{-1}(a_{i+1} - a_i) = \widehat{D}\hat{v}(\hat{a}_{i+1} - \hat{a}_i).$$

As a consequence, the value of the (scaled) tangential derivative of v at a vertex equals to the value of tangential derivative of $\hat{v}$ at the associated vertex. We get similar expressions for the other dof. Finally, the space of local ansatz functions is transformed to

$$\mathcal{P}_K = Q_3(K) = \{p : K \to \mathbb{R} : p = \hat{p} \circ F_K^{-1}, \hat{p} \in Q_3(\widehat{K})\}.$$

Using Theorem 1.5 (Sobolev embedding), we are able to define point values of $D^\alpha v$, $|\alpha| \leq 2$, provided that $v \in H^4(\Omega)$. Thus, we can define the interpolation operator $\Pi_K : H^4(K) \to \mathcal{P}_K$ locally by

$$\widetilde{N}_i(\Pi_K v) = \widetilde{N}_i(v), \quad v \in H^4(K), i = 0, \ldots, 15.$$

Since $\Pi_K p = p$ for all polynomials $p \in P_3(K)$ we have the interpolation estimates

$$|v - \Pi_K v|_{m,K} \leq Ch_K^{4-m}|v|_{4,K}, \quad v \in H^4(K), 0 \leq m \leq 4. \tag{5.5}$$

The global interpolation $\Pi_h : H^4(\Omega) \to V_h \subset C^1(\overline{\Omega})$ is defined by

$$\Pi_h v|_K = \Pi_K(v|_K) \quad \text{for all } K \in \mathcal{T}_h.$$

Theorem 5.3. *Let the solution u of problem (5.2) belong to $H_0^2(\Omega) \cap H^4(\Omega)$ and u_h denote the solution of the discrete problem (5.4) with the Bogner–Fox–Schmit finite element space. Then, $\|u - u_h\|_2 \leq Ch^2|u|_4$.*

Proof. Apply Theorem 5.2 and the interpolation error estimates (5.5). ■

Remark 5.2. We could also use other $C^1(\overline{\Omega})$ elements like the Argyris triangle or the Bell's triangle with 21 and 18 dof, respectively. Note, however, that both elements are not affine equivalent and in order to obtain an analog to the interpolation error estimate (5.5), one has to modify the concept of affine equivalent elements. For details, see Ciarlet (2002).

Exercise 5.1. Let K be a triangle with vertices a_i, $i = 0, 1, 2$, and midpoints $a_{ij} = (a_i + a_j)/2$, $0 \leq i < j \leq 2$. Show that the set of dof

$$\Sigma_K := \{v(a_i), D^2v(a_i), i = 0, 1, 2, v(a_{ij}), 0 \leq i < j \leq 2\}$$

is $\mathcal{P}_K$ unisolvent, where $\mathcal{P}_K = P_4(K)$. Does this element yield the inclusion $V_h \subset C(\overline{\Omega})$ and $V_h \subset C^1(\overline{\Omega})$, respectively?

The Hsieh–Clough–Tocher (HCT) triangle

Let $\Omega \subset \mathbb{R}^2$ be decomposed into triangles K. We denote the vertices of a single triangle K by a_i, $i = 0, 1, 2$. Each triangle is subdivided into three sub-triangles where K_j, $j = 0, 1, 2$, are created by connecting an inner point a_K with the vertices a_{j+1} and a_{j+2} (counted modulo 3). The midpoint of an edge opposite to a_i is noted as b_i, $i = 0,1,2$, as indicated in Figure 5.3. As a local approximation space we chose C^1 continuous, piecewise polynomials of degree three, more precisely

$$\mathcal{P}_K = \{v \in C^1(K) : v|_{K_i} \in P_3(K_i),\ i = 0,1,2\}.$$

We define the set of dof Σ_K by the nodal functionals

$$N_i(v) := v(a_i), \qquad N_{3+i}(v) := \frac{\partial v}{\partial x_1}(a_i), \qquad N_{6+i}(v) := \frac{\partial v}{\partial x_2}(a_i),$$
$$N_{9+i}(v) := \frac{\partial v}{\partial n}(b_i), \quad i = 0,1,2.$$

Since $\dim P_3(K_i) = 10$, we need 30 equations to define the three polynomials $v|_{K_i}$, $i = 0,1,2$. The nodal functionals define 6 function values and 12 derivatives at the vertices. Moreover, 3 normal derivatives at the midpoint of edges complete the given 21 dof. For the C^1 continuity over the inner edges of the subdomains we add six equations to guarantee the C^1 continuity at a_K and three equations for the continuity of the normal derivative in the midpoints of inner edges of the subdomains. Indeed, the restrictions of functions $v \in P_3(K_{i+1})$ and $w \in P_3(K_{i+2})$ (counted modulo 3) on the inner edge $[a_K, a_i]$ are cubic functions in one variable for which the function values and the tangential derivatives coincide, that is, $v|_{[a_K,a_i]} = w|_{[a_K,a_i]}$ is the unique Hermite interpolation. Consequently, v and the tangential derivatives are continuous across $[a_K, a_i]$. In order to prove the continuity of the normal derivative across $[a_K, a_i]$ we note that $n \cdot \nabla v \in P_2([a_K, a_i])$ is continuous at a_K, a_i, and $(a_K + a_i)/2$ and thus uniquely defined on any inner edge. Finally, the C^1 continuity of the generated finite element space across edges of K can be shown in the same way.

Lemma 5.4. *The set of dof Σ_K in the HCT element is $\mathcal{P}_K$ unisolvent. The generated finite element space V_h belongs to $C^1(\overline{\Omega})$. We get $V_h \subset H_0^2(\Omega)$ by setting all dof on the boundary $\partial\Omega$ equal to zero.*

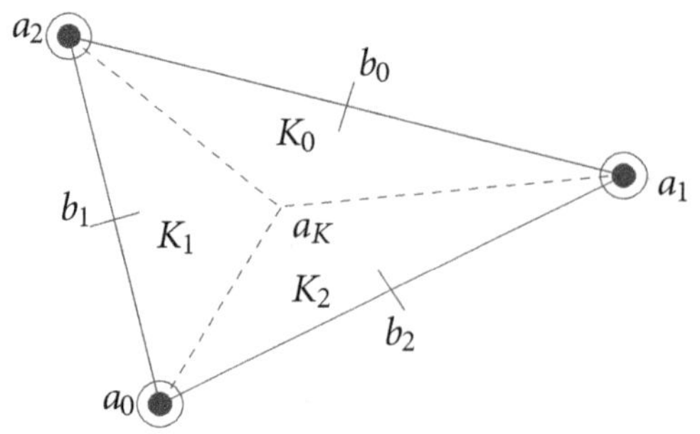

FIGURE 5.3 Degrees of freedom of the HCT triangle. Values of the function, its first derivatives at the vertices, and its normal derivatives at the midpoints.

Proof. See Section 6.1.3 of Ciarlet (2002) for the first two statements. We show $V_h \subset H_0^2(\Omega)$ provided all dof on the boundary are set to be zero. Let E be an edge on the boundary. Since $u_h|_E \in P_3(E)$ and vanishes with its tangential derivatives at the endpoints of E, we conclude $u_h = 0$ on E. Note that $(\partial u_h/\partial n)|_E \in P_2(E)$ vanishes at the mid and endpoints of E. Then, $\partial u_h/\partial n = 0$ on the boundary edge E. ■

Remark 5.3. The HTC element is not affine equivalent due to the presence of the normal derivatives at b_i, $i=0,1,2$, as dof. An additional reason for being not affine is that a_K may be allowed to vary inside of K. Nevertheless, the estimate (5.5) for the interpolation error holds true, for proof see Theorem 6.1 of Ciarlet (2002).

Remark 5.4. If we require that the normal derivative vary linearly along the edges of K we obtain the reduced HCT element. This results in three additional constraints and the dimension of the associated local approximation space

$$\mathcal{P}_K = \left\{ v \in C^1(K) : v|_{K_i} \in P_3(K_i),\ \left.\frac{\partial v}{\partial n}\right|_{[a_i,a_{i+1}]} \in P_1([a_i,a_{i+1}]),\ i=0,1,2 \right\}$$

reduces to 9. The set of dof become the function values and the first derivatives at the vertices of K. For the HCT element it holds $P_3 \subset \mathcal{P}_K$ but for the reduced HCT element we only have $P_2 \subset \mathcal{P}_K$ and the estimate for the interpolation error becomes

$$|v - \Pi_K v|_{m,K} \le C h_K^{3-m} |v|_{3,K}, \quad v \in H^3(K),\ 0 \le m \le 3.$$

A C^1 tetrahedral finite element

A three-dimensional conforming finite element based on ninth degree polynomials has been constructed by Ženíšek (1974a). It has $\dim P_9(K) = 220$ dof. Recently, a composite element based on piecewise polynomials of degree 5 with 45 dof has been proposed by Walkington (2014). In the following, we explain the main features of this element.

Let $\Omega \subset \mathbb{R}^3$ be decomposed into tetrahedrons K. We denote the vertices of a single tetrahedron K by a_i, $i=0,1,2,3$. Then, each tetrahedron is subdivided into four sub-tetrahedrons where K_j, $j=0,1,2,3$, are created by connecting the barycentre a_K with three vertices a_{j+1}, a_{j+2}, a_{j+3} of K (indices are counted modulo 4). If $f \subset \partial K$ is a triangular face, then its centroid is denoted by b_f and the outer normal by n_f, see Figure 5.4. The local approximation space is given by piecewise polynomials of degree 5 with normal derivatives on the faces restricted to polynomials of degree 3, more precisely,

$$\mathcal{P}_K = \left\{ v \in C^1(K) \cap C^4(a_K) : v|_{K_i} \in P_5(K_i),\ \left.\frac{\partial v}{\partial n_f}\right|_f \in P_3(f) \text{ for all } f \subset \partial K \right\},$$

where $v \in C^4(a_K)$ means that v and its derivatives up to order 4 are continuous at a_K. The set of dof Σ_K consists of

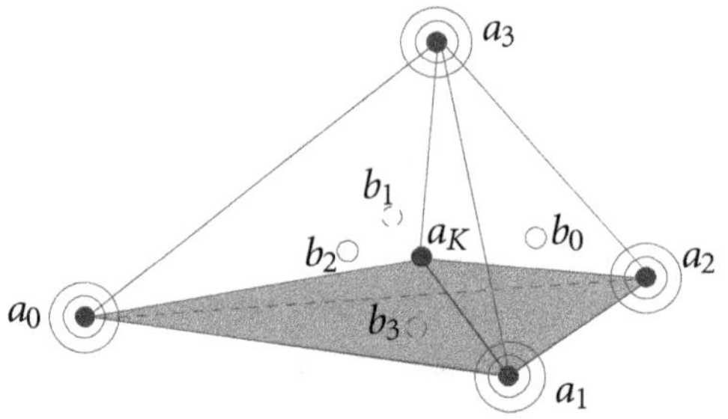

FIGURE 5.4 Degrees of freedom of the composite tetrahedron. Values of the function, its first and second derivatives at the vertices, and its normal derivatives at the centroids of faces.

(i) the function values and its derivative up to order 2 at the vertices (40 dof)

$$N_i^\alpha(v) := D^\alpha v(a_i), \quad 0 \le |\alpha| \le 2, \quad i = 0,1,2,3, \tag{5.6}$$

(ii) the normal derivatives at the centroids of the faces (4 dof)

$$N_f(v) := n_f \cdot \nabla v(b_f), \quad f \subset \partial K, \tag{5.7}$$

(iii) the function value at the barycenter of the tetrahedron (1 dof)

$$N(v) := v(a_K). \tag{5.8}$$

Lemma 5.5. *The set of dof Σ_K in the composite tetrahedron is $\mathcal{P}_K$ unisolvent. The generated finite element space V_h belongs to $C^1(\overline{\Omega})$.*

Proof. See Lemma 2.3 of Walkington (2014). ■

As usual we consider the canonical interpolation operator Π_K defined by

$$N_i^\alpha(\Pi_K v) = N_i^\alpha(v), \qquad N_f(\Pi_K v) = N_f(v), \qquad N(\Pi_K v) = N(v),$$
$$\text{for all } 0 \le |\alpha| \le 2, i = 0,1,2,3, f \subset \partial K.$$

Theorem 1.5 (Sobolev embedding) tells us that $v \in W^{m,p}(\Omega) \subset C^2(\overline{\Omega})$ for $2 + 3/p < m$ such that the interpolant $\Pi_K : W^{m,p}(K) \to \mathcal{P}_K$ is well defined.

Remark 5.5. We can replace the set of dof associated with the vertices of the tetrahedron by

$$\widetilde{N}_i^\alpha(v) := v(a_i), \quad |\alpha| = 0, \qquad \widetilde{N}_i^\alpha(v) := \nabla v(a_i) \cdot (a_j - a_i), \quad |\alpha| = 1, j \ne i,$$
$$\widetilde{N}_i^\alpha(v) := (a_k - a_i)^T D^2 v(a_i)(a_j - a_i), \quad |\alpha| = 2, j, k \ne i, 0 \le j \le k \le 3.$$

Nevertheless, as for the HTC element, the face dof $N_f(v)$ are not invariant under affine maps. Thus, we cannot apply the classical concept of affine equivalent finite elements to get estimates for the interpolation error.

Theorem 5.4. *Let $v \in W^{m,p}(K)$ with $2+3/p \leq m \leq 5$. Then, the canonical interpolation $\Pi_K : W^{m,p}(\Omega) \to \mathcal{P}_K$ is well defined and satisfies*

$$|v - \Pi_K v|_{l,p,K} \leq C h_K^{m-l} |v|_{m,p,K}, \qquad 0 \leq l \leq m.$$

Proof. Let $F_K : \widehat{K} \to K$ be the bijective affine mapping from the reference cell $\widehat{K}$ onto a cell $K \in \mathcal{T}_h$. The interpolation Π_K defines an interpolation $\widehat{\Pi}_K$ on the reference cell by

$$(\widehat{\Pi}_K \hat{v})(\hat{x}) := \left(\Pi_K \left(\hat{v} \circ F_K^{-1}\right)\right)(F_K(\hat{x})),$$

where $\widehat{\Pi}_K$ cannot depend on K for an affine equivalent family of elements. As a consequence, the uniform continuity of $\widehat{\Pi}_K : W^{m,p}(\widehat{K}) \to W^{l,p}(\widehat{K})$ with respect to $K \in \mathcal{T}_h$ can be established and the estimate follows by applying the Bramble-Hilbert lemma. Note that this classical argument with minor modification holds true provided that $\widehat{\Pi}_K$ is invariant under the translation and scaling of K and depends continuously on the Jacobian DF_K. For more details, see Theorem 2.5 of Walkington (2014). ■

Finally, we discuss the construction of the finite element space $V_h \cap H_0^2(\Omega)$ by setting appropriate dof equal to zero. The starting point is the observation that $v = 0$ on a face $f \subset \partial K \cap \partial\Omega$ of a cell K implies that the tangential derivatives of v vanish on f. Furthermore, we have to take into consideration the boundary condition $n_f \cdot \nabla v = 0$ on f. Thus, we require

$$D^{\alpha} v(a_i) = 0, \quad 0 \leq \alpha \leq 1, \qquad n_f \cdot \nabla v(b_j) = 0$$

at a vertex a_i on the boundary and at the centroid of the face $f \subset \partial\Omega$. It remains to set the second derivatives at a_i. We distinguish two cases:

(i) all boundary edges $E \subset \partial\Omega$ meeting at a_i are situated in the same plane (a_i is an inner point of a boundary face of the polyhedron Ω) and
(ii) there are three boundary edges $E \subset \partial\Omega$ meeting at a_i building a nondegenerated parallelepiped (a_i is a vertex or an inner point of an boundary edge of the polyhedron Ω).

Let τ_1 and τ_2 be the two unit vectors spanning the boundary face f with the normal n_f in the first case. Since $v = 0$ and $n_f \cdot \nabla v = 0$ on f, we conclude

$$D^2 v(\tau_j, \tau_k) = 0, \qquad D^2 v(n_f, \tau_j) = 0, \quad \text{on } f, \ 1 \leq j \leq k \leq 2.$$

In the second case, we have two boundary faces spanned by τ_1, τ_2 and τ_2, τ_3, respectively. Therefore, we conclude that all second order derivatives have to vanish

$$D^2 v(\tau_j, \tau_k) = 0, \quad \text{on } f, \ 1 \leq j \leq k \leq 3.$$

We complete our setting of dof in the first case by

$$\text{(i)} \quad D^2 v(a_i)(\tau_j, \tau_k) = 0, \qquad D^2 v(a_i)(n_f, \tau_j) = 0, \quad 1 \leq j \leq k \leq 2,$$

letting $D^2 v(a_i)(n_f, n_f)$ free and in the second case by

$$\text{(ii)} \quad D^2 v(a_i)(\tau_j, \tau_k) = 0, \quad 1 \leq j \leq k \leq 3.$$

We still have to show that the setting of dof above guarantees $u_h = 0$ and $n_f \cdot \nabla u_h = 0$ on f. Let E be an edge of f. We know that $w := n_f \cdot \nabla u_h|_f \in P_3(f)$, therefore $z := w|_E \in P_3(E)$. The setting above implies that z and its tangential derivatives vanish at the two endpoints of E. Thus, $z = 0$ on E. Repeating the arguments for another edge shows that $w = 0$ on ∂f. Recall that $w \in P_3(f)$ and vanishes at the barycenter of f. This gives $w = 0$ on f. Now we prove $u_h = 0$ on f. Consider f as Argyris triangle. Then, all dof at the three vertices of f (u_h and its tangential derivatives upto second order) vanish. It remains to show that the conormal derivative (normal to $E \subset \partial f$ and tangential to f) in the midpoints of the three edges vanish. For this we use the observation that on an edge $E = [a_i, a_j]$ with midpoint $a_{ij} = (a_i + a_j)/2$ it holds

$$P_3(E) = \{p \in P_4(E) : \Lambda(p) = 0\},$$

where

$$\Lambda(p) := \frac{1}{2}(p(a_i) + p(a_j)) - p(a_{ij}) + \frac{1}{8}(Dp(a_i) - Dp(a_j))(a_j - a_i).$$

Let E be the intersection of the two faces f and f' of K. Since $n_f \cdot \nabla u_h \in P_3(E)$, we can express $n_f \cdot \nabla u_h(a_{ij})$ as a linear combination of first and mixed second order derivatives at the endpoints of E. Thus, $n_f \cdot \nabla u_h(a_{ij}) = 0$. The same argument applied to $n_{f'} \cdot \nabla u_h \in P_3(E)$ leads to $n_{f'} \cdot \nabla u_h(a_{ij}) = 0$. Therefore, any first derivative of u_h in a direction perpendicular to E vanishes at a_{ij}, in particular, the conormal derivative mentioned above.

5.4. Nonconforming finite element methods

In order to avoid C^1elements with a large number of dof locally in each cell, we consider now the case $V_h \not\subset V = H_0^2(\Omega)$. Then, the bilinear form $a : H_0^2(\Omega) \times H_0^2(\Omega) \to \mathbb{R}$ is, in general, not defined on the finite element space V_h. Therefore, in a first step, we extend the bilinear form onto $(V + V_h) \times (V + V_h)$. A natural choice would be to compute the bilinear form cellwise

$$a_h^*(u_h, v_h) := \sum_{K \in \mathcal{T}_h} \int_K \Delta u_h \Delta v_h \, dx.$$

Of course $a_h^* : (V + V_h) \times (V + V_h) \to \mathbb{R}$ is an extension of $a : V \times V \to \mathbb{R}$ satisfying

$$a_h^*(u, v) = a(u, v) \quad \text{for all } u, v \in V. \tag{5.9}$$

However, for a piecewise linear functions $w_h \in V_h \not\subset V$ we get $\Delta w_h|_K = 0$ and thus $a_h^*(w_h, w_h) = 0$ without being $w_h = 0$. This shows that, in general, a_h^* is not coercive on a nonconforming space V_h. The reason is that we cannot apply Lemma 5.1 for functions from the discrete space V_h.

Now the idea is to reformulate the continuous bilinear form by means of Lemma 5.1 and then using some cellwise computed version of it which guarantees the coercivity on the discrete finite element space. We define

$$a_h(u_h, v_h) := \sum_{K\in\mathcal{T}_h} \int_K \Delta u_h \Delta v_h \, dx$$

$$+ (1-\sigma) \sum_{K\in\mathcal{T}_h} \int_K \sum_{i,j=1}^{d} \left(\frac{\partial^2 u_h}{\partial x_i \partial x_j} \frac{\partial^2 v_h}{\partial x_i \partial x_j} - \frac{\partial^2 u_h}{\partial x_i^2} \frac{\partial^2 v_h}{\partial x_j^2} \right) dx.$$

Note that the second term vanishes for $u_h, v_h \in H_0^2(\Omega)$ according to Lemma 5.1, thus a_h is an extension of a on $V + V_h$ satisfying (5.9). Now

$$\Delta u_h \Delta v_h = \sum_{i,j=1}^{d} \frac{\partial^2 u_h}{\partial x_i^2} \frac{\partial^2 v_h}{\partial x_j^2}$$

consequently, the bilinear form a_h can be written as

$$a_h(u_h, v_h) = \sum_{K\in\mathcal{T}_h} \int_K \left\{ \sigma \Delta u_h \Delta v_h + (1-\sigma) \sum_{i,j=1}^{d} \frac{\partial^2 u_h}{\partial x_i \partial x_j} \frac{\partial^2 v_h}{\partial x_i \partial x_j} \right\} dx. \tag{5.10}$$

Theorem 5.5. *Let $0 \le \sigma < 1$ and let the broken seminorm $|\cdot|_{2,h}$ be a norm on $V + V_h$. Then, the bilinear form $a_h : (V + V_h) \times (V + V_h) \to \mathbb{R}$ is coercive, that is,*

$$a_h(v, v) \ge (1-\sigma)|v|_{2,h}^2 \quad \textit{for all } v \in V + V_h.$$

Proof. Setting $u_h = v_h = v \in V + V_h$ in Equation (5.10), we obtain immediately

$$a_h(v, v) = \sum_{K\in\mathcal{T}_h} [\sigma \|\Delta v\|_{0,K}^2 + (1-\sigma)|v|_{2,K}^2] \ge (1-\sigma)|v|_{2,h}^2.$$

■

The continuity of the bilinear form is shown by applying Cauchy–Schwarz inequality (in case of sums and integrals), for $u, v \in V + V_h$ we have

$$|a_h(u, v)| \le \sum_{K\in\mathcal{T}_h} \left[\sigma \|\Delta u\|_{0,K} \|\Delta v\|_{0,K} + (1-\sigma)|u|_{2,K}|v|_{2,K}\right] \le C\|u\|_{2,h}\|v\|_{2,h}.$$

The discrete problem reads as

$$\text{Find } u_h \in V_h \text{ such that } a_h(u_h, v_h) = F(v_h) \text{ for all } v_h \in V_h. \tag{5.11}$$

Theorem 5.6. *Let $0 \le \sigma < 1$ and let the broken seminorm $|\cdot|_{2,h}$ be a norm on $V + V_h$. Then, there is a unique solution $u_h \in V_h$ of the discrete problem (5.11) satisfying*

$$\|u - u_h\|_{2,h} \le C \left(\inf_{v_h\in V_h} \|u - v_h\|_{2,h} + \sup_{w_h\in V_h} \frac{|a_h(u, w_h) - F(w_h)|}{\|w_h\|_{2,h}} \right).$$

Proof. The existence of a unique solution of the discrete problem follows from the Lax–Milgram theorem. Consider now the error estimate. Let $v_h \in V_h$ arbitrary. Then, the triangle inequality

$$\|u-u_h\|_{2,h} \leq \|u-v_h\|_{2,h} + \|v_h-u_h\|_{2,h}$$

and Theorem 5.5 give

$$\begin{aligned}(1-\sigma)|v_h-u_h|^2_{2,h} &\leq a_h(v_h-u_h, v_h-u_h)\\ &= a_h(u-u_h, v_h-u_h) + a_h(v_h-u, v_h-u_h)\\ &= a_h(u, v_h-u_h) - F(v_h-u_h) + a_h(v_h-u, v_h-u_h)\\ &\leq C\left(\sup_{w_h\in V_h}\frac{|a_h(u,w_h)-F(w_h)|}{\|w_h\|_{2,h}} + \|u-v_h\|_{2,h}\right)\|v_h-u_h\|_{2,h}.\end{aligned}$$

Using the equivalence of the seminorm $|\cdot|_{2,h}$ to the norm $\|\cdot\|_{2,h}$ and combining the inequalities above, we get for all $v_h \in V_h$

$$\|u-u_h\|_{2,h} \leq C\left(\|u-v_h\|_{2,h} + \sup_{w_h\in V_h}\frac{|a_h(u,w_h)-F(w_h)|}{\|w_h\|_{2,h}}\right).$$

Taking the infimum over all $v_h \in V_h$ the statement follows. ■

Remark 5.6. As shown for the conforming case, the discrete problem (5.11) is equivalent to a linear algebraic system of equations. Let Φ_i, $i = 1,\ldots,n_{\text{glob}}$, be a basis of V_h, $U = (U_j)$ the unknown coefficient vector in the basis representation of u_h, $a_{ij} = a_h(\Phi_j, \Phi_i)$ the matrix entries of A, and $f_i = F(\Phi_i)$ the entries of vector b. Then, $AU = b$.

The rectangular Adini element

As a first example of a nonconforming finite element method for solving the biharmonic equation, the rectangular Adini element is considered. Let $\Omega \subset \mathbb{R}^2$ be decomposed into axiparallel rectangles K. We denote the vertices of a single rectangle K by a_i, $i = 0,\ldots,3$ as indicated in Figure 5.5. An enrichment of the space of polynomials of degree less than or equal to three is taken as the local approximation space, more precisely

$$\mathcal{P}_K = P_3(K) \oplus \text{span}\{x_1x_2^3, x_1^3x_2\}.$$

The dimension is given by $\dim \mathcal{P}_K = 10+2 = 12$. The set of dof Σ_K is defined by the 12 nodal functionals

$$N_i(v) := v(a_i), \qquad N_{4+i}(v) := \frac{\partial v}{\partial x_1}(a_i), \qquad N_{8+i}(v) := \frac{\partial v}{\partial x_2}(a_i), \quad \text{for } i = 0,\ldots,3.$$

Lemma 5.6. *The set of dof Σ_K in Adini's element is $\mathcal{P}_K$ unisolvent.*

Proof. We have to show that $v \in \mathcal{P}_K$ with $N_j(v) = 0$, $j = 0,\ldots,11$, implies $v = 0$. The restriction of v on the segment $[a_1, a_2]$ is a polynomial with respect to x_1 of degree less than or equal three.

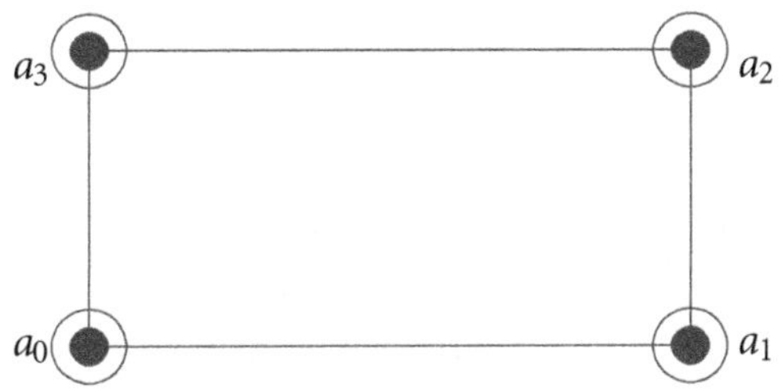

FIGURE 5.5 Degrees of freedom of the Adini's rectangular element. Values of the function and its first derivatives at the vertices.

This polynomial satisfies

$$v(a_1) = N_1(v) = v(a_2) = N_2(v) = \frac{\partial v}{\partial x_1}(a_1) = N_5(v) = \frac{\partial v}{\partial x_1}(a_2) = N_6(v) = 0.$$

From the 1d Hermite interpolation we conclude $v|_{[a_1,a_2]} = 0$. The same argument can be used to show that v vanishes on the other edges of $K = [\alpha,\beta] \times [\gamma,\delta]$. As a consequence, v admits a representation

$$v(x_1,x_2) = (x_1 - \alpha)(\beta - x_1)(x_2 - \gamma)(\delta - x_2)p(x_1,x_2)$$

with some polynomial p. Since, however, the function $(x_1,x_2) \mapsto x_1^2 x_2^2$ does not belong to $\mathcal{P}_K$, the polynomial p has to be identically zero. ■

The local basis functions φ_j defined by $N_i(\varphi_j) = \delta_{ij}$, with $\delta_{ij} = 0$ for $i \neq j$ and $\delta_{ii} = 1$ are given on the reference cell $\widehat{K} = (-1,+1)^2$ by

$$\varphi_0(x_1,x_2) = \frac{(1+x_1)(1+x_2)}{4}\left(1 + \frac{x_1+x_2}{2} - \frac{x_1^2+x_2^2}{2}\right),$$

$$\varphi_1(x_1,x_2) = \frac{(1-x_1)(1+x_2)}{4}\left(1 + \frac{x_2-x_1}{2} - \frac{x_1^2+x_2^2}{2}\right),$$

$$\varphi_2(x_1,x_2) = \frac{(1-x_1)(1-x_2)}{4}\left(1 - \frac{x_1+x_2}{2} - \frac{x_1^2+x_2^2}{2}\right),$$

$$\varphi_3(x_1,x_2) = \frac{(1+x_1)(1-x_2)}{4}\left(1 + \frac{x_1-x_2}{2} - \frac{x_1^2+x_2^2}{2}\right),$$

$$\varphi_4(x_1,x_2) = \frac{(1+x_2)(1+x_1)^2(x_1-1)}{8},$$

$$\varphi_5(x_1,x_2) = \frac{(1+x_2)(1-x_1)^2(x_1+1)}{8},$$

$$\varphi_6(x_1,x_2) = \frac{(1-x_2)(1-x_1)^2(x_1+1)}{8},$$

$$\varphi_7(x_1,x_2) = \frac{(1-x_2)(1+x_1)^2(x_1-1)}{8},$$

$$\varphi_8(x_1,x_2) = \frac{(1+x_1)(1+x_2)^2(x_2-1)}{8},$$

$$\varphi_9(x_1,x_2) = \frac{(1-x_1)(1+x_2)^2(x_2-1)}{8},$$

$$\varphi_{10}(x_1,x_2) = \frac{(1-x_1)(1-x_2)^2(x_2+1)}{8},$$

$$\varphi_{11}(x_1,x_2) = \frac{(1+x_1)(1-x_2)^2(x_2+1)}{8}.$$

We construct a finite element space V_h based on the Adini element. A function $v_h \in V_h$ is defined locally by its dof in all vertices of the decomposition of Ω into rectangles

$$V_h := \{v_h : \Omega \to \mathbb{R} : v_h|_K \in \mathcal{P}_K, D^\alpha v_h \text{ continuous at the vertices,}$$
$$D^\alpha v_h = 0 \text{ at boundary vertices, } |\alpha| \leq 1\}.$$

Note that the restriction of v_h onto an edge is a polynomial of degree less than or equal to three and depends on the two function values and the two values for the tangential derivatives at the endpoint of the edge. Therefore, $V_h \subset C(\overline{\Omega})$. We also see, however, that $V_h \not\subset C^1(\overline{\Omega})$. Indeed, along the edge $x_1 = 1$, we have

$$\frac{\partial \varphi_9}{\partial x_1}(1,x_2) = (1+x_2)^2(1-x_2)$$

but the nodal functionals along this edge

$$N_0(\varphi_9) = N_3(\varphi_9) = N_4(\varphi_9) = N_7(\varphi_9) = N_8(\varphi_9) = N_{11}(\varphi_9) = 0$$

vanish and allow $v = 0$ in the neighbouring cell, that is, the normal derivative is not continuous over this edge.

Lemma 5.7. *On the space V_h of Adini elements $v_h \mapsto |v_h|_{2,h}$ is a norm.*

Proof. Since $v_h \mapsto |v_h|_{2,h}$ is a seminorm we have only to show that $|v_h|_{2,h} = 0$ implies $v_h = 0$. Let $v_h \in V_h$ with $|v_h|_{2,h} = 0$. Then, the first derivatives are constant on each $K \in \mathcal{T}_h$. The first derivatives are continuous in the vertices, thus they are constant on $\overline{\Omega}$. The first derivatives at boundary nodes are zero and we conclude that they are identically zero. It follows that $v_h|_K$ is constant on each $K \in \mathcal{T}_h$. The continuity of v_h on $\overline{\Omega}$ and the homogenuous boundary conditions imply that this constant is equal to zero. ■

The family of Adini's rectangle can be modified to build an affine equivalent family by replacing its dof by

$$\widetilde{N}_i(v) = v(a_i), \qquad \widetilde{N}_{4+i}(v) = Dv(a_i)(a_{i-1} - a_i), \qquad \widetilde{N}_{8+i}(v) = Dv(a_i)(a_{i+1} - a_i),$$

where $i = 0, \ldots, 3$ and the index of a_i is counted modulo 4. Thus, assuming $u \in H_0^2(\Omega) \cap H^3(\Omega)$ the approximation error is bounded by

$$\inf_{v_h \in V_h} \|u - v_h\|_{2,h} \leq C \left(\sum_{K \in \mathcal{T}_h} \|u - \Pi_K u\|_{2,K}^2 \right)^{1/2} \leq Ch|u|_3.$$

Based on Theorem 5.6 it remains to estimate the consistency error

$$\sup_{w_h \in V_h} \frac{|a_h(u, w_h) - F(w_h)|}{\|w_h\|_{2,h}}.$$

Let $u \in H_0^2(\Omega) \cap H^3(\Omega)$ be a solution of problem (5.2). We show that

$$F(w_h) = -\int_\Omega \nabla(\Delta u) \cdot \nabla w_h \, \mathrm{d}x \qquad \text{for all } w_h \in V_h.$$

We recall that $V_h \subset C(\overline{\Omega}) \cap H_0^1(\Omega)$. Thus, there is a sequence $(w_h^k)_{k \in \mathbb{N}} \in C_0^\infty(\Omega)$ converging to w_h in $H_0^1(\Omega)$. Then, integrating by parts we obtain

$$\begin{aligned} F(w_h) &= \lim_{k \to \infty} F(w_h^k) = \lim_{k \to \infty} a(u, w_h^k) = \lim_{k \to \infty} \int_\Omega \Delta u \Delta w_h^k \, \mathrm{d}x \\ &= \lim_{k \to \infty} \left[-\int_\Omega \nabla(\Delta u) \cdot \nabla w_h^k \, \mathrm{d}x \right] = -\int_\Omega \nabla(\Delta u) \cdot \nabla w_h \, \mathrm{d}x. \end{aligned}$$

Now, integrating elementwise by parts

$$\begin{aligned} &\int_K \Delta u \Delta w_h \, \mathrm{d}x + (1 - \sigma) \int_K \sum_{i,j=1}^{2} \left(\frac{\partial^2 u}{\partial x_i \partial x_j} \frac{\partial^2 w_h}{\partial x_i \partial x_j} - \frac{\partial^2 u}{\partial x_i^2} \frac{\partial^2 w_h}{\partial x_j^2} \right) \mathrm{d}x \\ &= -\int_K \nabla(\Delta u) \cdot \nabla w_h \, \mathrm{d}x + \int_{\partial K} \Delta u \frac{\partial w_h}{\partial n_K} \, \mathrm{d}\gamma \\ &\quad + (1 - \sigma) \int_{\partial K} \left(-\frac{\partial^2 u}{\partial \tau_K^2} \frac{\partial w_h}{\partial n_K} + \frac{\partial^2 u}{\partial \tau_K \partial n_K} \frac{\partial w_h}{\partial \tau_K} \right) \mathrm{d}\gamma \end{aligned}$$

and sum up over all cells K we get

$$a_h(u, w_h) = -\int_\Omega \nabla(\Delta u)\cdot\nabla w_h\,\mathrm{d}x + \sum_{K\in\mathcal{T}_h}\int_{\partial K}\Delta u\frac{\partial w_h}{\partial n_K}\,\mathrm{d}\gamma$$
$$+(1-\sigma)\sum_{K\in\mathcal{T}_h}\int_{\partial K}\left(-\frac{\partial^2 u}{\partial \tau_K^2}\frac{\partial w_h}{\partial n_K} + \frac{\partial^2 u}{\partial \tau_K\partial n_K}\frac{\partial w_h}{\partial \tau_K}\right)\mathrm{d}\gamma\,.$$

Here, we used the notation n_K for the outer normal along ∂K and τ_K for the (counter clockwise rotated n_K) unit vector tangent to ∂K, respectively. For an inner edge $E = \partial K \cap \partial K'$ we observe that $\tau_K = -\tau_{K'}$, thus the two integrals over E in the sum cancel. The tangential derivative of w_h along an edge $E \subset \Gamma$ vanishes due to $V_h \subset C(\overline{\Omega})$, consequently we end up with a new representation of the numerator within the consistency error

$$a_h(u, w_h) - F(w_h) = \sum_{K\in\mathcal{T}_h}\int_{\partial K}\left(\Delta u - (1-\sigma)\frac{\partial^2 u}{\partial \tau_K^2}\right)\frac{\partial w_h}{\partial n_K}\,\mathrm{d}\gamma\,. \tag{5.12}$$

Let $E_i = [a_i, a_{i+1}]$, $i = 0,\ldots,3$, denote the edges of K where the indices are counted modulo 4. Then, (5.12) can be written as

$$a_h(u, w_h) - F(w_h)$$
$$= \sum_{K\in\mathcal{T}_h}\left(\int_{E_0}\left(\Delta u - (1-\sigma)\frac{\partial^2 u}{\partial x_1^2}\right)\frac{\partial w_h}{\partial x_2}\,\mathrm{d}\gamma - \int_{E_2}\left(\Delta u - (1-\sigma)\frac{\partial^2 u}{\partial x_1^2}\right)\frac{\partial w_h}{\partial x_2}\,\mathrm{d}\gamma\right)$$
$$+ \sum_{K\in\mathcal{T}_h}\left(\int_{E_3}\left(\Delta u - (1-\sigma)\frac{\partial^2 u}{\partial x_2^2}\right)\frac{\partial w_h}{\partial x_1}\,\mathrm{d}\gamma - \int_{E_1}\left(\Delta u - (1-\sigma)\frac{\partial^2 u}{\partial x_2^2}\right)\frac{\partial w_h}{\partial x_1}\,\mathrm{d}\gamma\right).$$

Replacing the first derivatives of w_h by a continuous function vanishing along boundary edges the above expression vanishes. Indeed, in the sum integrals over inner edges appear twice with opposite sign (and thus cancel) and the tangential derivative over boundary edges vanish. This observation allows us to subtract a continuous piecewise bilinear function without changing the value of the functional $w_h \mapsto a_h(u, w_h) - F(w_h)$. Taking into consideration that the first derivatives of w_h are continuous in the vertices of the decomposition, we can define the Q_1 interpolation Π_h^1, locally by

$$\Pi_K^1\frac{\partial w_h}{\partial x_2}(a_i) = \frac{\partial w_h}{\partial x_2}(a_i), \qquad \Pi_K^1\frac{\partial w_h}{\partial x_1}(a_i) = \frac{\partial w_h}{\partial x_1}(a_i), \quad i = 0,\ldots,3.$$

Now (5.12) is rewritten as

$$a_h(u, w_h) - F(w_h) = \sum_{K\in\mathcal{T}_h}\left[D_1^K\left(u, \frac{\partial w_h}{\partial x_2}\right) + D_2^K\left(u, \frac{\partial w_h}{\partial x_1}\right)\right],$$

where $D_1^K : H^3(K) \times \frac{\partial}{\partial x_2}\mathcal{P}_K \to \mathbb{R}$ and $D_2^K : H^3(K) \times \frac{\partial}{\partial x_1}\mathcal{P}_K \to \mathbb{R}$ are given by

$$D_1^K(u, z_h) = \left(\int_{E_0} - \int_{E_2}\right)\left(\Delta u - (1-\sigma)\frac{\partial^2 u}{\partial x_1^2}\right)(z_h - \Pi_K^1 z_h)\, \mathrm{d}\gamma\,,$$

$$D_2^K(u, z_h) = \left(\int_{E_3} - \int_{E_1}\right)\left(\Delta u - (1-\sigma)\frac{\partial^2 u}{\partial x_2^2}\right)(z_h - \Pi_K^1 z_h)\, \mathrm{d}\gamma\,.$$

In the following, we only consider $D_1^K(u, z_h)$ since the estimation of $D_2^K(u, z_h)$ is similar. The term to be estimated is a bilinear form $d : H^1(K) \times \frac{\partial}{\partial x_2}\mathcal{P}_K \to \mathbb{R}$, given by

$$d(\varphi, z_h) = \left(\int_{E_0} - \int_{E_2}\right)\varphi\,(z_h - \Pi_K^1 z_h)\, \mathrm{d}\gamma\,.$$

Transforming it to the reference cell $\widehat{K} = (0,1)^2$ with the edges $\widehat{E}_0, \ldots, \widehat{E}_3$ and using the continuity of the trace operator we obtain

$$\left|\frac{1}{h_1} d(\varphi, z_h)\right| = \left|\hat{d}(\hat{\varphi}, \hat{z})\right| \le C\|\hat{\varphi}\|_{1,\widehat{K}}\,\|\hat{z}\|_{1,\widehat{K}},$$

that is, the bilinear form $\hat{d}$ is continuous on $H^1(\widehat{K}) \times \frac{\partial}{\partial \hat{x}_2}P_{\widehat{K}}$. Since

$$\begin{aligned} \hat{d}(\hat{\varphi}, \hat{q}) &= 0 \quad \text{for all } \hat{\varphi} \in H^1(\widehat{K}),\, \hat{q} \in P_0(\widehat{K}), \\ \hat{d}(\hat{p}, \hat{z}) &= 0 \quad \text{for all } \hat{p} \in P_0(\widehat{K}),\, \hat{z} \in \frac{\partial}{\partial \hat{x}_2}P_{\widehat{K}}, \end{aligned} \tag{5.13}$$

the generalization of Bramble–Hilbert lemma shows that even

$$\left|\hat{d}(\hat{\varphi}, \hat{z})\right| \le C|\hat{\varphi}|_{1,\widehat{K}}\,|\hat{z}|_{1,\widehat{K}}.$$

The second property of $\hat{d}$ in Equation (5.13) needs to be explained in some more detail. It holds true iff

$$\int_{\widehat{E}_0} (\hat{z} - \widehat{\Pi}^1\hat{z})\,\mathrm{d}\gamma = \int_{\widehat{E}_2} (\hat{z} - \widehat{\Pi}^1\hat{z})\,\mathrm{d}\gamma \quad \text{for all } \hat{z} \in \frac{\partial}{\partial \hat{x}_2}P_{\widehat{K}}.$$

Now a function $\hat{z} \in \frac{\partial}{\partial \hat{x}_2}P_{\widehat{K}}$ can be represented as

$$\hat{z} = A(\hat{x}_2) + B(\hat{x}_2)\hat{x}_1 + C\hat{x}_1^2 + D\hat{x}_1^3,$$

where A and B are polynomials of degree less than or equal to two with respect to $\hat{x}_2$. Subtracting the $Q_1(\widehat{K})$ interpolation, we have

$$(\hat{z} - \widehat{\Pi}^1\hat{z})|_{\widehat{E}_0} = C\hat{x}_1(\hat{x}_1 - 1) + D\hat{x}_1(\hat{x}_1^2 - 1) = (\hat{z} - \widehat{\Pi}^1\hat{z})|_{\widehat{E}_2}.$$

Transforming back to the cell K and replacing φ and z_h by the corresponding values, we end up with

$$\left|D_1^K\left(u, \frac{\partial w_h}{\partial x_2}\right)\right| = |d(\varphi, z_h)| \leq Ch_K|\varphi|_{1,K}\,|z_h|_{1,K} \leq Ch_K|u|_{3,K}\,|w_h|_{2,K}.$$

Applying Cauchy–Schwarz inequality, we get

$$\left|\sum_{K\in\mathcal{T}_h} D_1^K\left(u, \frac{\partial w_h}{\partial x_2}\right)\right| \leq Ch\left(\sum_{K\in\mathcal{T}_h} |u|_{3,K}^2\right)^{1/2}\left(\sum_{K\in\mathcal{T}_h} |w_h|_{2,K}^2\right)^{1/2} \leq Ch|u|_3\,\|w_h\|_{2,h}.$$

The second sum over D_2^K can be estimated in the same way.

Theorem 5.7. *Let the solution u of problem (5.2) belong to $H_0^2(\Omega) \cap H^3(\Omega)$ and u_h denote the solution of the discrete problem (5.11) with the Adini's finite element. Then, $\|u - u_h\|_{2,h} \leq Ch|u|_3$.*

Proof. Collect the estimates for the approximation and consistency error, respectively, and apply Lemma 5.7 and Theorem 5.6. ■

The triangular Morley element

In the previous subsection we saw that the finite element space V_h based on the Adini element belongs to $C(\overline{\Omega})$ but not to $C^1(\overline{\Omega})$. We will now show that even finite element spaces which do not belong to $C(\overline{\Omega})$ can be used to approximate the solution of the biharmonic equation.

Let $\Omega \subset \mathbb{R}^2$ be decomposed into triangles $K \in \mathcal{T}_h$. We denote the vertices and midpoints of edges of a single triangle K by a_i and b_i, $i = 0, 1, 2$, respectively, as indicated in Figure 5.6. For the local approximation space we choose $\mathcal{P}_K = P_2(K)$ with $\dim \mathcal{P}_K = 6$. We define six degrees of freedom by the nodal functionals

$$N_i(v) := v(a_i), \qquad N_{i+3}(v) := \frac{\partial v}{\partial n}(b_i), \quad i = 0, 1, 2.$$

Lemma 5.8. *The set of dof Σ_K in the Morley element is $\mathcal{P}_K$ unisolvent.*

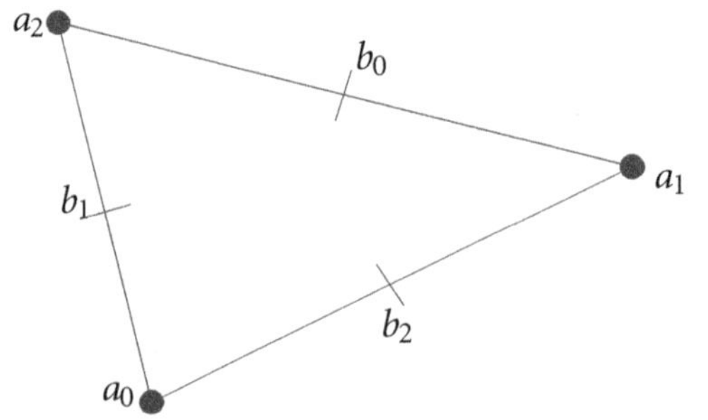

FIGURE 5.6 Degrees of freedom of the Morley triangle. Values of the function at the vertices and first normal derivatives at the midpoint of edges.

Proof. We show that $v \in \mathcal{P}_K$ with $N_i(v) = 0$, $i = 0,\dots,5$, implies $v = 0$. As usual we use the convention that the indices are considered modulo 3, that is, $a_{i+3} = a_i$, $b_{i+3} = b_i$, for all i. The first derivative of $v \in P_2(K)$ belongs to $P_1(K)$, thus

$$\frac{\partial v}{\partial \tau}(b_i) = \frac{1}{|E_i|}\int_{E_i} \frac{\partial v}{\partial \tau}\,\mathrm{d}\gamma = \frac{1}{|E_i|}(v(a_{i+2}) - v(a_{i+1})) = 0,$$

$$\frac{\partial v}{\partial n}(b_i) = |E_i|\,N_i(v) = 0, \quad i = 0,1,2,$$

where τ denotes the tangent unit vector obtained by rotating the normal, n, counter clockwise. Since the first derivatives of v vanish at the three midpoints of edges and $Dv \in P_1^2(K)$ we conclude $Dv = 0$ on K which means $v = \text{const}$ on K, but v vanishes at the vertices which gives $v = 0$ on K. ■

The local basis functions φ_j, defined by $N_i(\varphi_j) = \delta_{ij}$, are given on the reference triangle $\widehat{K}$ with the vertices $a_0 = (0,0)$, $a_1 = (1,0)$, $a_2 = (0,1)$ by

$$\begin{aligned}
\varphi_0(x_1,x_2) &= (1 - x_1 - x_2)^2 + x_1(1 - x_1) + x_2(1 - x_2),\\
\varphi_1(x_1,x_2) &= x_1^2 - \frac{1}{2}(x_1 + x_2)(x_1 + x_2 - 1),\\
\varphi_2(x_1,x_2) &= x_2^2 - \frac{1}{2}(x_1 + x_2)(x_1 + x_2 - 1),\\
\varphi_3(x_1,x_2) &= \frac{1}{\sqrt{2}}(x_1 + x_2)(x_1 + x_2 - 1),\\
\varphi_4(x_1,x_2) &= x_1(x_1 - 1),\\
\varphi_5(x_1,x_2) &= x_2(x_2 - 1).
\end{aligned}$$

We construct a finite element space V_h based on the Morley element. A function $v_h \in V_h$ is defined locally by its dof in all vertices and midpoints of the edges of the decomposition of Ω into triangles, that is,

$$\begin{aligned}
V_h := \{v_h : \Omega \to \mathbb{R} :\ & v_h|_K \in \mathcal{P}_K,\ v_h \text{ continuous at the inner vertices},\\
& \partial_n v_h \text{ continuous at the midpoint of inner edges},\ v_h(a_k) = 0,\\
& \partial_n v_h(b_k) = 0 \text{ for } a_k, b_k \in \Gamma\}.
\end{aligned}$$

We see that $V_h \not\subset C(\overline{\Omega})$. Indeed, consider the local basis function φ_1 on the reference triangle $\widehat{K}$ along the edge $x_1 = 0$, we have

$$\varphi_1(0,x_2) = \frac{1}{2}x_2(1 - x_2)$$

but all nodal functional along this edge vanish,

$$N_0(\varphi_1) = N_2(\varphi_1) = N_4(\varphi_1) = 0$$

and allow $v = 0$ in the neighbouring cell.

Exercise 5.2. Consider the triangulation of Ω as indicated in Figure 5.7. Show that the function

$$v_h(x_1,x_2) := \begin{cases} x_1(1-x_1) - x_2(1-x_2) & \text{in } K_1 \\ x_1(1+x_1) + x_2(1-x_2) & \text{in } K_2 \\ -x_1(1+x_1) + x_2(1+x_2) & \text{in } K_3 \\ -x_1(1-x_1) - x_2(1+x_2) & \text{in } K_4 \end{cases}$$

belongs to V_h but is not continuous on Ω.

Although the space V_h of Morley elements does not belong to $C^1(\overline{\Omega})$ it satisfied a weakened continuity property, as shown in the next lemma.

Lemma 5.9. *The mean values along the edges of the first derivatives of a function in V_h are equal on both sides of an inner edge and equal to zero on a boundary edge.*

Proof. Let $E_i = [a_{i+1}, a_{i+2}]$ and τ be the tangential unit vector directed from a_{i+1} to a_{i+2}, $i = 0,1,2$, where the indices are counted modulo 3. Then, the mean value of the tangential derivative over the edge satisfies

$$\int_{E_i} \frac{\partial v_h}{\partial \tau}\, d\gamma = v_h(a_{i+2}) - v_h(a_{i+1}), \quad i = 0,1,2.$$

The continuity of v_h at the vertices shows that the mean value along both sides of E_i of the tangential derivative are equal. In case that E_i is a boundary edge, the mean value vanishes due to $v_h(a_{i+1}) = v_h(a_{i+2}) = 0$. The midpoint rule is exact for polynomials of degree less than or equal to one, thus

$$\int_{E_i} \frac{\partial v_h}{\partial n}\, d\gamma = |E_i| \frac{\partial v_h}{\partial n}(b_i) \quad \text{for all } v_h \in V_h,\ i = 0,1,2.$$

From the continuity of the normal derivatives at the midpoint of inner edges and its vanishing at boundary edges we obtain the stated result. ■

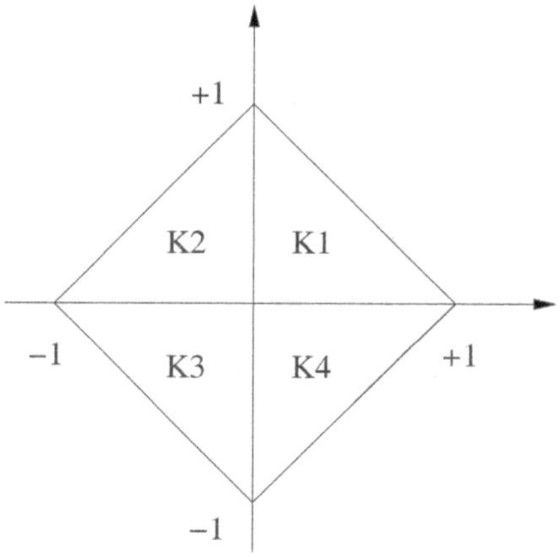

FIGURE 5.7 Triangulation of Ω for a finite element space generated by the Morley element.

Lemma 5.10. *On the space V_h of Morley elements $v_h \mapsto |v_h|_{2,h}$ is a norm.*

Proof. Since $v_h \mapsto |v_h|_{2,h}$ is a seminorm, we have only to show that $|v_h|_{2,h} = 0$ implies $v_h = 0$. Let $v_h \in V_h$ with $|v_h|_{2,h} = 0$. Then, the first order derivatives of v_h are constant on each cell $K \in \mathcal{T}_h$. Since the mean value of these derivatives are continuous at the inner cell edges and equal to zero along the boundary edges, the first order derivatives are equal to zero. This implies that v_h is constant on each cell $K \in \mathcal{T}_h$. Since v_h is continuous at the vertices of the cells and equal to zero at boundary vertices, we conclude $v_h = 0$. ∎

Taking into consideration that for all $v \in \mathcal{P}_K$

$$\frac{\partial v}{\partial n}(b_i) = \frac{1}{|E_i|}\int_{E_i} \frac{\partial v}{\partial n}\,\mathrm{d}\gamma, \quad i = 0,1,2,$$

we could replace the set of dof by

$$N_i^*(v) = v(a_i), \qquad N_{i+3}^*(v) = \frac{1}{|E_i|}\int_{E_i} \frac{\partial v}{\partial n}\,\mathrm{d}\gamma, \quad i = 0,1,2.$$

Note that this replacement changes the domain of definition in the canonical interpolation Π_K defined by $N_i^*(\Pi_K v) = N_i^*(v)$, $i = 0,\ldots,5$. The normal derivatives need not be continuous; it is enough if they are integrable over the edges. Unfortunately, both families of Morley triangles are not affine equivalent, however, we can replace the dof using the nodal functionals

$$\widetilde{N}_i(v) = v(a_i), \qquad \widetilde{N}_{i+3}(v) = \frac{1}{|E_i|}\int_{E_i} Dv\,(a_i - b_i)\,\mathrm{d}\gamma, \quad i = 0,1,2$$

to get an affine equivalent family of elements. It turns out that the canonical interpolations based on N_i^* and $\widetilde{N}_i$, $i = 0,\ldots,5$, are equal. Let us define the canonical local interpolations Π_K^* and $\widetilde{\Pi}_K$ by

$$N_i^*(\Pi_K^* v) = N_i^*(v), \qquad \widetilde{N}_i(\widetilde{\Pi}_K v) = \widetilde{N}_i(v), \quad i = 0,\ldots,5.$$

We have

$$N_i^*(\widetilde{\Pi}_K v) = \widetilde{\Pi}_K v(a_i) = v(a_i) = \Pi_K^* v(a_i) = N_i^*(\Pi_K^* v), \quad i = 0,1,2.$$

From the representation

$$a_i - b_i = ((a_i - b_i)\cdot n)\,n + ((a_i - b_i)\cdot\tau)\,\tau$$

we get for $\alpha_i := (a_i - b_i)\cdot n$, $\beta_i := (a_i - b_i)\cdot\tau$, $i = 0,1,2$,

$$\begin{aligned}\frac{1}{|E_i|}\int_{E_i} Dv\,(a_i - b_i)\,\mathrm{d}\gamma &= \frac{\alpha_i}{|E_i|}\int_{E_i}\frac{\partial v}{\partial n}\,\mathrm{d}\gamma + \frac{\beta_i}{|E_i|}\int_{E_i}\frac{\partial v}{\partial \tau}\,\mathrm{d}\gamma \\ &= \frac{\alpha_i}{|E_i|}\int_{E_i}\frac{\partial v}{\partial n}\,\mathrm{d}\gamma + \frac{\beta_i}{|E_i|}\,(v(a_{i+2}) - v(a_{i+1})).\end{aligned}$$

Since $(\widetilde{\Pi}_K v - v)(a_j) = 0$, we conclude

$$\alpha_i N_{i+3}^*(\widetilde{\Pi}_K v - v) = \widetilde{N}_{i+3}(\widetilde{\Pi}_K v - v) = 0.$$

Note that $\alpha_i \neq 0$, thus, we have shown

$$N^*_{i+3}(\widetilde{\Pi}_K v) = N^*_{i+3}(v) = N^*_{i+3}\left(\Pi^*_K v\right), \quad i = 0,1,2.$$

Now, $\widetilde{\Pi}_K v = \Pi^*_K v$ follows from the $\mathcal{P}_K$ unisolvence of the set of dof N^*_i, $i = 0,\dots,5$.

The interpolation operators $\widetilde{\Pi}_K = \Pi^*_K$ equal the identity on the subspace $\mathcal{P}_K = P_2(K)$. Thus, assuming $u \in H^2_0(\Omega) \cap H^3(\Omega)$ the approximation error is bounded by

$$\inf_{v_h \in V_h} \|u - v_h\|_{2,h} \leq C \left(\sum_{K \in \mathcal{T}_h} \|u - \widetilde{\Pi}_K u\|^2_{0,K} \right)^{1/2} \leq Ch|u|_3.$$

Based on Theorem 5.6 it remains to estimate the consistency error

$$\sup_{w_h \in V_h} \frac{|a_h(u,w_h) - F(w_h)|}{\|w_h\|_{2,h}}.$$

In case of the Adini element we used Equation (5.7) for the numerator which holds true for $V_h \subset C(\overline{\Omega})$, however, in case of the Morley element we have $V_h \not\subset C(\overline{\Omega})$. Assuming $u \in H^4(\Omega)$ and $f \in L^2(\Omega)$ we obtain

$$F(w_h) = \int_\Omega f w_h \, dx = \sum_{K \in \mathcal{T}_h} \left\{ \sigma \int_K \Delta\Delta u \, w_h \, dx + (1-\sigma) \int_K \Delta\Delta u \, w_h \, dx \right\}.$$

Elementwise integration by parts yields

$$\int_K \Delta\Delta u \, w_h \, dx = \int_K \Delta u \Delta w_h \, dx + \int_{\partial K} \frac{\partial \Delta u}{\partial n_K} w_h \, d\gamma - \int_{\partial K} \Delta u \frac{\partial w_h}{\partial n_K} \, d\gamma,$$

$$\int_K \Delta\Delta u \, w_h \, dx = \int_K \sum_{i,j=1}^{2} \frac{\partial^2 u}{\partial x_i \partial x_j} \frac{\partial^2 w_h}{\partial x_i \partial x_j} \, dx + \int_{\partial K} \frac{\partial \Delta u}{\partial n_K} w_h \, d\gamma$$
$$+ \int_{\partial K} \left(\frac{\partial^2 u}{\partial \tau_K^2} - \Delta u \right) \frac{\partial w_h}{\partial n_K} \, d\gamma - \int_{\partial K} \frac{\partial^2 u}{\partial n_K \partial \tau_K} \frac{\partial w_h}{\partial \tau_K} \, d\gamma,$$

where we used the notation n_K for the outer normal along ∂K and τ_K for the (counter clockwise rotated n_K) unit vector tangent to ∂K, respectively. Multiply the expressions by σ and $(1-\sigma)$, and take into consideration the definition of the discrete bilinear form a_h given in (5.10), we obtain

$$F(w_h) = a_h(u,w_h) + R_1(u,w_h) + R_2(u,w_h) + R_3(u,w_h)$$

with

$$R_1(u,w_h) = \sum_{K\in\mathcal{T}_h} \int_{\partial K} \left((1-\sigma)\frac{\partial^2 u}{\partial \tau_K^2} - \Delta u \right) \frac{\partial w_h}{\partial n_K}\, \mathrm{d}\gamma\,,$$

$$R_2(u,w_h) = -\sum_{K\in\mathcal{T}_h} \int_{\partial K} (1-\sigma)\frac{\partial^2 u}{\partial n_K\,\partial \tau_K}\,\frac{\partial w_h}{\partial \tau_K}\, \mathrm{d}\gamma\,,$$

$$R_3(u,w_h) = \sum_{K\in\mathcal{T}_h} \int_{\partial K} \frac{\partial \Delta u}{\partial n_K}\, w_h\, \mathrm{d}\gamma\,.$$

Before estimating R_i, $i = 1,2,3$, let us mention that $R_2(u,w_h) = R_3(u,w_h) = 0$ for $w_h \in C(\overline{\Omega})$ and $u \in H^4(\Omega)$. Indeed, for any boundary edge $E \subset \Gamma$ we have $w_h = 0$ from which $\partial w_h/\partial \tau_K = 0$ follows. Further, for an inner edge $E = \partial K \cap \partial K'$ the contribution associated with the integral over E appears twice with opposite sign since on E the outer normal of K equals the inner normal of K'.

In order to estimate R_1 we first use the property of the finite element space generated by the Morley element that the mean values along edges of the first derivatives of functions of V_h are equal on both sides of an inner edge and equal to zero on a boundary edge (see Lemma 5.9). Therefore, R_1 can be represented as

$$R_1(u,w_h) = \sum_{K\in\mathcal{T}_h} \int_{\partial K} \left((1-\sigma)\frac{\partial^2 u}{\partial \tau_K^2} - \Delta u \right) \left(\frac{\partial w_h}{\partial n_K} - \Pi_0 \frac{\partial w_h}{\partial n_K} \right) \mathrm{d}\gamma\,,$$

where $\Pi_0(\partial w_h/\partial n_K)$ denotes the mean value of $\partial w_h/\partial n_K$ over the edge E which is well defined in Lemma 5.9. We mention that the continuous bilinear form $D : H^1(K) \times H^1(K) \to \mathbb{R}$ given by

$$D(\varphi,\psi) := \int_E \varphi(\psi - \Pi_0\psi)\,\mathrm{d}\gamma$$

has the properties

$$D(\varphi,q) = 0 \quad \text{for all } \varphi \in H^1(K),\ q \in P_0(K),$$
$$D(p,\psi) = 0 \quad \text{for all } p \in P_0(K),\ \psi \in H^1(K).$$

Using this observation on a reference cell, applying the generalization of Bramble–Hilbert lemma and summing up the local estimates, we can show

$$|R_1(u,w_h)| \le Ch|u|_3\, \|w_h\|_{2,h}, \quad u \in H^3(\Omega),\ w_h \in V_h,$$

for details we refer to Lascaux and Lesaint (1975). The same inequality can be derived for $R_2(u,w_h)$. In order to estimate $R_3(u,w_h)$ we introduce the continuous piecewise linear interpolation Π_1 which is well defined on $H^2(\Omega) + V_h$. Since $\Pi_1 w_h$ is continuous over the inner edges and zero along the boundary edges, we can write

$$R_3(u,w_h) = \sum_{K\in\mathcal{T}_h} \int_{\partial K} \frac{\partial \Delta u}{\partial n_K}(w_h - \Pi_1 w_h)\,\mathrm{d}\gamma\,, \quad u \in H^4(\Omega),\ w_h \in V_h.$$

Standard local interpolation estimates lead to

$$|R_3(u,w_h)| \le Ch(|u|_3 + h|u|_4)\,\|w_h\|_{2,h}, \qquad u \in H^4(\Omega),\ w_h \in V_h.$$

Theorem 5.8. *Let the solution u of problem (5.2) belong to $H_0^2(\Omega)\cap H^4(\Omega)$ and u_h denote the solution of the discrete problem (5.11) with the Morley finite element space. Then,*

$$\|u-u_h\|_{2,h} \le Ch(|u|_3 + h|u|_4).$$

Proof. Combine the approximation error estimate with the estimates of the consistency error and apply Theorem 5.6. ■

A nonconforming tetrahedral element

In the following subsection we describe a nonconforming tetrahedral element proposed by Ming and Xu (2007). Let $\Omega \subset \mathbb{R}^3$ be decomposed into tetrahedrons $K \in \mathcal{T}_h$. We denote the vertices and centroids of faces of a single tetrahedron K by a_i and b_i, $i=0,\ldots,3$, respectively. We choose the local approximation space $\mathcal{P}_K = P_3(K)$ with $\dim \mathcal{P}_K = 20$. We define 20 dof by the nodal functionals (see Figure 5.8)

$$N_i^\alpha(v) := D^\alpha v(a_i), \qquad N_i(v) := \frac{\partial v}{\partial n}(b_i) \quad 0 \le |\alpha| \le 1,\ i=0,\ldots,3.$$

Remark 5.7. Note that the three, first derivative dof at the vertex a_i, $i=0,\ldots,3$ can be replaced by the first derivatives in the direction of the edges $[a_j,a_i]$, $j \ne i$, that is, $Dv(a_i)(a_j-a_i)$, without changing the finite element space. Nevertheless, the associated family of elements is not affine equivalent due to the normal derivative dof.

Lemma 5.11. *The set of dof Σ_K in the Ming–Xu element is $\mathcal{P}_K$ unisolvent.*

Proof. The basis functions φ_i, ψ_i, $i=0,\ldots,3$ and φ_{ij}, $0 \le i \ne j \le 3$, are explicitly given in Ming and Xu (2007) for which the canonical interpolation $\Pi_K : C^1(K) \to \mathcal{P}_K$ reads

$$\Pi_K v := \sum_{i=0}^{3} v(a_i)\,\varphi_i + \sum_{i=0}^{3} \frac{\partial v}{\partial n}(b_i)\,\psi_i + \sum_{0\le i\ne j\le 3} Dv(a_i)(a_j-a_i)\,\varphi_{ij}.$$

■

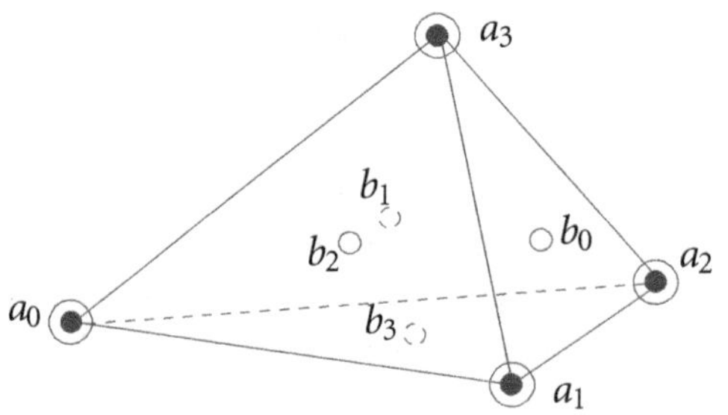

FIGURE 5.8 Degrees of freedom of the nonconforming tetrahedral element. Values of the function, its first derivatives at the vertices, and its normal derivatives at the barycenters of the faces.

Remark 5.8. A reduced element with 16 dof can be constructed by removing the normal derivative dof. Let f be a face associated with the three vertices a_i, a_j, a_k and the outer normal, n. Then, the normal derivative of dof at the centroid $a_{ijk} = (a_i + a_j + a_k)/3$ is replaced by

$$\frac{\partial v}{\partial n}(a_{ijk}) = \frac{1}{3} \sum_{l=i,j,k} \frac{\partial v}{\partial n}(a_l)$$

leading to the local approximation space $\mathcal{P}_K^{\text{red}}$ of dimension $\dim \mathcal{P}_K^{\text{red}} = 16$. Note that $P_2(K) \subset \mathcal{P}_K^{\text{red}}$.

We construct finite element spaces V_h and V_h^{red} based on the nonconforming tetrahedral and the reduced element, respectively. A function $v_h \in V_h$ is defined locally by its dof at all vertices and centroids of the faces of the decomposition of Ω into tetrahedrons, that is,

$$\begin{aligned} V_h := \{v_h : \Omega \to \mathbb{R} : v_h|_K \in \mathcal{P}_K, D^\alpha v_h \text{ continuous at the inner vertices,} \\ \partial_n v_h \text{ continuous at the centroids of inner faces, } D^\alpha v_h(a_k) = 0, \\ \partial_n v_h(b_k) = 0 \text{ for } a_k, b_k \in \Gamma, 0 \le |\alpha| \le 1\}. \end{aligned}$$

Similarly the reduced finite element space is defined as

$$\begin{aligned} V_h^{\text{red}} := \{v_h : \Omega \to \mathbb{R} : v_h|_K \in \mathcal{P}_K^{\text{red}}, D^\alpha v_h \text{ continuous at the inner vertices,} \\ D^\alpha v_h(a_k) = 0, \text{ for } a_k \in \Gamma, 0 \le |\alpha| \le 1\}. \end{aligned}$$

As in the Morley triangle we have the following result.

Lemma 5.12. *The mean values over faces of the first derivatives of a function of $v_h \in V_h$ are equal on both sides of an inner face and equal to zero on a boundary face. Moreover $v_h \mapsto |v_h|_{2,h}$ is a norm on V_h.*

Proof. For the first statement see Lemma 3.1 of Ming and Xu (2007). Since $v_h \mapsto |v_h|_{2,h}$ is a seminorm, we have to show that $|v_h|_{2,h} = 0$ implies $v_h = 0$. The arguments are same as in the proof of Lemma 5.10. ∎

Finally, a careful study of the approximation and consistency error leads to the following theorem.

Theorem 5.9. *Let the solution u of problem (5.2) belong to $H_0^2(\Omega) \cap H^4(\Omega)$ and u_h denote the solution of the discrete problem (5.11) with the nonconforming tetrahedral or the reduced finite element space. Then,*

$$\|u - u_h\|_{2,h} \le Ch(|u|_3 + h|u|_4).$$

CHAPTER 6

PARABOLIC PROBLEMS

In the previous chapter, the finite element discretization of stationary partial differential equations for a scalar function $u : \Omega \to \mathbb{R}$ depending on the position $x \in \Omega$ in space has been considered. In practice, the solution u of the differential equations may also dependent on time. Therefore, a temporal discretization, in addition to a finite element discretization in space, needs to be applied to approximate the solution of time-dependent partial differential equations. In this chapter, we discuss the most commonly used time discretization schemes in the context of finite elements. In particular, temporal discretization schemes for scalar parabolic equations are discussed in detail. In addition, splitting schemes that handle high-dimensional differential equations are briefly discussed.

6.1. Conservation of energy

Let ω be a small control volume. The principle of conservation of energy states that the rate of change of internal energy equals the sum of net heat flux and produced heat, in formulas

$$\frac{\mathrm{d}}{\mathrm{d}t}\int_\omega \rho c_p T \,\mathrm{d}x = \int_\omega f \,\mathrm{d}x - \int_{\partial\omega} \mathbf{J}\cdot\mathbf{n}\,\mathrm{d}s,$$

where T is the temperature, ρ the density, c_p the specific heat of the material, f the heat source intensity, $\mathbf{J}$ the total heat flux and $\mathbf{n}$ the outer unit normal with respect to ω. Transforming the boundary integral into a volume integral by the Gauss theorem, we get

$$\int_\omega \left(\rho c_p \frac{\partial T}{\partial t} + \operatorname{div}\mathbf{J}\right)\mathrm{d}x = \int_\omega f \,\mathrm{d}x.$$

The flux can be diffusive, convective, or both, that is, $\mathbf{J} = \mathbf{J}_{\text{diff}} + \mathbf{J}_{\text{conv}}$. Fourier's law states that the diffusive heat flux is proportional to the negative local temperature gradient, that is,

$$\mathbf{J}_{\text{diff}} = -k\nabla T,$$

where k is the thermal conductivity. For the convective heat flux we have

$$\mathrm{J}_{\mathrm{conv}} = \rho c_p \mathbf{u} T,$$

where $\mathbf{u}$ is the velocity of the fluid in ω. Hence, we get

$$\int_{\omega} \left(\rho c_p \frac{\partial T}{\partial t} + \operatorname{div}\left(\rho c_p \mathbf{u} T - k \nabla T \right) \right) \mathrm{d}x = \int_{\omega} f \,\mathrm{d}x.$$

Since ω is an arbitrary control volume, we obtain the differential equation

$$\rho c_p \frac{\partial T}{\partial t} + \operatorname{div}\left(\rho c_p \mathbf{u} T - k \nabla T \right) = f \quad \text{in } \Omega \times (0, \mathrm{I}],$$

where $\Omega \subset \mathbb{R}^3$ is an open bounded domain and I in the time interval $(0,\mathrm{I}]$ is a given final time. For an incompressible fluid it holds $\operatorname{div} \mathbf{u} = 0$ in Ω and the equation can be written as

$$\rho c_p \left(\frac{\partial T}{\partial t} + \mathbf{u} \cdot \nabla T \right) - k \Delta T = f \quad \text{in } \Omega \times (0, \mathrm{I}]. \tag{6.1}$$

As in the case of an elliptic differential equation we have to specify boundary conditions on the boundary Γ of the computational domain (see Section 2.1):

- Dirichlet boundary condition (prescribed temperature)

$$T = T_{\mathrm{D}} \quad \text{on } \Gamma \times (0, \mathrm{I}],$$

- Neumann boundary condition (prescribed heat flux)

$$k \frac{\partial T}{\partial \mathbf{n}} = j \quad \text{on } \Gamma \times (0, \mathrm{I}],$$

- Robin boundary condition (heat flux proportional to heat transfer)

$$k \frac{\partial T}{\partial \mathbf{n}} = \alpha (T_{\infty} - T) \quad \text{on } \Gamma \times (0, \mathrm{I}].$$

Finally, the initial condition

$$T(\cdot, 0) = T_0 \quad \text{in } \Omega$$

prescribes a given temperature field at time $t = 0$.

6.2. A general parabolic problem of second order

Let $\Omega \subset \mathbb{R}^d$ be a bounded domain with Lipschitz continuous boundary and $(0,\mathrm{I}]$ be a given time interval with final time I. Equation (6.1) is a special case of the more general second order scalar parabolic equation of the form

$$\frac{\partial u}{\partial t} - \sum_{i,j=1}^{d} \frac{\partial}{\partial x_j} \left(a_{ij} \frac{\partial u}{\partial x_i} \right) + \sum_{i=1}^{d} b_i \frac{\partial u}{\partial x_i} + c u = f \quad \text{in } \Omega \times (0, \mathrm{I}]. \tag{6.2}$$

Here, $a_{ij}(x,t)$ is the diffusion, $b_i(x,t)$ the advection, $c(x,t)$ the reaction and $f(x,t)$ the source term, and these terms are assumed to be sufficiently smooth. The differential Equation (6.2) is parabolic at the point $(x,t) \in \Omega \times (0,I]$ if the matrix $a_{ij}(x,t)$ is positive definite. As in the elliptic case, often the slightly stronger assumption of being uniformly parabolic is made, that is, there exists a positive constant α_0 such that

$$\sum_{i,j=1}^{d} a_{ij}(x,t)\xi_i\xi_j \geq \alpha_0 \sum_{i=1}^{d} \xi_i^2 \quad \text{for all } (x,t) \in \overline{\Omega} \times [0,I],\ \xi \in \mathbb{R}^d.$$

To show that Equation (6.1) is a special case of Equation (6.2), we set in Equation (6.2)

$$u := T, \quad a_{ij} := \frac{k}{\rho c_p}\delta_{ij}, \quad \mathbf{b} := \mathbf{u}, \quad c := 0, \quad f := \frac{1}{\rho c_p} f$$

and multiply the equation with ρc_p.

6.3. Weak formulation of initial boundary value problems

Unsteady processes (for example, the temperature in a room during the day) can be considered from different view points. Let us assume that the process takes place in a domain $\Omega \subset \mathbb{R}^d$ over a time interval $[0,I] \subset \mathbb{R}$. Then we can study the process quantity u (temperature) depending on space and time, that is, we consider the mapping

$$(x,t) \in \Omega \times [0,I] \mapsto u(x,t) \in \mathbb{R}.$$

It turns out, however, that a different point of view is more appropriate for the mathematical treatment of unsteady processes. Instead of considering a single temperature value at a space–time point $(x,t) \in \Omega \times [0,I]$ we investigate the time-dependent mapping

$$\mathrm{t} \in [0,I] \mapsto \mathrm{u}(\mathrm{t}) \in \mathrm{V},$$

that is, the temperature field, $\mathrm{u}(\mathrm{t}) : \Omega \to \mathbb{R}$, at a certain time t belongs to some Banach space V. In the first case, the space and time variable are equally acting whereas in the latter the focus is on time-dependent mappings into Banach spaces, that is, $\mathrm{u}(\mathrm{t}) = u(\cdot,t) \in \mathrm{V}$.

Definition 6.1. Let $V = H_0^1(\Omega)$. We denote by $L^2(0,I;V)$, the set of all measurable functions $\mathrm{u} : [0,I] \to V$ with

$$\|\mathrm{u}\|_{L^2(0,I;V)} := \left(\int_0^I \|\mathrm{u}(t)\|_V^2 \,\mathrm{d}t \right)^{1/2} < \infty.$$

Definition 6.2. A *Gelfand triple* is a pair of imbeddings $V \subset H = H^* \subset V^*$, where V is a separable and reflexive Banach space, H is a separable Hilbert space and V is dense in H and continuously embedded in H. The notation $H = H^*$ has to be understood in the sense that according to Theorem 1.2 (Riesz representation theorem) the dual space of H can be identified with H.

A typical example of a Gelfand triple $V \subset H \subset V^*$ in the context of second order parabolic equations is

$$V = H_0^1(\Omega), \qquad H = L^2(\Omega), \qquad V^* = H^{-1}(\Omega).$$

The time derivative has to be understood in the sense of distributions.

Definition 6.3. The function $\mathrm{u} \in L^2(0,\mathrm{I};V)$ admits a generalized time derivative $\mathrm{w} = \mathrm{u}' \in L^2(0,\mathrm{I};V^*)$ if

$$\int_0^{\mathrm{I}} \varphi'(t)\mathrm{u}(t)\,\mathrm{d}t = -\int_0^{\mathrm{I}} \varphi(t)\mathrm{w}(t)\,\mathrm{d}t \quad \text{for all } \varphi \in C_0^\infty(0,\mathrm{I}).$$

We set $W^1(0,\mathrm{I};V,H) := \{\mathrm{u} \in L^2(0,\mathrm{I};V) : \mathrm{u}' \in L^2(0,\mathrm{I};V^*)\}$ with the norm

$$\|\mathrm{u}\|_{W^1} = \|\mathrm{u}\|_{L^2(0,\mathrm{I};V)} + \|\mathrm{u}'\|_{L^2(0,\mathrm{I};V^*)}.$$

The introduction of the solution space $W^1(0,\mathrm{I};V,H)$ has two important consequences.

Lemma 6.1. *Let* u *and* v *belong to* $W^1(0,\mathrm{I};V,H)$. *Then,*

(i) *the mapping* $\mathrm{u} : [0,\mathrm{I}] \to H$ *is continuous,*
(ii) *it holds the integration by parts formula*

$$\int_s^t \langle \mathrm{u}'(\tau),\mathrm{v}(\tau)\rangle\,\mathrm{d}\tau = (\mathrm{u}(t),\mathrm{v}(t)) - (\mathrm{u}(s),\mathrm{v}(s)) - \int_s^t \langle \mathrm{v}'(\tau),\mathrm{u}(\tau)\rangle\,\mathrm{d}\tau$$

for $0 \le s \le t \le \mathrm{I}$.

Proof. See Gajewski et al. (1974), Thomée (2006) and Zeidler (1990). ∎

Now, we are ready to formulate the initial boundary value problem for Equation (6.2) in a weak sense. As usual we multiply Equation (6.2) by a test function $v \in V$, integrate over Ω and apply the integration by parts formula to the term with the highest space derivatives. Then, for given $u_0 \in L^2(\Omega)$ and $\mathrm{f} \in L^2(0,\mathrm{I};V^*)$ the weak formulation of Equation (6.2) completed by an initial condition and by homogeneous Dirichlet boundary condition reads

Find $\mathrm{u} \in W^1(0,\mathrm{I};V,H)$ such that

$$\langle \mathrm{u}'(t),v\rangle + a(t,\mathrm{u}(t),v) = \langle f(t),v\rangle \quad \text{for a.e. } t \in (0,\mathrm{I}), \text{ for all } v \in V, \tag{6.3}$$

$$(\mathrm{u}(0),v) = (u_0,v) \quad \text{for all } v \in L^2(\Omega), \tag{6.4}$$

where $a : (0,\mathrm{I}) \times V \times V \to \mathbb{R}$ is given by

$$a(t,u,v) = \int_\Omega \left(\sum_{i,j=1}^d a_{ij}(x,t)\frac{\partial u}{\partial x_i}\frac{\partial v}{\partial x_j} + \sum_{i=1}^d b_i(x,t)\frac{\partial u}{\partial x_i}v + c(x,t)\,uv \right) \mathrm{d}x.$$

Concerning the solvability of problem (6.3), (6.4) we assume that the mapping $a: (0,\mathrm{I})\times V \times V \to \mathbb{R}$ satisfies:

(A1) the function $t \mapsto a(t,u,v)$ is measurable for all $u, v \in V$,
(A2) $a(t,\cdot,\cdot)$ is bilinear and continuous on V for a.e. $t \in (0,\mathrm{I})$,
(A3) there are $\alpha > 0$ and $\gamma \in \mathbb{R}$ such that

$$a(t,v,v) \geq \alpha \|v\|_V^2 - \gamma \|v\|_H^2 \quad \text{for a.e. } t \in (0,\mathrm{I}), \text{ for all } v \in V.$$

Note that in contrast to elliptic equations, the bilinear form $a(t,\cdot,\cdot)$ is not supposed to be coercive, but satisfies the weaker condition (A3), which is called Gårding's inequality. Nevertheless, up to a change of variable it is always possible to modify condition (A3) in such a way that $\gamma = 0$, that is the bilinear form $a(t,\cdot,\cdot)$ is coercive on V. The transformation $u = \tilde{u}\exp(\gamma t)$, $\gamma > 0$, leads to an equivalent problem for $\tilde{u}$ with the bilinear form

$$\tilde{a}(t,u,v) = a(t,u,v) + \gamma\,(u,v)$$

for which

$$\tilde{a}(t,v,v) \geq \alpha \|v\|_V^2 - \gamma \|v\|_H^2 + \gamma \|v\|_H^2 \geq \alpha \|v\|_V^2$$

holds. Consequently, we may assume (A3) with $\gamma = 0$ without changing generality.

Theorem 6.1. *Under the assumptions (A1), (A2) and (A3), problem* (6.3), (6.4) *has a unique solution.*

Proof. See Lions and Magenes (1968). ■

The solution u that satisfies the variational form (6.3) is a *weak solution* of the parabolic Equation (6.2) with homogeneous Dirichlet boundary conditions. According to Lemma 6.1 (i), the initial condition $u(0) = u_0$ formulated in Equation (6.4) makes sense when considering $u \in W^1(0,\mathrm{I}; V,H)$ as a mapping $u : (0,\mathrm{I}) \to H$. Note that the solution u of the parabolic problem (6.3), (6.4) is considered as a mapping of t into the space V and H, respectively, instead of being a function of x and t in $\mathbb{R}$.

6.4. Semidiscretization by finite elements

The following example can be considered as the simplest case of an initial boundary value problem for a parabolic equation. In the following, we use it as our model problem to discuss different time-stepping methods.

Consider Equation (6.2) with $a_{ij}(x,t) = \delta_{ij}$, $i,j = 1,\ldots,d$, $b_i(x,t) = 0$, $i = 1,\ldots,d$, $c(x,t) = 0$ and a source term f that is sufficiently smooth in $\Omega \times (0,\mathrm{I})$, completed by homogeneous Dirichlet boundary condition and the initial data $u_0 \in L^2(\Omega)$

$$
\begin{aligned}
\frac{\partial u}{\partial t} - \Delta u &= f \quad \text{in } \Omega \times (0,\mathrm{I}], \\
u &= 0 \quad \text{on } \Gamma \times (0,\mathrm{I}], \\
u(\cdot,0) &= u_0 \quad \text{in } \Omega.
\end{aligned} \tag{6.5}
$$

The weak formulation of problem (6.5) leads to the special case of (6.3) and (6.4)

Find $\mathrm{u} \in W^1(0,\mathrm{I}; H_0^1(\Omega), L^2(\Omega))$ such that

$$
\begin{aligned}
\langle \mathrm{u}'(t), v\rangle + (\nabla \mathrm{u}(t), \nabla v) &= \langle \mathrm{f}(t), v\rangle \qquad \text{for a.e. } t \in (0,\mathrm{I}), \text{ for all } v \in H_0^1(\Omega), \\
(\mathrm{u}(0), v) &= (u_0, v) \qquad \text{for all } v \in L^2(\Omega).
\end{aligned}
$$

Using the standard Galerkin approach for a finite element space $V_h \subset V$ described in Section 2.3 we get the semidiscrete (continuous in time) problem as

Find $\mathrm{u}_h(t) \in V_h$ such that for a.e. $t \in (0,\mathrm{I})$, for all $v_h \in V_h$

$$
\langle \mathrm{u}_h'(t), v_h\rangle + (\nabla \mathrm{u}_h(t), \nabla v_h) = \langle \mathrm{f}(t), v_h\rangle, \tag{6.6}
$$

$$
(\mathrm{u}_h(0), v_h) = (u_0, v_h). \tag{6.7}
$$

We now show that the semidiscrete problem is equivalent to a system of ordinary differential equations. Let $\{\Phi_i\}$, $i = 1,\ldots,n_{\text{glob}}$, be a basis of V_h. We represent the discrete solution $\mathrm{u}_h(t) \in V_h$ in terms of the basis of V_h as

$$
\mathrm{u}_h(t) := \sum_{j=1}^{n_{\text{glob}}} U_j(t)\, \Phi_j, \qquad \nabla \mathrm{u}_h(t) := \sum_{j=1}^{n_{\text{glob}}} U_j(t)\, \nabla \Phi_j, \tag{6.8}
$$

where $U_j(t)$, $j = 1,\ldots,n_{\text{glob}}$, $t \in [0,\mathrm{I}]$, are the unknown coefficients to be determined. For expressing $\langle \mathrm{u}_h'(t), v_h\rangle$, we mention that Lemma 6.1 (integration by parts formula) implies

$$
\frac{\mathrm{d}}{\mathrm{d}t}(\mathrm{u}(t), v) = \langle \mathrm{u}'(t), v\rangle \quad \text{for all } v \in V.
$$

Therefore, we have

$$
\langle \mathrm{u}_h'(t), \Phi_i\rangle = \frac{\mathrm{d}}{\mathrm{d}t}(\mathrm{u}_h(t), \phi_i) = \sum_{j=1}^{n_{\text{glob}}} U_j'(t)(\Phi_j, \Phi_i).
$$

Setting $v_h = \Phi_i$ in Equations (6.6) and (6.7), and using the above expressions, we obtain the system of ordinary differential equations for the vector $U(t) = (U_j(t))$

$$
M\, U'(t) + A\, U(t) = F(t), \qquad M\, U(0) = G, \tag{6.9}
$$

where the matrices $M = (m_{ij})$, $A = (a_{ij})$, and the vectors $F(t) = (f_i(t))$ and $G = (g_i)$ are given by

$$
\begin{aligned}
m_{ij} &:= (\Phi_j, \Phi_i), \qquad & a_{ij} &:= (\nabla \Phi_j, \nabla \Phi_i) \quad i,j = 1,\ldots,n_{\text{glob}}, \\
f_i(t) &:= \langle \mathrm{f}(t), \Phi_i\rangle, \qquad & g_i &:= (u_0, \Phi_i) \quad i = 1,\ldots,n_{\text{glob}}.
\end{aligned}
$$

Here, M is called the mass matrix and A the stiffness matrix. The mass matrix is symmetric and positive definite. Indeed, for $\xi \in \mathbb{R}^{n_{\text{glob}}}$ we have

$$\xi^T M\xi = \sum_{i,j=1}^{n_{\text{glob}}} \xi_i m_{ij} \xi_j = \sum_{i,j=1}^{n_{\text{glob}}} \xi_i (\Phi_j, \Phi_i)\xi_j = \left(\sum_{j=1}^{n_{\text{glob}}} \xi_j \Phi_j, \sum_{i=1}^{n_{\text{glob}}} \xi_i \Phi_i \right) = \left\| \sum_{i=1}^{n_{\text{glob}}} \xi_i \Phi_i \right\|^2$$

from which $\xi^T M\xi > 0$ for all $0 \neq \xi \in \mathbb{R}^{n_{\text{glob}}}$ follows. Therefore, the inverse of M exists and problem (6.9) can be written as

$$U'(t) + M^{-1}AU(t) = M^{-1}F(t), \qquad U(0) = M^{-1}G. \tag{6.10}$$

The unique solvability of problem (6.10) depending on the regularity of F can be shown by means of standard theorems in the theory of ordinary differential equations. Next, we state a stability result.

Lemma 6.2. *Suppose* $u_0 \in L^2(\Omega)$ *and* $\mathrm{f} \in L^2(0, \mathrm{I}; H)$, *then the solution of the semidiscrete problem* (6.6), (6.7) *depends continuously on* u_0 *and* f, *that is,*

$$\|\mathrm{u}_h(t)\|_0 \leq \|u_0\|_0 + \int_0^t \|\mathrm{f}(\tau)\|_0 \, \mathrm{d}\tau, \quad t \in (0, \mathrm{I}].$$

Proof. Take $v = \mathrm{u}_h(t)$ in Equation (6.6) and apply Lemma 6.1 (integration by parts formula) to get

$$\frac{1}{2}\frac{\mathrm{d}}{\mathrm{d}t}\left(\|\mathrm{u}_h(t)\|_0^2\right) + \|\nabla \mathrm{u}_h(t)\|_0^2 \leq \|\mathrm{f}(t)\|_0 \, \|\mathrm{u}_h(t)\|_0.$$

Assuming that $\|\mathrm{u}_h(t)\|_0$ is differentiable, we conclude

$$\|\mathrm{u}_h(t)\|_0 \frac{\mathrm{d}}{\mathrm{d}t}\|\mathrm{u}_h(t)\|_0 = \frac{1}{2}\frac{\mathrm{d}}{\mathrm{d}t}\left(\|\mathrm{u}_h(t)\|_0^2\right) \leq \|\mathrm{f}(t)\|_0 \, \|\mathrm{u}_h(t)\|_0$$

and integration over $(0, t)$ gives the result. Next, we discuss a regularization of the term with the time derivative and show how difficulties can be avoided when $\|\mathrm{u}_h(t)\|_0$ is not differentiable for $\mathrm{u}_h(t) = 0$, see Thomée (2006). Let ε be positive. Taking into consideration that

$$\frac{1}{2}\frac{\mathrm{d}}{\mathrm{d}t}\left(\|\mathrm{u}_h(t)\|_0^2 + \varepsilon^2\right) = \frac{1}{2}\frac{\mathrm{d}}{\mathrm{d}t}\left(\|\mathrm{u}_h(t)\|_0^2\right) \leq \|\mathrm{f}(t)\|_0 \sqrt{\|\mathrm{u}_h(t)\|_0^2 + \varepsilon^2}$$

and

$$\frac{1}{2}\frac{\mathrm{d}}{\mathrm{d}t}\left(\|\mathrm{u}_h(t)\|_0^2 + \varepsilon^2\right) = \sqrt{\|\mathrm{u}_h(t)\|_0^2 + \varepsilon^2}\,\frac{\mathrm{d}}{\mathrm{d}t}\sqrt{\|\mathrm{u}_h(t)\|_0^2 + \varepsilon^2},$$

we obtain the inequality

$$\frac{\mathrm{d}}{\mathrm{d}t}\sqrt{\|\mathrm{u}_h(t)\|_0^2 + \varepsilon^2} \leq \|\mathrm{f}(t)\|_0.$$

Again integration over $(0,t)$ leads to

$$\sqrt{\|\mathrm{u}_h(t)\|_0^2+\varepsilon^2} \leq \sqrt{\|\mathrm{u}_h(0)\|_0^2+\varepsilon^2}+\int_0^t \|\mathrm{f}(\tau)\|_0 \,\mathrm{d}\tau$$

and the limit $\varepsilon \to 0$ proves the statement of the lemma. ■

The stability estimate in Lemma 6.2 implies the boundedness of a sequence of solutions of the semidiscrete problem for given fixed data. As a consequence, there is a subsequence of solutions u_h that converges weakly to a limit in $L^2(0,\mathrm{I};V)$. The existence of a weak solution is obtained by passing the weak limits. Finally, the uniqueness follows directly from the stability estimate of Lemma 6.2 by setting $u_0=0$ and $\mathrm{f}=0$. Since a solution of problem (6.6), (6.7) exists, is unique, and depends continuously on the data, problem (6.6), (6.7) is well-posed.

Now we derive an error estimate for the solution of the semidiscrete problem. Let V_h be the finite element space of continuous, piecewise polynomials of degree r. We introduce the Ritz projection $R_h : V \to V_h$ given by

$$(\nabla(u-R_hu),\nabla v_h)=0 \quad \text{for all } v_h \in V_h. \tag{6.11}$$

The Ritz projection $R_hu \in V_h$ is the finite element solution of the elliptic problem with the solution $u \in V$. We recall the error estimate

$$\|R_hu-u\|_0+h|R_hu-u|_1 \leq Ch^{s+1}\|u\|_{s+1}, \qquad u \in H^{s+1}(\Omega), 1 \leq s \leq r.$$

Theorem 6.2. *Let* u *and* u_h *be the solutions of the continuous problem* (6.5) *and the discrete problem* (6.6), (6.7), *respectively. Then, the error estimate*

$$\|\mathrm{u}_h(t)-\mathrm{u}(t)\|_0 \leq \|\mathrm{u}_h(0)-\mathrm{u}(0)\|_0+Ch^{r+1}\left(\|u_0\|_{r+1}+\int_0^t \|\mathrm{u}'(\tau)\|_{r+1}\,\mathrm{d}\tau\right) \quad \text{for } t \geq 0$$

holds true.

Proof. We split the error into two parts

$$\mathrm{u}_h-\mathrm{u}=(\mathrm{u}_h-R_h\mathrm{u})+(R_h\mathrm{u}-\mathrm{u})=\xi+\rho.$$

An upper bound for the second term follows directly from the estimates for the Ritz projection

$$\begin{aligned}\|\rho(t)\|_0 &\leq Ch^{r+1}\|\mathrm{u}(t)\|_{r+1}=Ch^{r+1}\left\|u_0+\int_0^t \mathrm{u}'(\tau)\,\mathrm{d}\tau\right\|_{r+1}\\ &\leq Ch^{r+1}\left(\|u_0\|_{r+1}+\int_0^t \|\mathrm{u}'(\tau)\|_{r+1}\,\mathrm{d}\tau\right).\end{aligned}$$

For $\xi(t) \in V_h$, we have the error equation

$$\begin{aligned}\langle \xi'(t),v_h\rangle+(\nabla\xi(t),\nabla v_h) &= \langle \mathrm{f}(t),v_h\rangle-\langle (R_h\mathrm{u})'(t),v_h\rangle-(\nabla R_h\mathrm{u}(t),\nabla v_h)\\ &= \langle \mathrm{u}'(t),v_h\rangle-\langle R_h(\mathrm{u}'(t)),v_h\rangle+(\nabla(\mathrm{u}(t)-R_h\mathrm{u}(t)),\nabla v_h)\\ &= \langle \mathrm{u}'(t)-R_h(\mathrm{u}'(t)),v_h\rangle,\end{aligned}$$

where we used the commutation property of the Ritz projection with the time derivative $(R_h\mathrm{u})'(t) = R_h(\mathrm{u}'(t))$. We see that $\xi(t) \in V_h$ is the solution of a semidiscrete problem to the right hand side $\mathrm{u}'(t) - R_h(\mathrm{u}'(t))$. Applying the stability estimate of Lemma 6.2 we have

$$\|\xi(t)\|_0 \le \|\xi(0)\|_0 + \int_0^t \|\mathrm{u}'(t) - R_h(\mathrm{u}'(t))\|_0 \, d\tau$$

with

$$\begin{aligned}\|\xi(0)\|_0 &= \|\mathrm{u}_h(0) - (R_h\mathrm{u})(0)\|_0 \le \|\mathrm{u}_h(0) - \mathrm{u}(0)\|_0 + \|\mathrm{u}(0) - (R_h\mathrm{u})(0)\|_0 \\ &\le \|\mathrm{u}_h(0) - \mathrm{u}(0)\|_0 + Ch^{r+1}\|u_0\|_{r+1}\end{aligned}$$

and

$$\|\mathrm{u}'(t) - R_h(\mathrm{u}'(t))\|_0 \le Ch^{r+1}\|\mathrm{u}'(t)\|_{r+1}.$$

Collecting all estimates we get the statement of the theorem. ■

6.5. Time discretization

As we saw in the previous section, semidiscretization of a linear parabolic initial-boundary value problem leads to an initial value problem of a system of ordinary differential equations (ODEs). One could think to solve the initial value problem (6.9) by some black-box ODE solver but the ODE systems generated by semidiscretization are *stiff systems.* For the heat Equation (6.5) in 1d and a semidiscretization by means of continuous, piecewise linear elements it turns out that the matrix $M^{-1}A$ has positive real eigenvalues, some of them are of moderate size, but others are very large typically $\mathcal{O}(1/h^2)$. In order to avoid the enforced use of very small time step sizes, one has to use a numerical method with certain stability properties to discretize the stiff systems. We start our study of suitable numerical methods for solving problem (6.9) or problem (6.10) by introducing the concept of A-*stability* and related notions.

A-stability and L-stability

In order to define different stability properties we consider for a given $\lambda \in \mathbb{C}$ with $\Re(\lambda) < 0$ and a given initial value $u_0 \in \mathbb{C}$, the test problem

Find a solution, $u : [0,\mathrm{I}] \to \mathbb{C}$ such that for all $t \in [0,\mathrm{I}]$

$$u'(t) = \lambda u(t), \qquad u(0) = u_0. \tag{6.12}$$

The solution of this problem is $u(t) = u_0 \exp(\lambda t)$ and $u \to 0$ as $t \to \infty$.

Let $0 = t_0 < t_1 < \cdots < t_{\mathcal{N}} = \mathrm{I}$ be an uniform decomposition of the considered time interval $[0,\mathrm{I}]$, and $\delta_t = t_n - t_{n-1}$, $1 \le n \le \mathcal{N}$, be the time step size. Using Taylor expansion at $t^* = \theta t_n + (1-\theta)t_{n-1}$ for $0 \le \theta \le 1$,

$$\frac{u(t_n)-u(t_{n-1})}{\delta_t} = u'(t^*) + \frac{(1-2\theta)}{2}\delta_t u''(t^*) + \mathcal{O}(\delta_t^2),$$
$$\theta u(t_n) + (1-\theta)u(t_{n-1}) = u(t^*) + \mathcal{O}(\delta_t^2)$$

and approximating $u(t_n)$ by u^n, $n = 0, 1, \ldots, \mathcal{N}$, the general θ-time stepping scheme applied to problem (6.12) becomes

$$\frac{u^n - u^{n-1}}{\delta_t} = \lambda(\theta u^n + (1-\theta)u^{n-1})).$$

With $z = \lambda\delta_t$, it reads

$$u^n - \theta z u^n = u^{n-1} + (1-\theta) z u^{n-1}, \quad 1 \le n \le \mathcal{N}.$$

Moreover, it can be written as

$$u^n = R(z)u^{n-1}, \qquad R(z) = \frac{1+(1-\theta)z}{1-\theta z}.$$

Here, $R(z)$ is the stability function of the θ-time-stepping scheme. Note, that for any one-step method we can find a stability function $R(z)$ satisfying the relation $u^n = R(z)\, u^{n-1}$, $1 \le n \le \mathcal{N}$. For the exact solution u of the test problem (6.12), the similar relationship

$$u(t_n) = u_0 \exp(\lambda(t_{n-1} + \delta_t)) = \exp(z)\, u(t_{n-1}), \quad 1 \le n \le \mathcal{N}$$

holds true. If the time step size δ_t tends to zero, we have $z \to 0$ and $R(z)$ is an approximation of $\exp(z)$. Indeed, for the θ-time-stepping scheme we have

$$R(z) = 1 + z + \theta z^2 + \mathcal{O}(z^3) = \exp(z) + \left(\theta - \frac{1}{2}\right) z^2 + \mathcal{O}(z^3) \quad \text{for } z \to 0.$$

We would like to pass certain properties of the solution of the test problem to the solution of the discrete scheme. First, we observe the boundedness of u in the left half-plane since

$$\mathbb{C}^- := \{z \in \mathbb{C} : \Re(z) \le 0\} = \{z \in \mathbb{C} : |\exp(z)| \le 1\}.$$

From the basic numerical methods, we know that $u^{\mathcal{N}} = [R(z)]^{\mathcal{N}} u_0$ is bounded iff $|R(z)| \le 1$. Since $z = \delta_t\lambda$ with $\Re(\lambda) \le 0$ and we aim to get boundedness of $(u^n)_{n\in\mathbb{N}}$ independent of the time step size $\delta_t > 0$, we require $|R(z)| \le 1$ for all $z \in \mathbb{C}^-$. This observation is the basis for the following definitions.

Definition 6.4. The region of absolute stability or the stability region of a time-stepping method with the stability function $R(z)$ is the set

$$\mathbb{D} = \{z \in \mathbb{C} : |R(z)| \le 1\}.$$

Definition 6.5. A time-stepping method is called A-stable if the region of absolute stability $\mathbb{D}$ contains the set $\mathbb{C}^-$, that is, if $\mathbb{D} \supseteq \mathbb{C}^-$.

In order to guarantee that independent of the time step size $u^{\mathcal{N}} \to 0$ as $t \to \infty$ we need the stronger requirement $|R(z)| < 1$.

Definition 6.6. A time-stepping method is called *strongly* A-stable, if it is A-stable and

$$\lim_{\Re(z)\to-\infty} |R(z)| < 1.$$

For moderate and larger time step sizes δ_t, the following property of the solution of the test problem (asymptotic behaviour in one step) is remarkable

$$\lim_{\delta_t\Re(\lambda)\to-\infty} u(t_{n-1}+\delta_t) = 0.$$

The next definition aims to transfer this property to the discrete case.

Definition 6.7. A time-stepping method is called L-stable, if it is A-stable and

$$\lim_{\Re(z)\to-\infty} |R(z)| = 0.$$

Even though the A-stable methods converge without any time step restriction and damp all decaying components, a strongly A-stable or L-stable method is necessary to damp the rapidly decaying components. As a consequence, strongly A-stable or L-stable methods are preferred for stiff problems with oscillating data.

We now examine the A-stability property of the θ-schemes. We start with the forward Euler method, that is, $\theta = 0$. The stability function of the forward Euler is

$$R(z) = 1 + z.$$

Thus, the region of stability of the forward Euler method is the unit circular domain center at $(-1,0)$ that is wholly inscribed in the left half-plane. Therefore, the forward Euler method is not A-stable. It can be shown that all explicit one-step methods, including the large class of explicit Runge–Kutta schemes, are not A-stable. Therefore, we focus only on implicit methods.

Let us first consider the implicit backward Euler method, that is, $\theta = 1$. The stability function of the backward Euler is

$$R(z) = \frac{1}{1-z}.$$

Hence, the region of stability of the backward Euler method is the exterior of the unit circular center at $(1,0)$, that is, the entire left half-plane and a major part of the right half-plane. Therefore, the backward Euler method is A-stable, in fact, it is L-stable due to $\lim_{\Re(z)\to-\infty} |R(z)| = 0$.

We next consider the Crank–Nicolson method, which is one of the most popular second order time-stepping method. It is the θ-scheme for $\theta = 1/2$. The stability function of the Crank–Nicolson method is

$$R(z) = \frac{1+\frac{1}{2}z}{1-\frac{1}{2}z}.$$

Hence, the stability region of the Crank–Nicolson method is left half-plane, that is, $\mathbb{D} = \mathbb{C}^-$. Thus, the Crank–Nicolson method is A-stable. However, the method is not strongly A-stable because $\lim_{\Re(z)\to-\infty} |R(z)| = 1$.

Time-stepping methods

Let V_h be the finite element space of continuous, piecewise polynomials of degree less than or equal to r. The weak formulation of the test problem (6.5) discretized in space leads to the semidiscrete form

Find $\mathrm{u}_h(t) \in V_h$ such that for all $v \in V_h$

$$\begin{aligned} \langle \mathrm{u}_h'(t), v\rangle + (\nabla \mathrm{u}_h(t), \nabla v) &= \langle f(t), v\rangle, \\ (\mathrm{u}_h(0), v) &= (u_0, v). \end{aligned} \tag{6.13}$$

Let u_h^n be an approximation of the solution $\mathrm{u}_h(t)$ of the semidiscrete problem at $t = t_n$. Applying the general θ-scheme, the discrete form of problem (6.5) in the interval (t_{n-1}, t_n) reads

For given $\mathrm{u}_h^{n-1}, f^{n-1}, f^n$ with u_h^0 the L^2-projection of u_0, find $\mathrm{u}_h^n \in V_h$ such that

$$\left(\frac{\mathrm{u}_h^n - \mathrm{u}_h^{n-1}}{\delta_t}, v\right) + \left(\theta \nabla \mathrm{u}_h^n + (1-\theta)\nabla \mathrm{u}_h^{n-1}, \nabla v\right) = \left(f^{n-1+\theta}, v\right) \tag{6.14}$$

for all $v \in V_h$, where $f^{n-1+\theta} = \theta f^n + (1-\theta) f^{n-1}$.

Backward Euler method

The backward Euler method is one of the simplest implicit and L-stable time discretization scheme. By choosing $\theta = 1$ in problem (6.14), we obtain the backward Euler scheme

For given u_h^{n-1}, f^n with u_h^0 the L^2-projection of u_0, find $\mathrm{u}_h^n \in V_h$ such that

$$\left(\frac{\mathrm{u}_h^n - \mathrm{u}_h^{n-1}}{\delta_t}, v\right) + \left(\nabla \mathrm{u}_h^n, \nabla v\right) = \left(f^n, v\right) \tag{6.15}$$

for all $v \in V_h$.

Lemma 6.3. (*Stability*) *The discrete solution of problem* (6.15) *depends continuously on* u_0 *and* f, *that is,*

$$\left\|\mathrm{u}_h^{\mathcal{N}}\right\|_0 \le \|u_0\|_0 + \sum_{n=1}^{\mathcal{N}} \delta_t \|f^n\|_0.$$

Proof. Set $v = u_h^n$ in problem (6.15) to get

$$\left(\mathrm{u}_h^n - \mathrm{u}_h^{n-1}, \mathrm{u}_h^n\right) + \delta_t |\mathrm{u}_h^n|_1^2 = \delta_t \left(f^n, \mathrm{u}_h^n\right).$$

The term $|\mathrm{u}_h^n|_1^2$ is nonnegative, further the Schwarz inequality yields

$$\begin{aligned} &\|\mathrm{u}_h^n\|_0^2 \le \left(\mathrm{u}_h^{n-1}, \mathrm{u}_h^n\right) + \delta_t \left(f^n, \mathrm{u}_h^n\right) \le \|\mathrm{u}_h^{n-1}\|_0 \|\mathrm{u}_h^n\|_0 + \delta_t \|f^n\|_0 \|\mathrm{u}_h^n\|_0, \\ &\|\mathrm{u}_h^n\|_0 \le \|\mathrm{u}_h^{n-1}\|_0 + \delta_t \|f^n\|_0. \end{aligned}$$

Summation over $n = 1,\ldots,\mathcal{N}$ provides the desired estimate. ■

In the following estimates we suppose enough regularity of the solution and the data such that the appearing terms exist.

Theorem 6.3. *Suppose* $\mathrm{u}_h^n \in V_h$, $1 \le n \le \mathcal{N}$, *and* u *be the solutions of the discrete problem* (6.15) *and the continuous problem* (6.5), *respectively, and* $\|\mathrm{u}_h^0 - u_0\|_0 \le Ch^{r+1}\|u_0\|_{r+1}$. *Then, we have*

$$\left\|\mathrm{u}_h^{\mathcal{N}} - \mathrm{u}(t_{\mathcal{N}})\right\|_0 \le Ch^{r+1}\left(\|u_0\|_{r+1} + \int_0^{t_{\mathcal{N}}} \left\|\mathrm{u}'(\tau)\right\|_{r+1} \mathrm{d}\tau\right) + \delta_t \int_0^{t_{\mathcal{N}}} \left\|\mathrm{u}''(\tau)\right\|_0 \mathrm{d}\tau.$$

Proof. Let the approximation error

$$\mathrm{u}_h^{\mathcal{N}} - \mathrm{u}(t_{\mathcal{N}}) = u_h^{\mathcal{N}} - R_h\mathrm{u}(t_{\mathcal{N}}) + R_h\mathrm{u}(t_{\mathcal{N}}) - \mathrm{u}(t_{\mathcal{N}}) = \xi^{\mathcal{N}} + \rho^{\mathcal{N}},$$

where $\xi^{\mathcal{N}} = u_h^{\mathcal{N}} - R_h\mathrm{u}(t_{\mathcal{N}})$ and $\rho^{\mathcal{N}} = R_h\mathrm{u}(t_{\mathcal{N}}) - \mathrm{u}(t_{\mathcal{N}})$. Moreover, the projection error estimate follows from the estimates of the Ritz projection, that is,

$$\begin{aligned}\|\rho(t_{\mathcal{N}})\|_0 &\le Ch^{r+1}\|\mathrm{u}(t_{\mathcal{N}})\|_{r+1} = Ch^{r+1}\left\|u_0 + \int_0^{t_{\mathcal{N}}} \mathrm{u}'(\tau)\,\mathrm{d}\tau\right\|_{r+1} \\ &\le Ch^{r+1}\left(\|u_0\|_{r+1} + \int_0^{t_{\mathcal{N}}} \|\mathrm{u}'(\tau)\|_{r+1}\mathrm{d}\tau\right).\end{aligned}$$

Next, for $1 \le n \le \mathcal{N}$, we have

$$\begin{aligned}(\xi^n - \xi^{n-1},\xi^n) + \delta_t(\nabla\xi^n,\nabla\xi^n) =& (\mathrm{u}_h^n - \mathrm{u}_h^{n-1},\xi^n) + \delta_t(\nabla \mathrm{u}_h^n,\nabla\xi^n) \\ &- \left(R_h\mathrm{u}(t_n) - R_h\mathrm{u}(t_{n-1}),\xi^n\right) - \delta_t\left(\nabla(R_h\mathrm{u}(t_n)),\nabla\xi^n\right).\end{aligned}$$

Using the Ritz projection (6.11), (6.15) and then (6.3), we get

$$(\xi^n - \xi^{n-1},\xi^n) + \delta_t|\xi^n|_1^2 = \delta_t(f^n,\xi^n) - \delta_t(\nabla\mathrm{u}(t_n),\nabla\xi^n) - (R_h\mathrm{u}(t_n) - R_h\mathrm{u}(t_{n-1}),\xi^n).$$

Since $|\xi^n|_1^2$ is nonnegative and further using (6.5), we obtain

$$\|\xi^n\|_0^2 \le (\xi^{n-1},\xi^n) + \delta_t\left(\frac{\partial \mathrm{u}(t_n)}{\partial t},\xi^n\right) - \left(R_h\mathrm{u}(t_n) - R_h\mathrm{u}(t_{n-1}),\xi^n\right).$$

Further, applying the Schwarz inequality yields

$$\begin{aligned}\|\xi^n\|_0 \le \|\xi^{n-1}\|_0 + &\left\|\delta_t\frac{\partial \mathrm{u}(t_n)}{\partial t} - (\mathrm{u}(t_n) - \mathrm{u}(t_{n-1}))\right\|_0 \\ &+ \|\mathrm{u}(t_n) - \mathrm{u}(t_{n-1}) - R_h(\mathrm{u}(t_n) - \mathrm{u}(t_{n-1}))\|_0.\end{aligned}$$

Now denoting

$$S_1^n = \delta_t\frac{\partial \mathrm{u}(t_n)}{\partial t} - (\mathrm{u}(t_n) - \mathrm{u}(t_{n-1})), \qquad S_2^n = \mathrm{u}(t_n) - \mathrm{u}(t_{n-1}) - R_h(\mathrm{u}(t_n) - \mathrm{u}(t_{n-1})),$$

we have

$$\|\xi^n\|_0 \le \|\xi^{n-1}\|_0 + \|S_1^n\|_0 + \|S_2^n\|_0.$$

Using the Taylor theorem with integral remainder

$$f(x) = f(a) + (x-a)f'(a) + \ldots + \frac{(x-a)^m}{m!} f^{(m)}(a) + \int_a^x \frac{(x-\tau)^m}{m!} f^{(m+1)}(\tau)\,\mathrm{d}\tau$$

for $m = 1$ and $m = 0$, respectively, we get

$$\|S_1^n\|_0 = \left\| \int_{t_{n-1}}^{t_n} (t_n - \tau)\mathrm{u}''(\tau)\,\mathrm{d}\tau \right\|_0 \le \delta_t \int_{t_{n-1}}^{t_n} \left\|\mathrm{u}''(\tau)\right\|_0 \mathrm{d}\tau,$$

$$\|S_2^n\|_0 = \|(\mathbb{I} - R_h)(\mathrm{u}(t_n) - \mathrm{u}(t_{n-1}))\|_0 = \left\| (\mathbb{I} - R_h) \int_{t_{n-1}}^{t_n} \mathrm{u}'(\tau)\,\mathrm{d}\tau \right\|_0$$

$$\le Ch^{r+1} \int_{t_{n-1}}^{t_n} \left\|\mathrm{u}'(\tau)\right\|_{r+1} \mathrm{d}\tau.$$

Using these estimates and applying summation over $n = 1, \ldots, \mathcal{N}$, we obtain

$$\|\xi^{\mathcal{N}}\|_0 \le \|\xi^0\|_0 + \delta_t \int_0^{t_{\mathcal{N}}} \left\|\mathrm{u}''(\tau)\right\|_0 \mathrm{d}\tau + Ch^{r+1} \int_0^{t_{\mathcal{N}}} \left\|\mathrm{u}'(\tau)\right\|_{r+1} \mathrm{d}\tau.$$

Finally, applying the estimate

$$\|\xi^0\|_0 = \left\|u_h^0 - R_h\mathrm{u}(0)\right\|_0 \le \left\|u_h^0 - u_0\right\|_0 + \|u_0 - R_h u_0\|_0 \le 2Ch^{r+1} \|u_0\|_{r+1},$$

we obtain the statement of the theorem. ∎

Next, using the definition of the discrete function (6.8) as in (6.9), the algebraic system of equations for the backward Euler becomes

$$(M + \delta_t A)\, U^n = MU^{n-1} + \delta_t F^n, \tag{6.16}$$

where the mass and stiffness matrix, and the load vector are defined as in (6.9).

Crank–Nicolson method

The backward Euler scheme is first order accurate in time. Nevertheless, a second order accurate in time can be obtained by symmetrically discretizing the equation in time at $t_{n-1/2}$. Choosing $\theta = 1/2$ in problem (6.14), the implicit Crank–Nicolson scheme in the time interval (t_{n-1}, t_n) reads

For given $\mathrm{u}_h^{n-1}, f^{n-1/2}$ with u_h^0 the L^2-projection of u_0, find $\mathrm{u}_h^n \in V_h$ such that

$$\left(\frac{\mathrm{u}_h^n - \mathrm{u}_h^{n-1}}{\delta_t}, v \right) + \frac{1}{2}\left(\nabla \mathrm{u}_h^n + \nabla \mathrm{u}_h^{n-1}, \nabla v\right) = (f^{n-1/2}, v) \tag{6.17}$$

for all $v \in V_h$. Since the Crank–Nicolson method approximates the differential equation at $t_{n-1/2}$, the source term is written as $(f^{n-1/2}, v)$.

Lemma 6.4. (*Stability*) *The discrete solution of problem* (6.17) *depends continuously on* u_0 *and* f, *that is,*

$$\left\| \mathrm{u}_h^{\mathcal{N}} \right\|_0 \leq \| u_0 \|_0 + \sum_{n=1}^{\mathcal{N}} \delta_t \| f^{n-1/2} \|_0.$$

Proof. Setting $v = \mathrm{u}_h^n + \mathrm{u}_h^{n-1}$ in (6.17), we get

$$\left(\mathrm{u}_h^n - \mathrm{u}_h^{n-1}, \mathrm{u}_h^n + \mathrm{u}_h^{n-1} \right) + \delta_t \frac{1}{2} |\mathrm{u}_h^n + \mathrm{u}_h^{n-1}|_1^2 = \delta_t \left(f^{n-1/2}, \mathrm{u}_h^n + \mathrm{u}_h^{n-1} \right).$$

The term $|\mathrm{u}_h^n + \mathrm{u}_h^{n-1}|_1^2$ is nonnegative, thus applying Schwarz inequality yields

$$\|\mathrm{u}_h^n\|_0^2 - \|\mathrm{u}_h^{n-1}\|_0^2 \leq \delta_t \|f^{n-1/2}\|_0 \|\mathrm{u}_h^n + \mathrm{u}_h^{n-1}\|_0 \leq \delta_t \|f^{n-1/2}\|_0 \left(\|\mathrm{u}_h^n\|_0 + \|\mathrm{u}_h^{n-1}\|_0 \right).$$

Cancelling $\|\mathrm{u}_h^n\|_0 + \|\mathrm{u}_h^{n-1}\|_0$ on both sides, and summing over $n = 1, \ldots, \mathcal{N}$, the lemma is proved. ∎

Theorem 6.4. *Suppose* $\mathrm{u}_h^n \in V_h$, $1 \leq n \leq \mathcal{N}$, *and* u *be the solutions of the discrete problem* (6.17) *and the continuous problem* (6.5), *respectively, and* $\left\| \mathrm{u}_h^0 - u_0 \right\|_0 \leq Ch^{r+1} \| u_0 \|_2$. *Then, we have*

$$\left\| \mathrm{u}_h^{\mathcal{N}} - \mathrm{u}(t_{\mathcal{N}}) \right\|_0 \leq Ch^{r+1} \left(\| u_0 \|_{r+1} + \int_0^{t_{\mathcal{N}}} \left\| \mathrm{u}'(\tau) \right\|_{r+1} d\tau \right)$$
$$+ C\delta_t^2 \int_0^{t_{\mathcal{N}}} \left(\left\| \mathrm{u}'''(\tau) \right\|_0 + \left\| \Delta \mathrm{u}''(\tau) \right\|_0 \right) d\tau.$$

Proof. Let the approximation error be split into

$$\mathrm{u}_h^{\mathcal{N}} - \mathrm{u}(t_{\mathcal{N}}) = u_h^{\mathcal{N}} - R_h \mathrm{u}(t_{\mathcal{N}}) + R_h \mathrm{u}(t_{\mathcal{N}}) - \mathrm{u}(t_{\mathcal{N}}) = \xi^{\mathcal{N}} + \rho^{\mathcal{N}},$$

where $\xi^{\mathcal{N}} = u_h^{\mathcal{N}} - R_h \mathrm{u}(t_{\mathcal{N}})$ and $\rho^{\mathcal{N}} = R_h \mathrm{u}(t_{\mathcal{N}}) - \mathrm{u}(t_{\mathcal{N}})$. An estimate for the projection error, $\rho^{\mathcal{N}}$, can be derived as in Theorem 6.3. Next, for $1 \leq n \leq \mathcal{N}$,

$$\left(\xi^n - \xi^{n-1}, v \right) + \frac{\delta_t}{2} \left(\nabla \xi^n + \nabla \xi^{n-1}, \nabla v \right)$$
$$= \left(u_h^n - u_h^{n-1}, v \right) + \frac{\delta_t}{2} \left(\nabla \mathrm{u}_h^n + \nabla \mathrm{u}_h^{n-1}, \nabla v \right) - \left(R_h \mathrm{u}(t_n) - R_h \mathrm{u}(t_{n-1}), v \right)$$
$$- \frac{\delta_t}{2} \left(\nabla \left(R_h \mathrm{u}(t_n) \right) + \nabla \left(R_h \mathrm{u}(t_{n-1}) \right), \nabla v \right)$$

for all $v \in V_h$. Using the Ritz projection (6.11), (6.17) and then (6.3), we obtain

$$\begin{aligned}
&(\xi^n - \xi^{n-1}, v) + \frac{\delta_t}{2}\left(\nabla \xi^n + \nabla \xi^{n-1}, \nabla v\right) \\
&= \delta_t\left(f^{n-1/2}, v\right) - \delta_t\left(\nabla \mathrm{u}\left(t_{n-1/2}\right), \nabla v\right) - \frac{\delta_t}{2}\left(\nabla\left(\mathrm{u}(t_n) + \mathrm{u}(t_{n-1}) - 2\mathrm{u}\left(t_{n-1/2}\right)\right), \nabla v\right) \\
&\qquad - \left(R_h \mathrm{u}(t_n) - R_h \mathrm{u}(t_{n-1}), v\right) \\
&= \left(\delta_t \frac{\partial u^{n-1/2}}{\partial t} + \frac{\delta_t}{2}\Delta\left(\mathrm{u}(t_n) + \mathrm{u}(t_{n-1}) - 2\mathrm{u}\left(t_{n-1/2}\right)\right) - R_h\left(\mathrm{u}(t_n) - \mathrm{u}(t_{n-1})\right), v\right).
\end{aligned}$$

Let $v = \xi^n + \xi^{n-1}$, then we have

$$\begin{aligned}
&\|\xi^n\|_0^2 - \|\xi^{n-1}\|_0^2 + \frac{\delta_t}{2}\left|\xi^n + \xi^{n-1}\right|_1^2 \\
&\qquad \le \left(\delta_t \frac{\partial u^{n-1/2}}{\partial t} - \left(\mathrm{u}(t_n) - \mathrm{u}(t_{n-1})\right), \xi^n + \xi^{n-1}\right) \\
&\qquad\quad + \left(\left(\mathrm{u}(t_n) - \mathrm{u}(t_{n-1})\right) - R_h\left(\mathrm{u}(t_n) - \mathrm{u}(t_{n-1})\right), \xi^n + \xi^{n-1}\right) \\
&\qquad\quad + \frac{\delta_t}{2}\left(\Delta\left(\mathrm{u}(t_n) + \mathrm{u}(t_{n-1}) - 2\mathrm{u}\left(t_{n-1/2}\right)\right), \xi^n + \xi^{n-1}\right).
\end{aligned}$$

The term $|\xi^n + \xi^{n-1}|_1^2$ is nonnegative, applying the Schwarz inequality and canceling the common factor $(\|\xi^n\|_0 + \|\xi^{n-1}\|_0)$ on both sides yield

$$\begin{aligned}
\|\xi^n\|_0 &\le \|\xi^{n-1}\|_0 + \left\|\delta_t \frac{\partial u^{n-1/2}}{\partial t} - \left(\mathrm{u}(t_n) - \mathrm{u}(t_{n-1})\right)\right\|_0 \\
&\quad + \left\|\mathrm{u}(t_n) - \mathrm{u}(t_{n-1}) - R_h\left(\mathrm{u}(t_n) - \mathrm{u}(t_{n-1})\right)\right\|_0 \\
&\quad + \delta_t\left\|\Delta\left(\frac{1}{2}\left(\mathrm{u}(t_n) + \mathrm{u}(t_{n-1})\right) - \mathrm{u}\left(t_{n-1/2}\right)\right)\right\|_0 \\
&= \|\xi^{n-1}\|_0 + \|S_1^n\|_0 + \|S_2^n\|_0 + \delta_t\|S_3^n\|_0.
\end{aligned}$$

Applying the Taylor theorem with integral remainder, we get

$$\|S_1^n\|_0 \le \frac{1}{2}\left\|\int_{t_{n-1}}^{t_n} (\tau - t_{n-1/2})^2 \mathrm{u}'''(\tau)\, d\tau\right\|_0 \le C\delta_t^2 \int_{t_{n-1}}^{t_n} \left\|\mathrm{u}'''(\tau)\right\|_0 d\tau.$$

Further, the estimate of S_2^n term is the same as in Theorem 6.3. Moreover,

$$\delta_t\|S_3^n\|_0 \le C\delta_t^2 \int_{t_{n-1}}^{t_n} \left\|\Delta \mathrm{u}''(\tau)\right\|_0 d\tau.$$

Hence, we obtain

$$\|\xi^n\|_0 \leq \|\xi^{n-1}\|_0 + C\delta_t^2 \int_{t_{n-1}}^{t_n} \left(\|u'''(\tau)\|_0 + \|\Delta u''(\tau)\|_0\right) d\tau$$

$$+ Ch^{r+1} \int_{t_{n-1}}^{t_n} \|u'(\tau)\|_{r+1} \, d\tau.$$

Finally, summing over $n = 1, \ldots, \mathcal{N}$ and bounding $\|\xi^0\|_0$ as in Theorem 6.3, provide the desired estimate. ■

Next, using the definition of the discrete function (6.8) as in (6.9), the algebraic system of equations for the Crank–Nicolson scheme becomes

$$\left(M + \frac{\delta_t}{2} A\right) U^n = \left(M - \frac{\delta_t}{2} A\right) U^{n-1} + \delta_t F^{n-1/2}, \tag{6.18}$$

where the mass and stiffness matrices are defined as in (6.16).

Fractional step θ-scheme

The backward Euler scheme is strongly A-stable but first order accurate in time and have strong numerical dissipation. Even though the Crank–Nicolson method is second order in time, it is neither strongly A-stable nor L-stable, and therefore it induces oscillation in the numerical solution when the initial or boundary data contain noise. Another popular method that is second order accurate in time and also strongly A-stable is the fractional step θ-scheme. This scheme was first proposed by Bristeau et al. (1987) as an operator splitting method to solve the incompressible Navier–Stokes equation by splitting the two essential effects incompressibility and nonlinearity. We split each time interval (t_{n-1}, t_n) into three subintervals as $(t_{n-1}, t_{n-1+\theta})$, $(t_{n-1+\theta}, t_{n-\theta})$ and $(t_{n-\theta}, t_n)$ with $t_{n-1+\theta} = t_{n-1} + \delta_t\theta$, and $t_{n-\theta} = t_n - \delta_t\theta$. The time step sizes of the three substeps are $\theta\delta_t$, $(1-2\theta)\delta_t$ and $\theta\delta_t$. In the first and third step, a general α-scheme and in the second step a general β-scheme with $\beta = 1 - \alpha$ are used. Note that a general θ-scheme with time step size δ_t has been introduced in (6.14). In the context of the fractional step θ-scheme, we have to perform one after another a general α-scheme with step size $\theta\delta_t$, a general β-scheme with step size $(1-2\theta)\delta_t$, and a general α-scheme with step size $\theta\delta_t$. The application of the fractional step θ-scheme to the ordinary differential equation (6.12) results in

For given u^{n-1} with $u^0 = u_0$, find u^n such that

$$\frac{u^{n-1+\theta} - u^{n-1}}{\theta\delta_t} - \lambda\left[\alpha u^{n-1+\theta} + (1-\alpha)u^{n-1}\right] = 0, \tag{6.19}$$

$$\frac{u^{n-\theta} - u^{n-1+\theta}}{(1-2\theta)\delta_t} - \lambda\left[\beta u^{n-\theta} + (1-\beta)u^{n-1+\theta}\right] = 0, \tag{6.20}$$

$$\frac{u^n - u^{n-\theta}}{\theta\delta_t} - \lambda\left[\alpha u^n + (1-\alpha)u^{n-\theta}\right] = 0. \tag{6.21}$$

Moreover, with $z = \lambda\delta_t$,

$$u^{\mathcal{N}} = [R(z)]^{\mathcal{N}} u_0, \qquad R(z) = \frac{(1+\beta\theta z)^2(1+\alpha(1-2\theta)z)}{(1-\alpha\theta z)^2(1-\beta(1-2\theta)z)}.$$

Here, $R(z)$ is the stability function of the fractional step θ-scheme. It holds

$$\begin{aligned} R(z) &= 1 + z + \frac{z^2}{2} + z^2(1-2\alpha)\left(\theta^2 - 2\theta + \frac{1}{2}\right) + \mathcal{O}(z^3) \\ &= \exp(z) + z^2(1-2\alpha)\left(\theta^2 - 2\theta + \frac{1}{2}\right) + \mathcal{O}(z^3) \quad \text{for } z \to 0. \end{aligned}$$

We see that two parameter choices, $\alpha = 1/2$ and $\theta = 1 - \sqrt{2}/2$, lead to a scheme of second order, however, $\alpha = 1/2$ implies $\beta = 1/2$, that is, in the three subintervals of the fractional step θ-scheme we would apply just the Crank–Nicolson scheme (which is, of course, of second order). The alternative is to chose $\theta = 1 - \sqrt{2}/2$ and to select the parameter α to improve the stability properties. Since

$$\lim_{\Re(z)\to-\infty} |R(z)| = \frac{\beta}{\alpha} = \frac{1-\alpha}{\alpha},$$

the scheme will be strongly A-stable only when $\beta < \alpha$, and thus a value for α is chosen in the interval $(0.5, 1]$.

Comparing with the one step θ-schemes discussed earlier, the fractional step θ-scheme, Equations (6.19)–(6.21), uses three different values for θ in each subinterval in the time interval (t_{n-1}, t_n).

Applying the fractional step θ-scheme to the semidiscrete form (6.6), the fully discrete form of (6.6) in the interval (t_{n-1}, t_n) reads

Step 1. For given $\mathrm{u}^{n-1}, f^{n-1}, f^{n-1+\theta} \in L^2(\Omega)$ with $\mathrm{u}^0 = u_0$, find $\mathrm{u}_h^{n-1+\theta} \in V_h$ such that for all $\phi_i \in V_h$

$$\left(\frac{\mathrm{u}_h^{n-1+\theta} - \mathrm{u}_h^{n-1}}{\theta\delta_t}, \phi_i\right) + \left(\alpha\nabla\mathrm{u}_h^{n-1+\theta} + \beta\nabla\mathrm{u}_h^{n-1}, \nabla\phi_i\right) = \left(\alpha f^{n-1+\theta} + \beta f^{n-1}, \phi_i\right).$$

Step 2. For given $\mathrm{u}^{n-1+\theta}, f^{n-1+\theta}, f^{n-\theta} \in L^2(\Omega)$, find $\mathrm{u}_h^{n-\theta} \in V_h$ such that for all $\phi_i \in V_h$

$$\left(\frac{\mathrm{u}_h^{n-\theta} - \mathrm{u}_h^{n-1+\theta}}{(1-2\theta)\delta_t}, \phi_i\right) + \left(\beta\nabla\mathrm{u}_h^{n-\theta} + \alpha\nabla\mathrm{u}_h^{n-1+\theta}, \nabla\phi_i\right) = \left(\beta f^{n-\theta} + \alpha f^{n-1+\theta}, \phi_i\right).$$

Step 3. For given $\mathrm{u}^{n-\theta}, f^{n-\theta}, f^n \in L^2(\Omega)$, find $\mathrm{u}_h^n \in V_h$ such that for all $\phi_i \in V_h$

$$\left(\frac{\mathrm{u}_h^{n} - \mathrm{u}_h^{n-\theta}}{\theta\delta_t}, \phi_i\right) + \left(\alpha\nabla\mathrm{u}_h^{n} + \beta\nabla\mathrm{u}_h^{n-\theta}, \nabla\phi_i\right) = \left(\alpha f^{n} + \beta f^{n-\theta}, \phi_i\right).$$

Note that the scheme becomes second order in time for

$$\theta = 1 - \frac{\sqrt{2}}{2} = 0.29289\ldots$$

and is strongly A-stable for any $\alpha \in (1/2, 1]$. The choice

$$\alpha = \frac{1-2\theta}{1-\theta} = 0.58579\ldots$$

implies $\alpha\theta = \beta(1-2\theta)$, thus in each step of the algorithm above we have to solve a linear algebraic equation with the same coefficient matrix. Finally, we mention that free oscillations are well preserved for the fractional step θ-scheme in contrast to the backward Euler scheme, indeed for the stability functions we have

$$|R^{\mathrm{FS}}(i\delta_t)| = 0.9998\ldots \quad \text{and} \quad |R^{\mathrm{BE}}(i\delta_t)| = 0.7809\ldots$$

for $\delta_t = 0.8$.

Galerkin and Galerkin–Petrov time-stepping methods

Suppose the mesh size becomes small, then the ODE (6.6) often becomes more stiff, and it necessities implicit schemes for time discretization, as the region of absolute stability of explicit schemes will be small. The implicit schemes, backward Euler and fractional step θ-scheme are strongly A-stable but only first and second order, respectively, in time. Popular higher order differential methods such as implicit Runge–Kutta methods that are often used for the solution of the ordinary differential equations can be used for higher order temporal discretization of the semidiscrete problem too. However, these methods are not so popular for temporal discretization of partial differential equations due to storage and memory requirements, and several function evaluations. Alternatively, Galerkin methods can be used for higher order temporal discretizations of partial differential equations. The ansatz space in time consists of piecewise polynomials of degree less than or equal to k and can either be discontinuous or continuous. One way to fix the $(k+1)$ dof in each time step is to impose the continuity of the ansatz function to the previous time step and the orthogonality of the residual to a test space of dimension k. This approach leads to different ansatz and test spaces, we call this class continuous Galerkin–Petrov (cGP(k)) method. Alternatively, the continuity requirement is relaxed to a weak one and incorporated into the variational formulation. This allows us to use equal ansatz and test spaces, thus we get a discontinuous Galerkin (dG(k)) in time method.

Discontinuous Galerkin time-stepping methods

In the following, we present the basic concepts and construction of dG(k) and cGP(k) methods for the temporal discretization. In order to explain the basis steps of Galerkin temporal discretization in a simple way, we consider an initial value problem for a system of ordinary differential equations

$$\frac{\mathrm{du}(t)}{\mathrm{d}t} + B\mathrm{u}(t) = f(t) \quad \text{in } (0,\mathrm{I}], \ \ \mathrm{u}(0) = \mathrm{u}_0, \tag{6.22}$$

where $\mathrm{u} : [0,\mathrm{I}] \to \mathbb{R}^{\mathrm{d}}$ is the unknown vector-valued function, $B \in \mathbb{R}^{d\times d}$ is a given nonsingular matrix, and $f : [0,\mathrm{I}] \to \mathbb{R}^{\mathrm{d}}$ a given source term. Recall that $0 = t_0 < t_1 < \cdots < t_{\mathcal{N}} = \mathrm{I}$ is a decomposition of the time interval $[0,\mathrm{I}]$ with $\delta_t = t_n - t_{n-1}$, $1 \le n \le \mathcal{N}$. To describe a dG(k) time marching process, let $\mathcal{I}_n := (t_{n-1}, t_n]$ be a left open and right closed time interval, whereas $\mathcal{I}^{\delta t} := \{\mathcal{I}_1, \ldots, \mathcal{I}_{\mathcal{N}}\}$ is a time mesh and $\mathcal{I} = [0,\mathrm{I}]$ is the given time interval. Now, we define

$$\mathbb{P}_k^{\mathrm{disc}}(\mathcal{I}^{\delta t}) := \left\{ w(t) : \mathcal{I} \to \mathbb{R}^d; \quad w(t)|_{\mathcal{I}_n} = \sum_{m=0}^{k} w_m^n t^m, \quad \text{for } n = 1, \ldots, \mathcal{N} \right\},$$

the space of vector-valued discontinuous, piecewise polynomials in time of order $k \ge 0$ over $\mathcal{I}$. Since the functions of $\mathbb{P}_k^{\mathrm{disc}}$ are allowed to be discontinuous at discrete points t_n, $1 \le n \le \mathcal{N}$, we denote for a function $w \in \mathbb{P}_k^{\mathrm{disc}}$

$$w_+^n := \lim_{t \to t_n + 0} w(t), \qquad w_-^n := \lim_{t \to t_n - 0} w(t), \qquad [w]_n = w_+^n - w_-^n.$$

Note that the initial condition $\mathrm{u}(0) = \mathrm{u}_0$ need to be imposed separately, since $0 \notin \mathcal{I}_1$. Let $\varphi : [0,\mathrm{I}] \to \mathbb{R}^{\mathrm{d}}$ be a smooth function in time with $\varphi(t_{\mathcal{N}}) = 0$. Upon multiplying the differential equation in (6.22) with φ and integrate over $\mathcal{I}$, we get

$$\int_0^{t_{\mathcal{N}}} \frac{\mathrm{du}}{\mathrm{d}t} \cdot \varphi \, \mathrm{d}t + \int_0^{t_{\mathcal{N}}} B\mathrm{u} \cdot \varphi \, \mathrm{d}t = \int_0^{t_{\mathcal{N}}} f \cdot \varphi \, \mathrm{d}t,$$

where the dot notation is used to indicate the inner product in $\mathbb{R}^d$. Now, applying integration by parts to the term containing the time derivative results in

$$-\int_0^{t_{\mathcal{N}}} \mathrm{u} \cdot \frac{\mathrm{d}\varphi}{\mathrm{d}t} \mathrm{d}t + \int_0^{t_{\mathcal{N}}} B\mathrm{u} \cdot \varphi \, \mathrm{d}t = \mathrm{u}_0 \cdot \varphi(0) + \int_0^{t_{\mathcal{N}}} f \cdot \varphi \, \mathrm{d}t.$$

Let $\mathrm{U} \in \mathbb{P}_k^{\mathrm{disc}}$ be an approximation of u. Replace u by U in the above equation, and apply integration by parts in each subinterval again to the term containing the time derivative to get

$$\begin{aligned}
-\int_0^{t_{\mathcal{N}}} \mathrm{U} \cdot \frac{\mathrm{d}\varphi}{\mathrm{d}t} \mathrm{d}t &= -\sum_{n=1}^{\mathcal{N}} \left[\mathrm{U} \cdot \varphi|_{t_{n-1}}^{t_n} - \int_{t_{n-1}}^{t_n} \frac{\mathrm{dU}}{\mathrm{d}t} \cdot \varphi \, \mathrm{d}t \right] \\
&= \sum_{n=1}^{\mathcal{N}} \int_{\mathcal{I}_n} \frac{\mathrm{dU}}{\mathrm{d}t} \cdot \varphi \, \mathrm{d}t - \sum_{n=1}^{\mathcal{N}} \left[\mathrm{U}_-^n \cdot \varphi^n - \mathrm{U}_+^{n-1} \cdot \varphi^{n-1} \right] \\
&= \sum_{n=1}^{\mathcal{N}} \int_{\mathcal{I}_n} \frac{\mathrm{dU}}{\mathrm{d}t} \cdot \varphi \, \mathrm{d}t + \sum_{n=1}^{\mathcal{N}-1} [\mathrm{U}]_n \cdot \varphi^n + \mathrm{U}_+^0 \cdot \varphi^0.
\end{aligned}$$

Note that in the last step we imposed $\varphi(t_{\mathcal{N}}) = 0$. Now replacing φ by $w \in \mathbb{P}_k^{\mathrm{disc}}$, the dG($k$) scheme reads

Find $U \in \mathbb{P}_k^{\mathrm{disc}}(\mathcal{I}^{\delta t})$ such that $U^0 = u_0$ and for all $w \in \mathbb{P}_k^{\mathrm{disc}}(\mathcal{I}^{\delta t})$

$$\sum_{n=1}^{\mathcal{N}} \int_{\mathcal{I}_n} \left(\frac{dU}{dt} \cdot w + BU \cdot w \right) dt + \sum_{n=1}^{\mathcal{N}-1} [U]_n \cdot w_+^n + U_+^0 \cdot w_+^0$$
$$= u_0 \cdot w_+^0 + \int_0^{t_{\mathcal{N}}} f \cdot w dt. \tag{6.23}$$

At the first glance, it seems that all $(k+1) \times \mathcal{N}$ dof are coupled in the system of algebraic equation. However, the function $w \in \mathbb{P}_k^{\mathrm{disc}}(\mathcal{I}^{\delta t})$ need not to be continuous at t_n, thus w can be chosen on the different time intervals independently. Indeed, assuming that $w = 0$ outside the interval $\mathcal{I}_n$, we obtain a time marching method

For all $n = 1, \ldots, \mathcal{N}$ find $U \in P_k(\mathcal{I}_n)$ such that for all $w \in P_k(\mathcal{I}_n)$

$$\int_{\mathcal{I}_n} \left(\frac{dU}{dt} \cdot w + BU \cdot w \right) dt + U_+^{n-1} \cdot w_+^{n-1} = U_-^{n-1} \cdot w_+^{n-1} + \int_{\mathcal{I}_n} f \cdot w dt.$$

We have the following convergence results.

Theorem 6.5. *Suppose* $U \in \mathbb{P}_k^{disc}(\mathcal{I}^{\delta t})$ *and* u *be the solutions of the discrete problem* (6.23) *and the continuous problem* (6.22), *respectively. Then,*

$$\|U^{\mathcal{N}} - u(t_{\mathcal{N}})\| \leq C(u)\,\delta_t^{k+1} \quad \textit{for } k \geq 0$$

and for $n = 1, \ldots, \mathcal{N}$

$$\sup_{t \in \mathcal{I}_n} \|U(t) - u(t)\| \leq \|U^n - u(t_n)\| + \|U^{n-1} - u(t_{n-1})\| + C(u)\,\delta_t^{k+1}.$$

Proof. See Theorems 12.1 and 12.2 of Thomée (2006). ■

Remark 6.1. In Theorem 6.5, $\|\cdot\|$ denotes the Euclidian norm in $\mathbb{R}^d$. We observe optimal error estimates locally at the nodes t_n but also globally in the maximum norm between the nodes. At the nodal points, one can even prove a superconvergent result of order $\mathcal{O}(2k+1)$, that is,

$$\|U^{\mathcal{N}} - u(t_{\mathcal{N}})\|_0 \leq C(u)\,\delta_t^{2k+1}$$

when $k \geq 1$, see Theorem 12.3 of Thomée (2006).

We next derive dG(k) time-stepping schemes for the model problem (6.6). Let $V = H_0^1(\Omega)$, and $V_h \subset V$ be a conforming finite element space. We define a discontinuous in time discrete space as

$$\mathbb{P}_k^{\mathrm{disc}}(\mathcal{I}^{\delta t}) := \{ v \in L^2(\mathcal{I}; V_h) : \quad v|_{\mathcal{I}_n} \in P_k(\mathcal{I}_n; V_h), \text{ for } n = 1, \ldots, \mathcal{N} \}, \tag{6.24}$$

the space of piecewise V_h valued polynomials in time of order less than or equal to $k \geq 0$ over $\mathcal{I}$. In particular, a function $U_h \in \mathbb{P}_k^{\mathrm{disc}}$ on the time interval $\mathcal{I}_n$ can be represented as

$$U_h(t) = \sum_{m=0}^{k} v_m^n \, \psi_m^{\mathrm{dG}}(t), \quad \forall\, t \in \mathcal{I}_n, \;\; n = 1, \ldots, \mathcal{N},$$

where $\mathrm{v}_m^n \in V_h$ and $\psi_m^{\mathrm{dG}} \in P_k(\mathcal{I}_n)$. Here, the $(k+1)$ time basis functions, $\psi_m^{\mathrm{dG}} \in P_k(\mathcal{I}_n)$, are linearly independent functions of the standard space of polynomials of order less than or equal to k defined on the time interval $\mathcal{I}_n$. Using the time-discrete space, the dG(k) time-discrete form of (6.6) reads

Find $\mathrm{U} \in \mathbb{P}_k^{\mathrm{disc}}(\mathcal{I}^{\delta t})$ such that $\mathrm{U}^0 = \mathrm{u}_{0h} = u_h(0)$ and for all $W \in \mathbb{P}_k^{\mathrm{disc}}(\mathcal{I}^{\delta t})$

$$\sum_{n=1}^{\mathcal{N}} \int_{\mathcal{I}_n} \left(\left(\frac{\mathrm{dU}}{\mathrm{d}t}, W \right) + (\nabla \mathrm{U}, \nabla W) \right) \mathrm{d}t + \sum_{n=1}^{\mathcal{N}-1} ([\mathrm{U}]_n, W_+^n) + (\mathrm{U}_+^0, W_+^0)$$

$$= (\mathrm{u}_{0h}, W_+^0) + \int_0^{t_{\mathcal{N}}} (\mathrm{f}, W)\, \mathrm{d}t. \tag{6.25}$$

Here, $(\cdot,\cdot)$ denote the inner product of the vector-valued $L^2(\Omega)$ space. Choosing a test function $W \in \mathbb{P}_k^{\mathrm{disc}}(\mathcal{I}^{\delta t})$ with $W|_{\mathcal{I}_n} = w \in P_k(\mathcal{I}_n)$ and $W = 0$ otherwise, we obtain the time marching method:

For all $n = 1, \dots, \mathcal{N}$ find $\mathrm{U} \in \mathbb{P}_k(\mathcal{I}_n)$ such that for all $\mathrm{w} \in \mathbb{P}_k(\mathcal{I}_n)$

$$\int_{\mathcal{I}_n} \left(\left(\frac{\mathrm{dU}}{\mathrm{d}t}, w \right) + (\nabla \mathrm{U}, \nabla w) \right) \mathrm{d}t + (\mathrm{U}_+^{n-1}, w_+^{n-1}) = (\mathrm{U}_-^{n-1}, w_+^{n-1}) + \int_{\mathcal{I}_n} (\mathrm{f}, w)\, \mathrm{d}t.$$

Recall from (6.8) that $\{\Phi_j\}$, $j = 1, \dots, n_{\mathrm{glob}}$ is a basis of V_h. Then the discrete solution $\mathrm{U} \in \mathbb{P}_k(\mathcal{I}_n)$ of the problem above can be represented in terms of spatial and temporal basis functions as

$$\mathrm{U}(x,t) = \sum_{m=0}^{k} \sum_{j=1}^{n_{\mathrm{glob}}} U_{j,m}^n \Phi_j(x)\, \psi_m^{\mathrm{dG}}(t), \quad \forall\ t \in \mathcal{I}_n.$$

Here, $U_{j,m}^n$ is an unknown space–time solution coefficient at the spatial nodal point j and temporal nodal point m in the time interval $\mathcal{I}_n$. Suppose $U_m^n = \mathrm{vec}\{U_{j,m}^n,\ j = 1, \dots, n_{\mathrm{glob}}\}$ is a vectorized form of the solution coefficients in space, and $U^n = \mathrm{vec}\{U_{j,m}^n,\ j = 1, \dots, n_{\mathrm{glob}}, m = 0, \dots, k\}$ is a vectorized form of the solution coefficients both in space and time. Now these function definitions result in a system of algebraic equations

$$\sum_{m=0}^{k} U_m^n \left(M \int_{\mathcal{I}_n} \frac{\mathrm{d}\psi_m^{\mathrm{dG}}}{\mathrm{d}t} \psi_\ell^{\mathrm{dG}}\, \mathrm{d}t + A \int_{\mathcal{I}_n} \psi_m^{\mathrm{dG}} \psi_\ell^{\mathrm{dG}}\, \mathrm{d}t + M\, \psi_{m,+}^{\mathrm{dG}} \psi_{\ell,+}^{\mathrm{dG}} \right)$$

$$= \sum_{m=0}^{k} M U_m^{n-1}\, \psi_{\ell,+}^{\mathrm{dG}} + \int_{\mathcal{I}_n} F \psi_\ell^{\mathrm{dG}}\, \mathrm{d}t, \tag{6.26}$$

for $\ell = 0, \dots, k$. Here, M, A and F are the mass matrix, stiffness matrix and load vector, respectively, as defined in (6.9).

In order to solve (6.26), the integrals over the time interval $\mathcal{I}_n = (t_{n-1}, t_n]$ need to be evaluated. The integrands on the left hand side integrals are polynomials in t of order $2k-1$ and $2k$, respectively. Since the Gauss–Radau quadrature formulas contain one end point in the interval as one of its quadrature points, Gauss–Radau quadrature formulas including the end point t_n

are chosen in general to evaluate the time integrals in the discontinuous time discrete algebraic system (6.26). It should be mentioned, that Gauss–Radau formulas with $(k+1)$ quadrature points $t_q \in (t_{n-1}, t_n]$ and positive weights w_q exist which are exact for polynomials of order $2k$, that is,

$$\int_{\mathcal{I}_n} g(t)\,\mathrm{d}t = \frac{\delta_t}{2}\sum_{q=1}^{k+1} w_q\, g(t_q) \quad \text{for all } g \in P_{2k}(\mathcal{I}_n).$$

For instance, in the case $k=1$, we use a two-points formula with $t_1 = (t_{n-1}+\delta_t/3)$ and $t_2 = t_n$ as quadrature points in the interval $(t_{n-1}, t_n]$, and $w_1 = 3/2$ and $w_2 = 1/2$ as their corresponding weights for the integrals with polynomial of order two.

dG(0) method Let us consider the case $k=0$, then $\psi_0^{\mathrm{dG}} \in P_0(\mathcal{I}_n)$ is a piecewise constant polynomial in time. Further, we have

$$\frac{\mathrm{d}\psi_0^{\mathrm{dG}}}{\mathrm{d}t} = 0, \qquad \psi_{0,+}^{\mathrm{dG}}|_{t_{n-1}} = \psi_0^{\mathrm{dG}}(t_n) = \psi_0^{\mathrm{dG}}(t) \quad \forall\, t \in \mathcal{I}_n.$$

Suppose, $\psi_0^{\mathrm{dG}} = 1$ in $\mathcal{I}_n$, then the local problem (6.26) in the interval $\mathcal{I}_n$ becomes
For given U^{n-1}, find U^n, where $U^n = U_0^n$, by solving the system

$$(M + \delta_t A)\, U_0^n = MU^{n-1} + \delta_t F,$$

which is same as the implicit Euler scheme (6.16). We have shown earlier that the implicit Euler method is L-stable, and in fact, it can be proven that dG(k), $k \geq 0$ is L-stable, see Hairer and Wanner (1996).

dG(1) method Next, let us consider the case $k=1$, that is, a piecewise linear polynomial in time with basis functions

$$\psi_0^{\mathrm{dG}} = 1, \qquad \psi_1^{\mathrm{dG}} = \frac{t - t_{n-1}}{\delta_t} \quad \forall\, t \in \mathcal{I}_n. \tag{6.27}$$

Using these linear basis functions in time, the discrete solution U can be represented in each time interval $\mathcal{I}_n$ as

$$\mathrm{U}(x,t) = \sum_{j=1}^{n_{\text{glob}}} \Phi_j(x)\left(U_{j,0}^n + \left(\frac{t - t_{n-1}}{\delta_t}\right) U_{j,1}^n\right) \quad \forall\ t \in \mathcal{I}_n$$

and hence, it holds

$$\mathrm{U}^{n-1} = \sum_{j=1}^{n_{\text{glob}}} \Phi_j(x)\, U_{j,0}^n, \qquad \mathrm{U}^n = \sum_{j=1}^{n_{\text{glob}}} \Phi_j(x)\left(U_{j,0}^n + U_{j,1}^n\right).$$

Now, for $\ell = 0$ in (6.26), that is, for the first test function $\psi_0^{\mathrm{dG}} = 1$, the local problem (6.26) becomes

$$\sum_{m=0}^{1} U_m^n \left(M \int_{\mathcal{I}_n} \frac{\mathrm{d}\psi_m^{\mathrm{dG}}}{\mathrm{d}t} \,\mathrm{d}t + A \int_{\mathcal{I}_n} \psi_m^{\mathrm{dG}} \,\mathrm{d}t + M \psi_{m,+}^{\mathrm{dG}} \right) = MU^{n-1} + \int_{\mathcal{I}_n} F \,\mathrm{d}t.$$

Further, using (6.27) and the two-point Gauss–Radau quadrature formula, we obtain

$$(M + \delta_t A)\, U_0^n + \left(M + \frac{\delta_t}{2} A \right) U_1^n = MU^{n-1} + \int_{\mathcal{I}_n} F \,\mathrm{d}t. \tag{6.28}$$

Next, consider $\ell = 1$ in (6.26), that is, the second test function $\psi_1^{\mathrm{dG}} = (t - t_{n-1})/\delta_t$. Note that, $\psi_1^{\mathrm{dG}}(t_{n-1}) = 0$ and further by using the two-point Gauss–Radau quadrature formula, we have

$$M \sum_{m=0}^{1} U_m^n \left(\int_{\mathcal{I}_n} \frac{\mathrm{d}\psi_m^{\mathrm{dG}}}{\mathrm{d}t} \left(\frac{t - t_{n-1}}{\delta_t} \right) \mathrm{d}t \right) = \frac{1}{2} M U_1^n,$$

$$A \sum_{m=0}^{1} U_m^n \left(\int_{\mathcal{I}_n} \psi_m^{\mathrm{dG}} \left(\frac{t - t_{n-1}}{\delta_t} \right) \mathrm{d}t \right) = \frac{\delta_t}{2} A U_0^n + \frac{\delta_t}{3} A U_1^n.$$

Hence, for the second test function ψ_1^{dG}, the local problem (6.26) becomes

$$\frac{1}{2} M U_1^n + \frac{1}{2} \delta_t A U_0^n + \frac{\delta_t}{3} A U_1^n = \frac{1}{\delta_t} \int_{\mathcal{I}_n} (t - t_{n-1})\, F \,\mathrm{d}t. \tag{6.29}$$

Finally, the dG(1) scheme for problem (6.22) in the time interval $\mathcal{I}_n$ reads
For given U^{n-1}, find U^n, where $U^n = U_0^n + U_1^n$, by solving the 2×2 block system

$$(M + \delta_t A)\, U_0^n + \left(M + \frac{\delta_t}{2} A \right) U_1^n = MU^{n-1} + \int_{\mathcal{I}_n} F \,\mathrm{d}t, \tag{6.30}$$

$$\frac{1}{2} \delta_t A U_0^n + \left(\frac{1}{2} M + \frac{\delta_t}{3} A \right) U_1^n = \frac{1}{\delta_t} \int_{\mathcal{I}_n} (t - t_{n-1})\, F \,\mathrm{d}t. \tag{6.31}$$

If the load vector F is a polynomial in t, the source term integrals in (6.30) and (6.31) can be exactly evaluated by applying a suitable quadrature rule. However, in general, a tailored (accurate enough but not more accurate as needed) quadrature formula is used to evaluate these integrals.

Continuous Galerkin–Petrov time-stepping methods

In order to construct a time-marching scheme, it would also be sufficient to use a discontinuous test function and a continuous ansatz function. This approach avoids the jumps in the solution terms in the time discretization and leads to the class of cGP(k) time-stepping methods.

Let us define

$$\mathbb{P}_k(\mathcal{I}^{\delta t}) := \{ \mathrm{v} \in C(\mathcal{I}; V_h) : \quad \mathrm{v}|_{\mathcal{I}_n} \in P_k(\mathcal{I}_n; V_h) \text{ for } n = 1, \ldots, \mathcal{N} \}, \tag{6.32}$$

the space of continuous, piecewise V_h valued polynomials in time of order less than or equal to $k \geq 0$ over $\mathcal{I}$. The functions $\mathrm{U}_h \in \mathbb{P}_k$ are continuous at discrete time points t_n, $n = 0, 1, \dots, \mathcal{N}$. Using the discontinuous and continuous time discrete spaces, the dG(k) time discrete form of (6.6) in the time interval $\mathcal{I}_n$ reads

For given f and u_h^{n-1}, find $\mathrm{u}_h(t) \in \mathbb{P}_k$, $k \geq 1$, such that for all $\mathrm{v} \in \mathbb{P}_{k-1}^{\mathrm{disc}}$

$$\int_{\mathcal{I}_n} \left(\frac{\mathrm{d}\mathrm{u}_h}{\mathrm{d}t}, \mathrm{v} \right) \mathrm{d}t + \int_{\mathcal{I}_n} (\nabla \mathrm{u}_h, \nabla \mathrm{v})\, \mathrm{d}t = \int_{\mathcal{I}_n} (\mathrm{f}, \mathrm{v})\, \mathrm{d}t. \tag{6.33}$$

Moreover, the system of algebraic equations in the time interval $\mathcal{I}_n$ becomes

$$\sum_{m=0}^{k} U_m^n \left(M \int_{\mathcal{I}_n} \frac{\mathrm{d}\phi_m}{\mathrm{d}t} \psi_\ell \, \mathrm{d}t + A \int_{\mathcal{I}_n} \phi_m \psi_\ell \, \mathrm{d}t \right) = \int_{\mathcal{I}_n} F \psi_\ell \, \mathrm{d}t \tag{6.34}$$

for $\ell = 0, \dots, k$. Here, M and A are the mass and stiffness matrices, respectively, as defined in (6.16). Further, $\psi_m^{\mathrm{dG}} \in P_k(\mathcal{I}_n)$ are time-basis functions.

Unlike Gauss–Radau quadrature, Gauss–Lobatto quadrature contains both end points of the interval as its quadrature points. Since the discrete solution is continuous at the end point of the interval, the Gauss–Lobatto quadrature formula is preferred in cGP(k) time-stepping methods. The $(k+1)$-point Gauss–Lobatto formula is exact for polynomials of order less than or equal to $2k-1$, thus the two integrals on the left hand side of (6.34) are exactly evaluated. For instance, the Gauss–Lobatto quadrature with $k = 1$ produce the Trapezoidal rule, which is exact for polynomials of order one.

cGP(1) method For the functions $\mathrm{v} \in \mathbb{P}_0^{\mathrm{disc}}$ and $\mathrm{U}_h \in \mathbb{P}_1$, let the constant test basis function for time be $\psi_0^{\mathrm{dG}} = 1$ and the linear solution basis functions for time in the interval $\mathcal{I}_n$ be

$$\psi_0(t) = \frac{t_n - t}{\delta_t}, \qquad \psi_1(t) = \frac{t - t_{n-1}}{\delta_t} \quad \forall\, t \in \mathcal{I}_n. \tag{6.35}$$

Note that t_n and t_{n-1}, which are the quadrature points of the two-point Gauss–Lobatto quadrature (Trapezoidal rule), are the nodal points of the chosen linear basis function. Hence, $U_0^n = U^{n-1}$ and $U_1^n = U^n$, and therefore we have only one unknown U_1^n and one equation for the test function, $\mathrm{v} = 1$. Moreover, the discrete solution of the problem (6.6) in each time interval $\mathcal{I}_n$ is defined by

$$\mathrm{U}_h(x,t) = \sum_{j=1}^{n_{\mathrm{glob}}} \Phi_j(x) \left(\left(\frac{t_n - t}{\delta_t} \right) U_{j,0}^n + \left(\frac{t - t_{n-1}}{\delta_t} \right) U_{j,1}^n \right) \quad \forall\ t \in \mathcal{I}_n,$$

where $\{\Phi_j\}$ is a basis of V_h. Now, for $\ell = 0$ in (6.33), that is, for the test function $\psi_0^{\mathrm{dG}} = 1$, the local problem (6.33) becomes

$$\sum_{m=0}^{1} U_m^n \left(M \int_{\mathcal{I}_n} \frac{\mathrm{d}\phi_m}{\mathrm{d}t} \mathrm{d}t + A \int_{\mathcal{I}_n} \phi_m \mathrm{d}t \right) = \int_{\mathcal{I}_n} F \mathrm{d}t.$$

Using (6.35) and the two-point Gauss–Lobatto quadrature formula, we obtain

$$MU_1^n + \frac{\delta_t}{2}\left(AU_0^n + AU_1^n\right) = MU_0^n + \int_{\mathcal{I}_n} F dt. \tag{6.36}$$

Since $U_0^n = U^{n-1}$ and $U_1^n = U^n$, the form (6.36) is the Crank–Nicolson scheme (6.18) that we discussed earlier. We have shown that the Crank–Nicolson scheme is only A-stable but not L-stable, and in general it holds true for cGP(k), $k \geq 1$, see Schieweck (2010). Compared to dG methods, the Crank–Nicolson scheme has "less advantageous smoothing properties", Thomée (2006), which might be a reason that they are less popular and should be avoided for the time discretization of the semidiscrete formulation of parabolic problems, see Grossmann et al. (2007).

6.6. Finite elements for high-dimensional parabolic problems

This section presents the numerical methods that can be used to solve high-dimensional (more than three dimensions in space) parabolic partial differential equations (PDEs). In addition to the space and time, PDEs in many applications might contain derivatives of the unknown function with respect to some of the properties of the problem. As a result, the PDEs are posed on high-dimensional domains. For example, the FENE Fokker–Planck equation for a bead–spring chain polymer model consisting $(M+1)$ beads and M springs is defined in $\mathbb{R}^{d(M+1)}$, where d is the dimension of the polymer flow domain. Another example of high-dimensional equation is the radiative transfer equation, which is used in many applications such as the furnace, solar energy conversion devices, nuclear reactors, astrophysics, bioconvection and propagation of light in tissues. Population balance equations that describe how populations of separate entities, for example, particles in particulate flows and cells in biology, develop with respect to specific properties over time is a further example of a high-dimensional parabolic equation.

In this section, numerical methods for the solution of high-dimensional parabolic PDEs in the finite element context will be presented.

High-dimensional equation

In order to explain the methodologies for solving a high-dimensional parabolic PDE, consider a simple differential equation

$$\begin{aligned} \frac{\partial u}{\partial t} + \mathbb{L}_x u + \mathbb{L}_\ell u &= f_x + f_\ell && \text{in } \Omega_x \times \Omega_\ell \times (0, \mathrm{I}], \\ u &= 0 && \text{on } \partial\Omega_x \times \partial\Omega_\ell \times (0, \mathrm{I}], \\ u|_{t=0} &= u_0 && \text{in } \Omega_x \times \Omega_\ell, \end{aligned} \tag{6.37}$$

where $\Omega_x \subset \mathbb{R}^2$, $\Omega_\ell \subset \mathbb{R}^2$. Here, t is the time in the given time interval $[0, \mathrm{I}]$, $u(x, \ell, t)$ is the unknown solution, $x = (x_1, x_2) \in \Omega_x$, $\ell = (\ell_1, \ell_2) \in \Omega_\ell$, $f_x, f_\ell \in L^2((0, \mathrm{I}] \times \Omega)$ are given source functions and $u_0 \in L^2(\Omega)$ is an initial value. Further, Ω is the Cartesian product of Ω_x and Ω_ℓ, $\mathbb{L}_x$ and $\mathbb{L}_\ell$ are scalar linear differential operators with respect to x and ℓ, respectively.

Tensor product finite elements for high-dimensional equation

Let $\Omega := \Omega_x \times \Omega_\ell \subset \mathbb{R}^4$ be the Cartesian product of Ω_x and Ω_ℓ, $\mathbb{L} = \mathbb{L}_x + \mathbb{L}_\ell$ and $f = f_x + f_\ell$. Then, the first equation in (6.37) becomes

$$\frac{\partial u}{\partial t} + \mathbb{L}u = f \quad \text{in } \Omega \times (0, \mathrm{I}]. \tag{6.38}$$

Following the steps given in Section 6.4, the weak form of (6.38) reads:

Find $\mathrm{u} \in W^1(0, \mathrm{I}; H_0^1(\Omega), L^2(\Omega))$ such that

$$\langle \mathrm{u}'(t), v\rangle + (\mathbb{L}\mathrm{u}(t), v) = \langle \mathrm{f}(t), v\rangle \quad \text{for a.e. } t \in (0, \mathrm{I}), \text{ for all } v \in H_0^1(\Omega),$$

$$(\mathrm{u}(0), v) = (u_0, v) \quad \text{for all } v \in L^2(\Omega).$$

Suppose $V_h^x \subset H_0^1(\Omega_x)$ and $V_h^\ell \subset H_0^1(\Omega_\ell)$ are conforming finite element spaces with basis functions $\{\Phi_i\}$, $i = 1, \ldots, \mathcal{M}$ and $\{\Psi_k\}$, $k = 1, \ldots, \mathcal{L}$, respectively. We then define a high-dimensional finite element space $V_h := V_h^x \otimes V_h^\ell$ such that for a function $\xi_h \in V_h$, we have

$$\xi_h(x, \ell) = \sum_{i=1}^{\mathcal{M}} \sum_{k=1}^{\mathcal{L}} \xi_{i,k} \Phi_i(x)\, \Psi_k(\ell). \tag{6.39}$$

Consequently, the finite element ansatz function, $\mathrm{u}_h^n \in V_h$ and an application of the operator on this function are given by

$$\mathrm{u}_h^n(x, \ell) = \sum_{j=1}^{\mathcal{M}} \sum_{l=1}^{\mathcal{L}} U_{j,l}^n \Phi_j(x) \Psi_l(\ell),$$

$$\mathbb{L}\mathrm{u}_h^n(x, \ell) = \sum_{j=1}^{\mathcal{M}} \sum_{l=1}^{\mathcal{L}} U_{j,l}^n (\Psi_l \mathbb{L}_x(\Phi_j) + \Phi_j \mathbb{L}_\ell(\Psi_l)),$$

where $U_{j,l}^n \in \mathbb{R}$ are the unknown solution coefficients. Using the standard Galerkin approach and the backward Euler method, the discrete form of (6.38) becomes

For given u_h^{n-1}, u_0 and f^n, find $\mathrm{u}_h^n \in V_h$ such that

$$\left(\mathrm{u}_h^n, v\right) + \delta_t \left(\mathbb{L}\mathrm{u}_h^n, v\right) = \left(\mathrm{u}_h^{n-1}, v\right) + \delta_t \left(\mathrm{f}^n, v\right),$$

$$\left(\mathrm{u}_h^0 - u_0, v\right) = 0$$

for all $v \in V_h$. Moreover, it results in the system of algebraic equations

$$(M + \delta_t A)\, \bar{U}^n = M \bar{U}^{n-1} + \delta_t F^n.$$

Here, $\bar{U}^n := \mathrm{vec}(U_{j,l}^n)$ is the vectorization of the solution matrix $[U_{j,l}^n]_{\mathcal{M} \times \mathcal{L}}$, M is the mass matrix and A the matrix generated by the differential operator $\mathbb{L}$. In particular, the matrices and the load vector are given by

$$M := M_x \otimes M_\ell, \qquad A := A_x \otimes M_\ell + M_x \otimes A_\ell, \qquad F^n_{i,k} := \int_\Omega f^n \Phi_i \Psi_k \,\mathrm{d}x\,\mathrm{d}\ell, \tag{6.40}$$

where $\otimes$ denotes the Kronecker product of two matrices, and

$$\begin{aligned} M_{x_{ij}} &= \int_{\Omega_x} \Phi_i \Phi_j \,\mathrm{d}x, & A_{x_{ij}} &= \int_{\Omega_x} \mathbb{L}_x \Phi_i \Phi_j \,\mathrm{d}x, \\ M_{\ell_{kl}} &= \int_{\Omega_\ell} \Psi_k \Psi_l \,\mathrm{d}\ell, & A_{\ell_{kl}} &= \int_{\Omega_\ell} \mathbb{L}_\ell \Psi_k \Psi_l \,\mathrm{d}\ell. \end{aligned} \tag{6.41}$$

For example, the $\mathcal{ML} \times \mathcal{ML}$ entries in the mass matrix are given by

$$M = M_x \otimes M_\ell := \begin{bmatrix} [\Phi_{1,1}]_{k,l} & \cdot \quad \cdot \quad \cdot & [\Phi_{1,\mathcal{M}}]_{k,l} \\ \cdot & \cdot & \cdot \\ \cdot & \cdot & \cdot \\ \cdot & \cdot & \cdot \\ [\Phi_{\mathcal{M},1}]_{k,l} & \cdot \quad \cdot \quad \cdot & [\Phi_{\mathcal{M},\mathcal{M}}]_{k,l} \end{bmatrix}_{\mathcal{ML}\times\mathcal{ML}},$$

where the entries in the $\mathcal{L} \times \mathcal{L}$ block matrix $[\Phi_{i,j}]_{k,l}$, $1 \le k, l \le \mathcal{L}$ can be evaluated by

$$[\Phi_{i,j}]_{k,l} := \int_{\Omega_x} \int_{\Omega_\ell} \Phi_i \Phi_j \Psi_k \Psi_\ell \,\mathrm{d}x\,\mathrm{d}\ell.$$

Remark 6.2. In general, one can construct a d-dimensional finite element function as in (6.39), and it will contain N^d dof when the number of dof in one dimension is N. Alternatively, one could use a sparse grid approach that involves only $\mathcal{O}(N(\log N)^{d-1})$ dof, for solving a d-dimensional problem, see Bungartz and Griebel (2004) for more details on sparse grid methods.

Operator-splitting techniques

A tensor product or sparse grid type finite element space can be constructed only when the computational domain is a Cartesian product of one-dimensional subdomains. However, this is not the case in many applications since the domains are more complex. Therefore, we need a different approach to handle high-dimensional problems and it will be discussed below.

Suppose that a complex high-dimensional domain can be defined as a Cartesian product of more than one complex low-dimensional subdomains as in the model problem (6.37). Further, assume that the differential operators $\mathbb{L}_x$ and $\mathbb{L}_\ell$ do not have derivatives or mixed derivatives with respect to ℓ and x, respectively. Then, we can apply an operator-splitting technique to solve the high-dimensional problem (6.37). We refer to McLachlan and Quispel (2002) and the references therein for a detailed discussion on operator splitting. In the following, we present the splitting techniques in the context of finite element method.

Lie–Trotter splitting

Lie–Trotter splitting is a first order sequential splitting, and it involves two steps in each time step (t^{n-1}, t^n). Applying Lie–Trotter splitting to the problem (6.37), results in two subequations in (t^{n-1}, t^n):
Step 1. For given f_x and $\tilde{u}^{n-1} = u^{n-1}$, find $\tilde{u}$ such that for all $\ell \in \Omega_\ell$

$$\frac{\partial \tilde{u}}{\partial t} + \mathbb{L}_x \tilde{u} = f_x \quad \text{in } \Omega_x \times (t^{n-1}, t^n). \tag{6.42}$$

Step 2. For given f_ℓ and $u^{n-1} = \tilde{u}^n$, find u such that for all $x \in \Omega_x$

$$\frac{\partial u}{\partial t} + \mathbb{L}_\ell u = f_\ell \quad \text{in } \Omega_\ell \times (t^{n-1}, t^n). \tag{6.43}$$

Note that the above x-direction Equation (6.42) has to be solved for all $\ell \in \Omega_\ell$ by considering ℓ as a parameter. The choice of ℓ is implementation specific, and it could be a quadrature point or a nodal point of finite elements, see the following sections for more details. Similarly, the ℓ-direction Equation (6.43) has to be solved for all $x \in \Omega_x$ by considering x as a parameter. Again the choice of x is implementation specific.

Remark 6.3. Suppose the source function $f(x,\ell)$ is not separable as f_x and f_ℓ, then it can be incorporated in one of the splitting steps or as a convex combination in both steps. Nevertheless, there will be a splitting error and thus a higher order splitting scheme is preferred for higher order temporal discretization methods.

Remark 6.4. Note that different discretizations can be used for the equations (6.42) and (6.43), and this is another advantage of using splitting techniques. Further, the computational complexity will also be less when splitting techniques are used. Suppose the computational complexity of a solver for the problem (6.37) is of order two, then the computational complexity in the tensor product finite element technique will be of order $\mathcal{O}(\mathcal{M}^2\mathcal{L}^2)$. However, the computational complexity in the Lie–Trotter splitting (6.42), (6.43), will only be of order $\mathcal{O}(\mathcal{M}\mathcal{L}(\mathcal{M}+\mathcal{L}))$. Note that the complexity will the significantly less compared to the tensor product technique in higher dimensional problems, and when $\mathcal{M}$ and $\mathcal{L}$ are large.

Strang splitting

Strang splitting is second order accurate in time, and it involves three steps in each time step (t^{n-1}, t^n). Applying Strang splitting scheme to the problem (6.37) results in:

Step 1. For given f_x and $\tilde{u}^{n-1} = u^{n-1}$, find $\tilde{u}$ such that for all $\ell \in \Omega_\ell$

$$\frac{\partial \tilde{u}}{\partial t} + \mathbb{L}_x \tilde{u} = f_x \quad \text{in } \Omega_x \times (t^{n-1}, t^{n-\delta_t/2}). \tag{6.44}$$

Step 2. For given f_ℓ and $\hat{u}^{n-1} = \tilde{u}^{n-\delta_t/2}$, find $\hat{u}$ such that for all $x \in \Omega_x$

$$\frac{\partial \hat{u}}{\partial t} + \mathbb{L}_\ell \hat{u} = f_\ell \quad \text{in } \Omega_\ell \times (t^{n-1}, t^n). \tag{6.45}$$

Step 3. For given f_x and $u^{n-\delta_t/2} = \hat{u}^n$, find u such that for all $\ell \in \Omega_\ell$

$$\frac{\partial u}{\partial t} + \mathbb{L}_x u = f_x \quad \text{in } \Omega_x \times (t^{n-\delta_t/2}, t^n). \tag{6.46}$$

Remark 6.5. Even though the Lie–Trotter and Strang splitting schemes induce splitting errors, the error in time will be of order $\mathcal{O}(\delta_t)$ and $\mathcal{O}(\delta_t^2)$ that are optimal for Euler and Crank–Nicolson methods, respectively. Thus, the optimal order of convergence of the temporal discretization scheme will not get affected when an appropriate splitting scheme is used.

In the following, splitting algorithms in the context of finite element method are presented. Two algorithms for the Lie–Trotter splitting method, the first is based on the finite element nodal point and the next is based on the quadrature points that are used to evaluate integrals in finite element matrix assembling are given below.

Nodal–nodal splitting algorithm

Suppose that the nodal functionals of V_h^x and V_h^ℓ are defined using point values. Further, the finite element functions $\tilde{u}_h^n(\cdot, \ell_q) \in V_h^x$ and $\hat{u}_h^n(x_j, \cdot) \in V_h^\ell$ are defined as

$$\tilde{u}_h^n(x, \ell_q) := \sum_{j=1}^{\mathcal{M}} \tilde{U}_{q,j}^n \Phi_j(x), \qquad \hat{u}_h^n(x_j, \ell) := \sum_{q=1}^{\mathcal{L}} \hat{U}_{j,q}^n \Psi_q(\ell).$$

Here, $x_j \in \Omega_x$ and $\ell_q \in \Omega_\ell$, are the Cartesian coordinates that are necessary to evaluate the nodal functionals of the finite element spaces V_h^x and V_h^ℓ, respectively. Since nodal functionals are defined using point values, we have $1 \le p \le \mathcal{M}$ and $1 \le q \le \mathcal{L}$. Note that the ℓ-direction solution matrix $\hat{U}_{j,q}^n$ is the transpose of the x-direction solution $\tilde{U}_{q,j}^n$.

Applying now the backward Euler scheme for the split Equations (6.42) and (6.43), the fully discrete form in the time interval (t^{n-1}, t^n) read as

x-Direction For given f_x^n and $\tilde{u}_h^{n-1}(x, \ell_q) = u_h^{n-1}$, $q = 1, \ldots, \mathcal{L}$, find an array of solution $\tilde{u}_h^n(x, \ell_q) \in V_h^x$ such that for all $v_x \in V_h^x$

$$(\tilde{u}_h^n, v_x)_{\Omega_x} + \delta t(\mathbb{L}_x \tilde{u}_h^n, v_x)_{\Omega_x} = (\tilde{u}_h^{n-1}, v_x)_{\Omega_x} + \delta t(f_x^n, v_x)_{\Omega_x}. \tag{6.47}$$

After that define

$$\hat{u}_h^{n-1}(x_j, \ell) = \sum_{q=1}^{\mathcal{L}} \hat{U}_{j,q}^{n-1} \Psi_q(\ell) = \sum_{q=1}^{\mathcal{L}} \tilde{U}_{q,j}^n \Psi_q(\ell), \quad j = 1, \ldots, \mathcal{M}$$

and solve

ℓ-Direction For given f_ℓ^n and $\hat{u}_h^{n-1}(x_j, \ell)$, $j = 1, \ldots, \mathcal{M}$, find an array of solution $\hat{u}_h^n(x_j, \ell) \in V_h^\ell$ such that for all $v_\ell \in V_h^\ell$

$$(\hat{u}_h^n, v_\ell)_{\Omega_\ell} + \delta t(\mathbb{L}_\ell \hat{u}_h^n, v_\ell)_{\Omega_\ell} = (\hat{u}_h^{n-1}, v_\ell)_{\Omega_\ell} + \delta t(f_\ell^n, v_\ell)_{\Omega_\ell}. \tag{6.48}$$

Finally, define

$$\tilde{\mathrm{u}}_h^n(x,\ell_q) := \sum_{j=1}^{\mathcal{M}} \tilde{U}_{q,j}^n \Phi_j(x) = \sum_{j=1}^{\mathcal{M}} \hat{U}_{j,q}^n \Phi_j(x)$$

and proceed to the next time step. Moreover, the discrete solution at t_n can also be obtained by

$$\mathrm{u}_h^n(x,\ell) = \sum_{j=1}^{\mathcal{M}} \sum_{q=1}^{\mathcal{L}} U_{q,j}^n \Phi_j(x) \Psi_q(\ell), \tag{6.49}$$

where $U_{q,j}^n = \tilde{U}_{q,j}^n$.

Quadrature–Nodal splitting algorithm

In the quadrature-nodal splitting approach, the ℓ-direction problem (6.48) has to be solved for all quadrature points that are necessary to evaluate the integrals in the x-direction problem (6.47).

Let K_k, $k = 1,\ldots,N$, be the cells in Ω_x, and N_KQ be the number of quadrature points in each cell K_k. Then, the total number of quadrature points in Ω_x is $N\_Q = N * N\_KQ$. As defined earlier, let the nodal functionals of V_h^x and V_h^ℓ are defined using point values, and let $\mathcal{M}$ and $\mathcal{L}$ be the number of dof in V_h^x and V_h^ℓ, respectively. Using these definitions, the first step of the quadrature–nodal splitting approach in the time interval (t^{n-1}, t^n) read as

x-Direction For given f_x^n and $\tilde{\mathrm{u}}_h^{n-1}(x,\ell_q) = \mathrm{u}_h^{n-1}$, $q = 1,\ldots,\mathcal{L}$, find an array of solution $\tilde{\mathrm{u}}_h^n(x,\ell_q) \in V_h^x$ such that for all $v_x \in V_h^x$

$$(\tilde{\mathrm{u}}_h^n, v_x)_{\Omega_x} + \delta t(\mathbb{L}_x \tilde{\mathrm{u}}_h^n, v_x)_{\Omega_x} = (\tilde{\mathrm{u}}_h^{n-1}, v_x)_{\Omega_x} + \delta t(\mathrm{f}_x^n, \Phi_i)_{\Omega_x}. \tag{6.50}$$

Next, interpolate the solution $\tilde{\mathrm{u}}_h^n$ to all quadrature points x_m, $m = 1,\ldots,N\_Q$ that are necessary to evaluate the integrals in (6.50), that is, compute the values

$$U_{m,q}^n = \sum_{j=1}^{\mathcal{M}} \tilde{U}_{q,j}^n \Phi_j(x_m), \quad q = 1,\ldots,\mathcal{L}, \quad m = 1,\ldots,N\_Q.$$

After defining

$$\hat{\mathrm{u}}_h^{n-1}(x_m,\ell) = \sum_{q=1}^{\mathcal{L}} \hat{U}_{m,q}^{n-1} \Psi_q(\ell) = \sum_{q=1}^{\mathcal{L}} U_{m,q}^n \Psi_q(\ell), \quad m = 1,\ldots,N\_Q,$$

the second step in quadrature–nodal splitting reads

ℓ-Direction For given f_ℓ^n and $\hat{\mathrm{u}}_h^{n-1}(x_m,\ell)$, $m = 1,\ldots,N\_Q$, find an array of solution $\hat{\mathrm{u}}_h^n(x_m,\ell) \in V_h^\ell$ such that for all $v_\ell \in V_h^\ell$

$$(\hat{\mathrm{u}}_h^n, v_\ell)_{\Omega_\ell} + \delta t(\mathbb{L}_\ell \hat{\mathrm{u}}_h^n, v_\ell)_{\Omega_\ell} = (\hat{\mathrm{u}}_h^{n-1}, v_\ell)_{\Omega_\ell} + \delta t(\mathrm{f}_\ell^n, v_\ell)_{\Omega_\ell}. \tag{6.51}$$

Unlike the nodal–nodal splitting, transferring the solution from Ω_ℓ to Ω_x is not straightforward, as the solution $\hat{u}_h^n$ is available only on the quadrature points and not on the nodal points of V_h^x. However, the nodal functionals of V_h^x can be computed using L^2-projection, that is, for given $\hat{u}_h^n(x_m, \ell) \in V_h^\ell$, $m = 1, \ldots, N\_Q$, compute $\tilde{u}_h^n(x, \ell_q)$, $q = 1, \ldots, \mathcal{L}$, such that

$$\left(\tilde{u}_h^n(x, \ell_q) - \hat{u}_h^n, v_x\right)_{\Omega_x} = 0 \quad \forall v_x \in V_h^x.$$

In particular, for given nodal values, $\hat{U}^n_{m,q}$, $m = 1, \ldots, N\_Q$, solve a $\mathcal{M} \times \mathcal{M}$ system

$$\begin{aligned}
\sum_{k=1}^{N} \int_{K_k} \sum_{j=1}^{\mathcal{M}} \tilde{U}^n_{q,j} \Phi_j \Phi_i &= \sum_{k=1}^{N} \sum_{m=1+(k-1)*N_KQ}^{k*N_KQ} \omega_m \left(\sum_{j=1}^{\mathcal{M}} \tilde{U}^n_{q,j} \Phi_j(x_m) \right) \Phi_i(x_m) \\
&= \sum_{k=1}^{N} \sum_{m=1+(k-1)*N_KQ}^{k*N_KQ} \omega_m \hat{U}^n_{m,q} \, \Phi_i(x_m), \quad i = 1, \ldots, \mathcal{M},
\end{aligned}$$

where ω_m is the quadrature weight of the quadrature point x_m. Suppose the x-direction problem (6.50) requires only the mass matrix evaluation from the initial value, as it is the case in backward Euler method, then the explicit calculation of the nodal values, $\tilde{U}^n_{q,j}$, is not necessary. Since the right hand side vector can be calculated using

$$\begin{aligned}
(\tilde{u}_h^{n-1}, v_x)_{\Omega_x} &= \sum_{k=1}^{N} \int_{K_k} \sum_{j=1}^{\mathcal{M}} \tilde{U}^{n-1}_{q,j} \Phi_j \, v_x \\
&= \sum_{k=1}^{N} \sum_{m=1+(k-1)*N_KQ}^{k*N_KQ} \omega_m \left(\sum_{j=1}^{\mathcal{M}} \tilde{U}^{n-1}_{q,j} \Phi_j(x_m) \right) v_x(x_m) \\
&= \sum_{k=1}^{N} \sum_{m=1+(k-1)*N_KQ}^{k*N_KQ} \omega_m \hat{U}^{n-1}_{m,q} \, v_x(x_m).
\end{aligned}$$

Hence, the L^2-projection to transfer the solution from ℓ-direction to x-direction is not needed in the backward Euler method.

Remark 6.6. Since the number of quadrature points in a cell will be more than the number of nodal points in general and the computational complexity will increase due to the L^2-projection, the computational cost of the quadrature–nodal algorithm will be considerably more than the nodal–nodal algorithm. Nevertheless, both the nodal–nodal and quadrature-nodal algorithms will be of same order of accuracy in time (Ganesan and Tobiska (2013)).

CHAPTER 7

SYSTEMS IN SOLID MECHANICS

In the previous sections, we considered PDEs for a scalar quantity. In this chapter we study systems of equations which are of importance in elasticity.

7.1. Linear elasticity

The behaviour of an elastic body can be described by the following system of equations

$$-\operatorname{div}\mathbb{S}(\mathbf{u}) = \mathbf{f} \quad \text{in } \Omega \subset \mathbb{R}^3, \qquad \mathbb{S}(\mathbf{u})\cdot\mathbf{n} = \mathbf{g} \quad \text{on } \Gamma_N, \quad \mathbf{u} = \mathbf{0} \quad \text{on } \Gamma_D,$$

where $\mathbb{S}(\mathbf{u}) = \mathbb{S}_{ij}(\mathbf{u})$, $i,j = 1,\dots,3$, denotes the stress tensor depending on the displacements $\mathbf{u} = (u_1, u_2, u_3)$, $\mathbf{n} = (n_1, n_2, n_3)$ is the outer unit normal vector, $\mathbf{f} = (f_1, f_2, f_3)$ is a given body force and $\mathbf{g} = (g_1, g_2, g_3)$ is a given traction on Γ_N. We assume that $\Gamma_D \cup \Gamma_N = \partial\Omega$ with $\Gamma_D \cap \Gamma_N = \emptyset$ and $|\Gamma_D| > 0$. The stress tensor is defined by Hook's law

$$\mathbb{S}(\mathbf{u}) = 2\mu\epsilon(\mathbf{u}) + \lambda \operatorname{tr}\epsilon(\mathbf{u})\mathbb{I}$$

with the deformation tensor $\epsilon(\mathbf{u}) = \frac{1}{2}(\nabla\mathbf{u} + \nabla\mathbf{u}^T)$ and the identity tensor $\mathbb{I} = \delta_{ij}$, $i,j = 1,\dots,3$. The positive constants λ and μ are called Lamé constants, and tr is the trace operator given by

$$\operatorname{tr}\epsilon(\mathbf{u}) = \sum_{k=1}^{3} \epsilon_{kk}(\mathbf{u}) = \sum_{k=1}^{3} \frac{\partial u_k}{\partial x_k} = \operatorname{div}\mathbf{u}.$$

The problem consists in computing the displacement $\mathbf{u}$ for a given body force $\mathbf{f}$ in Ω and a given traction $\mathbf{g}$ on Γ_N. Since

$$\mathbb{S}(\mathbf{u}) = \mu(\nabla\mathbf{u} + \nabla\mathbf{u}^T) + \lambda \operatorname{div}\mathbf{u}\,\mathbb{I},$$

we get for $i = 1,\dots,3$

$$
\begin{aligned}
-(\operatorname{div}\mathbb{S}(\mathbf{u}))_i &= -\sum_{j=1}^{3}\frac{\partial \mathbb{S}_{ij}(\mathbf{u})}{\partial x_j}\\
&= -\mu\sum_{j=1}^{3}\frac{\partial^2 u_i}{\partial x_j^2} - \mu\frac{\partial}{\partial x_i}\sum_{j=1}^{3}\frac{\partial u_j}{\partial x_j} - \lambda\sum_{j=1}^{3}\frac{\partial \operatorname{div}\mathbf{u}}{\partial x_j}\delta_{ij}\\
&= -\mu\,\Delta u_i - (\mu+\lambda)\frac{\partial}{\partial x_i}\operatorname{div}\mathbf{u} = f_i.
\end{aligned}
$$

Hence, the strong form of the system of PDEs in short notation reads

$$-\mu\,\Delta\mathbf{u} - (\mu+\lambda)\operatorname{grad}\operatorname{div}\mathbf{u} = \mathbf{f} \quad \text{in } \Omega.$$

The weak formulation of the problem is obtained as usual by multiplying the vector-valued equation with a vector-valued test function $\mathbf{v} = (v_1, v_2, v_3)$ which vanishes on the Dirichlet part Γ_{D}, integrating over Ω and using integration by parts to reduce the highest order of derivatives:

$$
\begin{aligned}
-\int_\Omega \sum_{i=1}^{3}(\operatorname{div}\mathbb{S}(\mathbf{u}))_i v_i\,\mathrm{d}x &= \int_\Omega \sum_{i,j=1}^{3}\mathbb{S}_{ij}(\mathbf{u})\frac{\partial v_i}{\partial x_j}\,\mathrm{d}x - \int_{\partial\Omega}\sum_{i=1}^{3}\sum_{j=1}^{3}\mathbb{S}_{ij}(\mathbf{u})n_j v_i\,\mathrm{d}s\\
&= \int_\Omega \sum_{i,j=1}^{3}\mathbb{S}_{ij}(\mathbf{u})\epsilon_{ij}(\mathbf{v})\,\mathrm{d}x - \int_{\Gamma_N}\sum_{i=1}^{3} g_i v_i\,\mathrm{d}s.
\end{aligned}
$$

In the above derivation, we used the symmetry $\mathbb{S}_{ij}(\mathbf{u}) = \mathbb{S}_{ji}(\mathbf{u})$ of the stress tensor. Let us introduce the solution space as

$$V := H^1_{\Gamma_{\mathrm{D}}}(\Omega)^3 = \{\mathbf{v} \in H^1(\Omega)^3 : \mathbf{v}|_{\Gamma_{\mathrm{D}}} = 0\}.$$

Then, the weak formulation of the problem reads

$$\text{Find } \mathbf{u} \in V \text{ such that for all } \mathbf{v} \in V: \quad a(\mathbf{u},\mathbf{v}) = F(\mathbf{v}),$$

where the bilinear form $a : V \times V \to \mathbb{R}$ and the linear form $F : V \to \mathbb{R}$ are given by

$$
\begin{aligned}
a(\mathbf{u},\mathbf{v}) &:= 2\mu(\epsilon(\mathbf{u}),\epsilon(\mathbf{v})) + \lambda(\operatorname{div}\mathbf{u},\operatorname{div}\mathbf{v})\\
F(\mathbf{v}) &:= (\mathbf{f},\mathbf{v}) + \langle\mathbf{g},\mathbf{v}\rangle_{\Gamma_{\mathrm{N}}}.
\end{aligned}
$$

Here, we used the notation $(\cdot,\cdot)$ and $\langle\cdot,\cdot\rangle_{\Gamma_{\mathrm{N}}}$ for the inner product in $L^2(\Omega)$, $L^2(\Gamma_{\mathrm{N}})$ and its vector-valued and matrix-valued versions.

Remark 7.1. The necessary condition for minimizing the energy functional

$$\mathbb{E}(\mathbf{u}) = \frac{1}{2}a(\mathbf{u},\mathbf{u}) - F(\mathbf{u})$$

over $V = H^1_{\Gamma_D}(\Omega)^3$ is just the above weak formulation. The energy functional can be written in the form

$$\mathbb{E}(\mathbf{u}) = \int_\Omega \left[\frac{1}{2} \sum_{i,j=1}^{3} \mathbb{S}_{ij}(\mathbf{u})\,\epsilon_{ij}(\mathbf{u}) - \sum_{i=1}^{3} f_i u_i \right] \mathrm{d}x - \int_{\Gamma_N} \sum_{i=1}^{3} g_i u_i \,\mathrm{d}s.$$

Due to the convexity of the energy functional, the minimization problem reads

$$\text{Find } \mathbf{u} \in V \text{ with } \mathbb{E}(\mathbf{u}) = \inf_{\mathbf{v}\in V} \mathbb{E}(\mathbf{v})$$

is equivalent to the weak formulation of the linear elasticity problem.

In order to study the properties of the bilinear form a and the linear form F that are needed for applying the Lax–Milgram theorem, we define a norm and a seminorm in vector-valued or matrix-valued function spaces canonically by

$$\|\mathbf{v}\|_1 := \left(\sum_{i=1}^{3} \|v_i\|_1^2 \right)^{1/2}, \qquad |\mathbf{v}|_1 := \left(\sum_{i=1}^{3} |v_i|_1^2 \right)^{1/2}, \qquad \mathbf{v} = (v_1, v_2, v_3) \in H^1(\Omega)^3,$$

$$\|\epsilon(\mathbf{u})\|_0 := \left(\sum_{i,j=1}^{3} \|\epsilon_{ij}(\mathbf{u})\|_0^2 \right)^{1/2}, \qquad \epsilon(\mathbf{u}) \in L^2(\Omega)^{3\times 3}.$$

The continuity of $a: V \times V \to \mathbb{R}$ follows by means of Schwarz inequality

$$|a(\mathbf{u},\mathbf{v})| \le 2\mu \|\epsilon(\mathbf{u})\|_0 \,\|\epsilon(\mathbf{v})\|_0 + \lambda \|\operatorname{div}\mathbf{u}\|_0 \,\|\operatorname{div}\mathbf{v}\|_0 \le (2\mu + 3\lambda)\|\mathbf{u}\|_1 \,\|\mathbf{v}\|_1,$$

where we have used

$$\|\epsilon(\mathbf{u})\|_0^2 \le \frac{1}{4} \int_\Omega \sum_{i,j=1}^{3} \left(\frac{\partial u_i}{\partial x_j} + \frac{\partial u_j}{\partial x_i} \right)^2 \mathrm{d}x \le \int_\Omega \sum_{i,j=1}^{3} \left(\frac{\partial u_i}{\partial x_j} \right)^2 \mathrm{d}x = |\mathbf{u}|_1^2 \le \|\mathbf{u}\|_1^2,$$

$$\|\operatorname{div}\mathbf{u}\|_0^2 \le \int_\Omega \left(\sum_{i=1}^{3} \frac{\partial u_i}{\partial x_i} \right)^2 \mathrm{d}x \le 3 \int_\Omega \sum_{i=1}^{3} \left(\frac{\partial u_i}{\partial x_i} \right)^2 \mathrm{d}x \le 3\,\|\mathbf{u}\|_1^2.$$

For $\mathbf{g} \in L^2(\Gamma_N)^d$ we use the continuity of the trace operator and get for $\mathbf{v} \in V$

$$|F(\mathbf{v})| \le \|\mathbf{f}\|_0 \,\|\mathbf{v}\|_0 + \|\mathbf{g}\|_{0,\Gamma_N} \|\mathbf{v}\|_{0,\Gamma_N} \le (\|\mathbf{f}\|_0 + C\|\mathbf{g}\|_{0,\Gamma_N})\,\|\mathbf{v}\|_1,$$

that is, $F: V \to \mathbb{R}$ is continuous.

Lemma 7.1 (Korn's inequality). *Let $\Omega \subset \mathbb{R}^3$ be a bounded domain with piecewise smooth boundary. Suppose that $\Gamma_D \subset \partial\Omega$ has a positive measure $|\Gamma_D| > 0$. Then, there is a positive constant $C = C(\Omega, \Gamma_D)$ such that*

$$\int_\Omega \epsilon(\mathbf{v}) : \epsilon(\mathbf{v})\,\mathrm{d}x \ge C\|\mathbf{v}\|_1^2 \quad \textit{for all } \mathbf{v} \in H^1_{\Gamma_D}(\Omega)^3. \tag{7.1}$$

Proof. We prove (7.1) in the case $\Gamma_D = \partial\Omega$, $\Gamma_N = \emptyset$, where $H^1_{\Gamma_D}(\Omega) = H^1_0(\Omega)$. For the general case, we refer to Braess (2007), Brenner and Scott (2008). Since $C_0^\infty(\Omega)$ is dense in $H^1_0(\Omega)$ it is sufficient to consider only functions $\mathbf{v} \in C_0^\infty(\Omega)^3$.

We start with the expressions

$$2\epsilon(\mathbf{v}):\epsilon(\mathbf{v}) - \nabla\mathbf{v}:\nabla\mathbf{v} = \sum_{i,j=1}^{3}\left[\frac{1}{2}\left(\frac{\partial v_i}{\partial x_j} + \frac{\partial v_j}{\partial x_i}\right)^2 - \left(\frac{\partial v_i}{\partial x_j}\right)^2\right] = \sum_{i,j=1}^{3}\frac{\partial v_i}{\partial x_j}\frac{\partial v_j}{\partial x_i},$$

$$\begin{aligned}\operatorname{div}[(\mathbf{v}\cdot\nabla)\mathbf{v} - (\operatorname{div}\mathbf{v})\mathbf{v}] &= \sum_{j=1}^{3}\frac{\partial}{\partial x_j}\left[\sum_{i=1}^{3}\left(v_i\frac{\partial v_j}{\partial x_i}\right) - \left(\sum_{i=1}^{3}\frac{\partial v_i}{\partial x_i}\right)v_j\right] \\ &= \sum_{i,j=1}^{3}\left[\frac{\partial v_i}{\partial x_j}\frac{\partial v_j}{\partial x_i} + v_i\frac{\partial^2 v_j}{\partial x_j\partial x_i} - \frac{\partial^2 v_i}{\partial x_j\partial x_i}v_j\right] - (\operatorname{div}\mathbf{v})^2\end{aligned}$$

and conclude

$$2\epsilon(\mathbf{v}):\epsilon(\mathbf{v}) - \nabla\mathbf{v}:\nabla\mathbf{v} = \operatorname{div}[(\mathbf{v}\cdot\nabla)\mathbf{v} - (\operatorname{div}\mathbf{v})\mathbf{v}] + (\operatorname{div}\mathbf{v})^2.$$

Integration over Ω and the Gauss theorem with $\mathbf{v} = \mathbf{0}$ on $\partial\Omega$ imply

$$2\int_\Omega \epsilon(\mathbf{v}):\epsilon(\mathbf{v})\,dx - |\mathbf{v}|_1^2 = \int_{\partial\Omega}[(\mathbf{v}\cdot\nabla)\mathbf{v} - (\operatorname{div}\mathbf{v})\mathbf{v}]\cdot\mathbf{n}\,ds + \int_\Omega(\operatorname{div}\mathbf{v})^2\,dx \geq 0$$

from which

$$|\mathbf{v}|_1 \leq \sqrt{2}\|\epsilon(\mathbf{v})\|_0 \quad \text{for all } \mathbf{v} \in H^1_0(\Omega)^3$$

follows. Note that in the considered case, the constant $\sqrt{2}$ is independent of the domain and $|\cdot|_1$ and $\|\cdot\|_1$ are equivalent norms on $H^1_0(\Omega)^3$. ■

We use continuous finite elements to approximate each component of $\mathbf{u} \in V$. Let $k \geq 1$ and the finite element space be given by

$$V_h := \{\mathbf{v}_h \in H^1(\Omega)^3 : \mathbf{v}_h|_K \in P_k(K)^3, \mathbf{v}_h = \mathbf{0} \text{ on } \Gamma_D\}.$$

The discrete linear elasticity problem reads

$$\text{Find } \mathbf{u}_h \in V_h \text{ such that for all } \mathbf{v}_h \in V_h: \quad a(\mathbf{u}_h, \mathbf{v}_h) = F(\mathbf{v}_h). \tag{7.2}$$

Theorem 7.1. *There are unique solutions* $\mathbf{u} \in V = H^1_{\Gamma_D}(\Omega)^3$ *and* $\mathbf{u}_h \in V_h$ *of the continuous and discrete linear elasticity problem, respectively. Moreover, if the solution of the continuous problem belongs to* $H^{k+1}(\Omega)^3$, *we have the error bound*

$$\|\mathbf{u} - \mathbf{u}_h\|_1 \leq Ch^k|\mathbf{u}|_{k+1}.$$

Proof. As shown above, the bilinear form $a : V \times V \to \mathbb{R}$ and the linear form $F : V \to \mathbb{R}$ satisfy the assumptions of Lax–Milgram theorem and Cea's lemma. Thus, we have unique solutions $\mathbf{u}$ and $\mathbf{u}_h$ satisfying

$$\|\mathbf{u} - \mathbf{u}_h\|_1 \leq C \inf_{\mathbf{v}_h \in V_h}\|\mathbf{u} - \mathbf{v}_h\|_1.$$

Under the additional regularity of the solution we replace the error of the best approximation by the standard interpolation error estimate. ■

Remark 7.2. For almost incompressible material (for example, rubber), we have Lamé coefficients with $\lambda \gg \mu$. As the above analysis shows, the constant in Cea's lemma behaves like $(2\mu + 3\lambda)/\mu \gg 1$, therefore, large errors can be expected for such materials. This phenomenon is known as locking and can be observed in real computations. Special formulations are needed to circumvent locking effects, see Braess (2007), Brenner and Scott (2008).

Remark 7.3. The discrete linear elasticity problem (7.2) corresponds to a linear system of algebraic equations. Since $V_h = Y_h^3$ is a product space of a scalar-valued space $Y_h = \text{span}(\Phi_1, \dots, \Phi_N)$, a basis in V_h is given by

$$\{\mathbf{e}_i\Phi_j : i = 1,2,3, j = 1,\dots,N\},$$

with $\mathbf{e}_1 := (1,0,0)^T$, $\mathbf{e}_2 := (0,1,0)^T$, $\mathbf{e}_3 := (0,0,1)^T$. Representing the displacements with respect to this basis and setting $\mathbf{v}_h = \mathbf{e}_i\Phi_j$, $i = 1,2,3$, $j = 1,\dots,N$ in (7.2) we get the linear system of equations.

7.2. Mindlin–Reissner plate

In Section 5.1 we considered a flat plate subjected to a load in the transversal direction. Under the assumption that the thickness of the plate is very thin compared to the other two dimensions, a PDE of fourth order for the vertical deflection $u : \Omega \to \mathbb{R}$ has been used to model the deformation of the plate. Let us assume that we have a plate of finite constant thickness t, that is, $\Omega \times (-t/2, +t/2)$, where t is small but not very small to neglect. A more precise description would be the 3d linear elasticity model studied in Section 7.1. Then, instead of one scalar quantity defined on a two-dimensional domain one has to determine three quantities $u_1, u_2, u_3 : \Omega \times (-t/2, +t/2) \to \mathbb{R}$, the displacement field on a three-dimensional domain. The Mindlin–Reissner plate is a model standing in between the 2d Kirchhoff model of thin plates (biharmonic equation) and the 3d linear elasticity model.

The Mindlin–Reissner model assumes that the in-plane displacements u_1, u_2 and the transversal displacement u_3 have the forms

$$\begin{aligned}
u_1(x_1,x_2,x_3) &= -x_3\theta_1(x_1,x_2),\\
u_2(x_1,x_2,x_3) &= -x_3\theta_2(x_1,x_2),\\
u_3(x_1,x_2,x_3) &= w(x_1,x_2).
\end{aligned}$$

Now we derive a model for computing the rotation $\theta = (\theta_1, \theta_2)$ and the deflection w. Compared with the 3d linear elasticity model we still have to compute three quantities but the newly introduced variables live on the two-dimensional domain Ω and not on the three-dimensional domain $\Omega \times (-t/2, +t/2)$. For simplicity we assume $\Gamma_N = \emptyset$ in the following case of a clamped plate, and a scaled vertical load $\mathbf{f} = (0,0,t^2 f)$. With the assumptions on the displacements above we get for the energy given in Remark 7.1 by integrating over x_3

$$\widetilde{\mathbb{E}}(\theta, w) = t^{-3}\mathbb{E}(\mathbf{u}) = \frac{1}{12} a(\theta,\theta) + \int_\Omega \left[\frac{\mu}{2t^2} |\nabla w - \theta|^2 - f\, w \right] \mathrm{d}x.$$

Here, the bilinear form a is given by

$$a(\theta, \psi) := \int_\Omega \left[2\mu \sum_{i,j=1}^{2} \epsilon_{ij}(\theta)\epsilon_{ij}(\psi) + \lambda \operatorname{div}\theta \operatorname{div}\psi \right] dx, \tag{7.3}$$

where ∇ and div are applied on scalar- and vector-valued functions of two variables, respectively. Furthermore,

$$\epsilon_{ij}(\theta) = \frac{1}{2}\left(\frac{\partial \theta_i}{\partial x_j} + \frac{\partial \theta_j}{\partial x_i} \right), \quad i,j = 1,2.$$

The energy functional $\widetilde{\mathbb{E}}$ is a weighted sum of the bending energy, the shear energy, and the potential of the surface load where sometimes different correction factors are used, see Hansbo and Larson (2003), Hansbo et al. (2011), Hansbo and Larson (2014), Boffi et al. (2013). Since the mathematical method is independent of the used factors, we do not discuss the associated modelling aspects and consider the energy functional for $(\theta, w) \in H_0^1(\Omega)^2 \times H_0^1(\Omega)$

$$\mathbb{E}(\theta, w) := \frac{1}{2} a(\theta,\theta) + \frac{\kappa}{2t^2} \int_\Omega |\nabla w - \theta|^2 \,\mathrm{d}x - \int_\Omega f\, w \,\mathrm{d}x. \tag{7.4}$$

The material constants are given by $\kappa := Ek/(2(1+\nu))$, $\mu := E/(6(1+\nu))$ and $\lambda := \nu E/(12(1-\nu^2))$, where E and ν are Young's modulus and Poisson's ratio, respectively, and $k \approx 5/6$ is a shear correction factor. If the thickness of the plate t tends to zero we expect $\theta \longrightarrow \nabla w$. Since

$$\theta = \nabla w \qquad \Rightarrow \qquad \epsilon_{ij}(\theta) = \frac{\partial^2 w}{\partial x_i \partial x_j},$$

the minimizer (θ, w) of the functional (7.4) should tend to $(\nabla w, w)$ and be close to the minimizer of the functional

$$\mathbb{E}^*(w) := \mathbb{E}(\nabla w, w) = \frac{1}{2} a(\nabla w, \nabla w) - \int_\Omega f w \,\mathrm{d}x,$$

$$\mathbb{E}^*(w) = \int_\Omega \left[\mu \sum_{i,j=1}^{2} \left(\frac{\partial^2 w}{\partial x_i \partial x_j} \right)^2 + \frac{\lambda}{2} (\Delta w)^2 - f\, w \right] \mathrm{d}x$$

over $W := H_0^2(\Omega)$. Using Lemma 5.1, we can rewrite the functional $\mathbb{E}^*$ as

$$\mathbb{E}^*(w) = \int_\Omega \left[\frac{2\mu + \lambda}{2} (\Delta w)^2 - f\, w \right] \mathrm{d}x.$$

It corresponds to the energy functional for the biharmonic equation modulo, the factor $(2\mu + \lambda)/2$, see Remark 5.1. Thus, the model for the Mindlin–Reissner plate based on the energy functional (7.4) tends for $t \to 0$ to the Kirchhoff plate model of the biharmonic equation.

At the first glance, the Mindlin–Reissner plate seems to be simpler to solve than the biharmonic equation since for the former we have a system of second order PDEs to solve in $H_0^1(\Omega)^2 \times H_0^1(\Omega)$, where C^0 finite elements can be used. Instead, the latter leads to a fourth order equation in $H_0^2(\Omega)$ such that conforming methods require C^1 finite elements. The difficulty consists in a suitable matching of the approximation spaces used for the rotation θ and the deflection w, respectively. If the thickness t tends to zero, the difference $\nabla w - \theta$ must tend to zero, which does not hold for standard choices of finite element spaces and leads to a deterioration of the approximation known as shear locking. One idea to overcome shear locking is to use projections $R_h : H^1(\Omega)^2 \to \Gamma_h$ and work with the modified energy functional

$$\mathbb{E}_h(\theta, w) := \frac{1}{2}a(\theta,\theta) + \frac{\kappa}{2t^2}\int_\Omega |\nabla w - R_h\theta|^2\,\mathrm{d}x - \int_\Omega f\,w\,\mathrm{d}x. \tag{7.5}$$

This approach has widely been used to prove convergence and is related to the class of elements with mixed interpolated tensorial components (MITC), introduced in Bathe and Dvorkin (1985). For more details, we refer to Braess (2007), Boffi et al. (2013).

In the following, we present a different concept based on a combination of the dG(k) method to approximate the rotations and continuous piecewise quadratic finite elements for the deflections. This approach was proposed by Hansbo and Larson (2003) for simplicial elements and has been extended to quadrilateral meshes in Hansbo and Larson (2014). Then, as an alternative we discuss an approach by Arnold and Brezzi (1993) based on a modified mixed formulation.

Minimizing the functional defined in (7.4) and (7.3) leads to the variational problem
Find $(\theta, w) \in H_0^1(\Omega)^2 \times H_0^1(\Omega)$ such that for all $(\psi, z) \in H_0^1(\Omega)^2 \times H_0^1(\Omega)$

$$a(\theta,\psi) + \frac{\kappa}{t^2}\int_\Omega (\nabla w - \theta)(\nabla z - \psi)\,\mathrm{d}x = \int_\Omega fz\,\mathrm{d}x. \tag{7.6}$$

Let Ω be decomposed into triangles $K \in \mathcal{T}_h$ and the triangulation $\mathcal{T}_h$ be quasi-uniform. We use continuous, piecewise quadratic approximations for the transverse displacements

$$V_h := \{w_h \in H_0^1(\Omega) : w_h|_K \in P_2(K) \text{ for all } K \in \mathcal{T}_h\}$$

and discontinuous, piecewise linear approximations for the rotations

$$\Theta_h := \{\theta \in L^2(\Omega)^2 : \theta|_K \in P_1(K)^2 \text{ for all } K \in \mathcal{T}_h\}.$$

Note that the spaces satisfy the inclusion $\nabla V_h \subset \Theta_h$, such that in the limit $t \to 0$ functions in Θ_h may belong to ∇V_h. Let $\mathcal{E}_\mathrm{I}$ and $\mathcal{E}_\mathrm{B}$ denote the set of inner and boundary edges, respectively. Each edge $E \in \mathcal{E}_\mathrm{I} \cup \mathcal{E}_\mathrm{B}$ is associated with a fixed unit normal n which in case of a boundary edge corresponds to the exterior unit normal. The jump of a function v at an edge E will be denoted by $[v] = v^+ - v^-$ for $E \in \mathcal{E}_\mathrm{I}$ and $[v] = v^+$ for $E \in \mathcal{E}_\mathrm{B}$, and the average $\langle v\rangle = (v^+ + v^-)/2$ for $E \in \mathcal{E}_\mathrm{I}$ and $\langle v\rangle = v^+$ for $E \in \mathcal{E}_\mathrm{B}$, where $v^\pm = \lim_{\varepsilon\to+0} v(x \mp \varepsilon n)$ with $x \in E$. Our dG/cG (discontinuous Galerkin/continuous Galerkin) approach reads

Find $(\theta_h, w_h) \in \Theta_h \times V_h$ such that for all $(\psi_h, z_h) \in \Theta_h \times V_h$

$$A_h(\theta_h, \psi_h) + \frac{\kappa}{t^2}\int_\Omega (\nabla w_h - \theta_h)(\nabla z_h - \psi_h)\,\mathrm{d}x = \int_\Omega f z_h\,\mathrm{d}x. \tag{7.7}$$

The bilinear form $A_h : \Theta_h \times \Theta_h \to \mathbb{R}$ is given by

$$\begin{aligned} A_h(\theta_h, \psi_h) = &\sum_{K\in\mathcal{T}_h}\int_K [2\mu\epsilon(\theta_h) : \epsilon(\psi_h) + \lambda\,\mathrm{div}\theta_h\,\mathrm{div}\,\psi_h]\,\mathrm{d}x \\ &- \sum_{E\in\mathcal{E}_\mathrm{I}\cup\mathcal{E}_\mathrm{B}}\int_E (\langle n\cdot\mathbb{S}(\theta_h)\rangle\cdot[\psi_h] + \langle n\cdot\mathbb{S}(\psi_h)\rangle\cdot[\theta_h])\,\mathrm{d}\gamma \\ &+ (2\mu+3\lambda)\gamma_\mathrm{P}\sum_{E\in\mathcal{E}_\mathrm{I}\cup\mathcal{E}_\mathrm{B}}\int_E h_E^{-1}[\theta_h][\psi_h]\,\mathrm{d}\gamma, \end{aligned} \tag{7.8}$$

where γ_P is a sufficiently large positive constant, $\mathbb{S}(\theta) = 2\mu\epsilon(\theta) + \lambda\,\mathrm{div}\theta\,\mathbb{I}$ is the stress tensor and h_E defined as

$$h_E = \frac{|K^+| + |K^-|}{2|E|} \quad \text{for } E = \partial K^+ \cap \partial K^-.$$

The unique solvability of the discrete problem (7.7), (7.8) can be proven by Theorem 2.1 (Lax–Milgram). For this the mesh-dependent energy-like norm

$$|||\theta||| := \left(\sum_{K\in\mathcal{T}_h}\int_K \left[2\mu\epsilon(\theta) : \epsilon(\theta) + \lambda(\mathrm{div}\theta)^2\right]\mathrm{d}x \right.$$
$$\left. + \frac{1}{2\mu+3\lambda}\sum_{E\in\mathcal{E}_\mathrm{I}\cup\mathcal{E}_\mathrm{B}}\int_E h_E|\langle n\cdot\mathbb{S}(\theta)\rangle|^2\,\mathrm{d}\gamma + (2\mu+3\lambda)\sum_{E\in\mathcal{E}_\mathrm{I}\cup\mathcal{E}_\mathrm{B}}\int_E h_E^{-1}|[\theta]|^2\,\mathrm{d}\gamma\right)^{1/2}$$

on $\Theta_h + H_0^1(\Omega)^2$ is introduced. A proper sufficiently large choice of γ_P guarantees that the bilinear form $A_h : (\Theta_h \times V_h) \times (\Theta_h \times V_h) \to \mathbb{R}$ is coercive.

Theorem 7.2. *Let (θ, w) and (θ_h, w_h) be the solutions of the continuous and the discrete problems (7.6) and (7.7), respectively. Suppose the scaled shear stress ξ and its discrete counter part ξ_h are defined as*

$$\xi := \frac{\kappa^{1/2}(\nabla w - \theta)}{t^2} \quad \text{and} \quad \xi_h := \frac{\kappa^{1/2}(\nabla w_h - \theta_h)}{t^2}.$$

Then, if Ω is a convex polygon and $f \in L^2(\Omega)$ we have

$$|||\theta - \theta_h||| + t\|\xi - \xi_h\|_0 \le Ch(\|f\|_{H^{-1}(\Omega)} + t\|f\|_0)$$

uniformly in t.

Proof. Consider the mesh-dependent norm ($t > 0$)

$$(\theta_h, w_h) \mapsto |||\theta_h||| + \frac{\kappa^{1/2}}{t} \|\nabla w_h - \theta_h\|_0$$

on the product space $\Theta_h \times V_h$. The left hand side of (7.7) defines a continuous bilinear form which is coercive for γ_p sufficiently large. The right hand side of (7.7) is a continuous linear form. The unique solvability of the discrete problem follows from Theorem 2.1 (Lax–Milgram). Using the Galerkin orthogonality, the quasi-optimality of the method, and suitable interpolation operators, we conclude the theorem (for details see Hansbo and Larson (2003)). ■

We explain now an alternative approach based on a mixed formulation. Our starting point is the energy functional $\mathbb{E}(\theta, w)$ defined in (7.4). Now a part of the shear energy is added to the bending energy, that is,

$$\mathbb{E}(\theta, w) = \frac{1}{2} a_m(\theta, w; \theta, w) + \frac{\kappa}{2}\frac{1-t^2}{t^2}(\nabla w - \theta, \nabla w - \theta) - (f, w).$$

Here, the bilinear form $a_m : (H_0^1(\Omega)^2 \times H_0^1(\Omega)) \times (H_0^1(\Omega)^2 \times H_0^1(\Omega)) \to \mathbb{R}$ is given by

$$a_m(\theta, w; \psi, z) := a(\theta, \psi) + \kappa(\nabla w - \theta, \nabla z - \psi).$$

Minimization of the energy functional leads to the problem

Find $(\theta, w) \in H_0^1(\Omega)^2 \times H_0^1(\Omega)$ such that for all $(\psi, z) \in H_0^1(\Omega)^2 \times H_0^1(\Omega)$

$$a_m(\theta, w; \psi, z) + \kappa \frac{1-t^2}{t^2}(\nabla w - \theta, \nabla z - \psi) = (f, z).$$

We introduce the scaled shear term $\xi = \kappa(1-t^2)t^{-2}(\nabla w - \theta)$ and obtain the associated mixed formulation with penalty term

Find $(\theta, w, \xi) \in H_0^1(\Omega)^2 \times H_0^1(\Omega) \times L^2(\Omega)^2$ such that

$$a_m(\theta, w; \psi, z) + (\nabla z - \psi, \xi) = (f, z), \tag{7.9}$$

$$(\nabla w - \theta, \eta) - \frac{t^2}{\kappa(1-t^2)}(\xi, \eta) = 0 \tag{7.10}$$

for all $(\psi, z, \eta) \in H_0^1(\Omega)^2 \times H_0^1(\Omega) \times L^2(\Omega)^2$.

Note that in contrast to a (which is coercive on $H_0^1(\Omega)^2$ only), the modified bilinear form a_m is coercive on the product space $H_0^1(\Omega)^2 \times H_0^1(\Omega)$.

Lemma 7.2. *There is a positive constant α such that*

$$a_m(\theta, w; \theta, w) \geq \alpha(\|\theta\|_1^2 + \|w\|_1^2)$$

for all $(\theta, w) \in H_0^1(\Omega)^2 \times H_0^1(\Omega)$.

Proof. Due to Korn's inequality there is a positive constant α_1 such that

$$a(\theta, \theta) \geq \alpha_1 \|\theta\|_1^2$$

for all $\theta \in H_0^1(\Omega)^2$. Then, applying triangle's inequality

$$\|\nabla w\|_0^2 \le (\|\nabla w - \theta\|_0 + \|\theta\|_0)^2 \le 2\|\nabla w - \theta\|_0^2 + 2\|\theta\|_1^2$$

and Friedrich's inequality

$$\|w\|_1^2 \le \alpha_2 |w|_1^2 \le 2\alpha_2(\|\nabla w - \theta\|_0^2 + \|\theta\|_1^2)$$

we finally get the coercivity

$$\begin{aligned}\|\theta\|_1^2 + \|w\|_1^2 &\le (1+2\alpha_2)(\|\nabla w - \theta\|_0^2 + \|\theta\|_1^2)\\ &\le (1+2\alpha_2)(\kappa^{-1} + \alpha_1^{-1})(\kappa\|\nabla w - \theta\|_0^2 + \alpha_1\|\theta\|_1^2)\\ &\le (1+2\alpha_2)(\kappa^{-1} + \alpha_1^{-1})\, a_m(\theta, w; \theta, w)\end{aligned}$$

with the constant $\alpha = (1+2\alpha_2)^{-1}(\kappa^{-1} + \alpha_1^{-1})^{-1}$. ■

Let $\Theta_h \subset H_0^1(\Omega)^2$, $V_h \subset H_0^1(\Omega)$ and $\Gamma_h \subset L^2(\Omega)^2$ be finite element spaces for discretizing the variational formulation (7.9) and (7.10). Then, the discrete problem reads

Find $(\theta_h, w_h, \xi_h) \in \Theta_h \times V_h \times \Gamma_h$ such that for all $(\psi_h, z_h, \eta_h) \in \Theta_h \times V_h \times \Gamma_h$

$$a_m(\theta_h, w_h; \psi_h, z_h) + (\nabla z_h - \psi_h, \xi_h) = (f, z_h), \tag{7.11}$$

$$(\nabla w_h - \theta_h, \eta_h) - \frac{t^2}{\kappa(1-t^2)}(\xi_h, \eta_h) = 0. \tag{7.12}$$

As usual the discrete problem is associated with a linear system of algebraic equations for the coefficients in the basis representation. Let us denote them by $(\overline{\theta}_h, \overline{w}_h, \overline{\xi}_h)$. Then, the linear system has the structure

$$\begin{bmatrix} A & B^T \\ B & -C \end{bmatrix} \begin{bmatrix} \overline{\theta}_h \\ \overline{w}_h \\ \overline{\xi}_h \end{bmatrix} = \begin{bmatrix} 0 \\ F \\ 0 \end{bmatrix}. \tag{7.13}$$

Exercise 7.1. Prove that the structure of the discrete problem (7.11), (7.12) coincise with (7.13). Show that the matrices A and C are symmetric and positive definite.

As usual in mixed methods we cannot choose the finite element spaces independently from each other. For stability, a stability condition with respect to the pair $((\Theta_h \times V_h), \Gamma_h)$ has to be satisfied. To be more precise we introduce the bilinear form $b : (H_0^1(\Omega)^2 \times H_0^1(\Omega)) \times L^2(\Omega)^2 \to \mathbb{R}$ by

$$b(\theta, w; \eta) := (\nabla w - \theta, \eta)$$

and use the notation

$$|||\eta|||_t := \|\eta\|_{-1} + \|\operatorname{div} \eta\|_{-1} + t\|\eta\|_0.$$

Lemma 7.3. *Let a bounded linear operator* $\Pi_h : H_0^1(\Omega)^2 \times H_0^1(\Omega) \to \Theta_h \times V_h$ *exists which satisfies*

$$b(\Pi_h(\theta, w); \eta_h) = b(\theta, w; \eta_h) \quad \text{for all } \eta_h \in \Gamma_h. \tag{7.14}$$

Then, the mixed method (7.11), (7.12) *is quasi-optimal, that is,*

$$\|\theta - \theta_h\|_1 + \|w - w_h\|_1 + |||\xi - \xi_h|||_t \le C\inf(\|\theta - \psi_h\|_1 + \|w - z_h\|_1 + |||\xi - \eta_h|||_t),$$

where the infimum is taken over $(\psi_h, z_h, \eta_h) \in \Theta_h \times V_h \times \Gamma_h$.

Proof. See Arnold and Brezzi (1993). ∎

Next we provide a possible choice of spaces and some hints for constructing the interpolation operator mentioned in Lemma 7.3. Let Θ_h be the space of continuous, piecewise linear functions enriched by cubic bubble functions, V_h be the space of continuous, piecewise quadratic functions and Γ_h be the space of discontinuous, piecewise constant functions, that is,

$$\Theta_h := \{\theta \in H_0^1(\Omega)^2 : \theta|_K \in (P_1(K) \oplus \operatorname{span} \lambda_1\lambda_2\lambda_3)^2 \text{ for all } K \in \mathcal{T}_h\}, \tag{7.15}$$

$$V_h := \{w \in H_0^1(\Omega) : w|_K \in P_2(K) \text{ for all } K \in \mathcal{T}_h\}, \tag{7.16}$$

$$\Gamma_h := \{\eta \in L^2(\Omega)^2 : \eta|_K \in P_0(K)^2 \text{ for all } K \in \mathcal{T}_h\}. \tag{7.17}$$

Setting $\Pi_h(\theta, w) = (\Pi_h^1\theta, \Pi_h^2 w)$ the condition (7.14) is satisfied if

$$0 = (\Pi_h^1\theta - \theta, \eta_h) = \sum_{K\in\mathcal{T}_h} \eta_h|_K \cdot \int_K (\Pi_h^1\theta - \theta)\,\mathrm{d}x, \tag{7.18}$$

$$0 = (\nabla(\Pi_h^2 w - w), \eta_h) = \sum_{K\in\mathcal{T}_h} \eta_h|_K \sum_{E\subset\partial K} \int_E (\Pi_h^2 w - w) \cdot n_K\,\mathrm{d}\gamma \tag{7.19}$$

for all $\eta_h \in \Gamma_h$. Let $K \in \mathcal{T}_h$ be a triangle with the edges E_0, E_1, E_2 and the opposite vertices a_0, a_1, a_2. In Θ_h the nodal functionals

$$N_i^1(\theta) = \theta(a_i),\ i = 0,1,2, \qquad N_3^1(\theta) = \frac{1}{|K|}\int_K \theta\,\mathrm{d}x$$

are considered. Then, the set of dof associated with these nodal functionals is $(P_1(K) \oplus \operatorname{span} \lambda_1\lambda_2\lambda_3)^2$ unisolvent. Similarly, in V_h we introduce the nodal functionals

$$N_i^2(w) = w(a_i), \qquad N_{i+3}^2 = \frac{1}{|E_i|}\int_{E_i} w\,\mathrm{d}\gamma\,,\ i = 0,1,2.$$

We see that the set of dof associated with the nodal functionals N_i^2, $i = 0,\ldots,5$ is $P_2(K)$ unisolvent. Then, the canonical interpolations Π_h^1, Π_h^2 generated from the nodal functionals N_i^1, $i = 0,\ldots,3$ and N_i^2, $i = 0,\ldots,5$, satisfy Equations (7.18) and (7.19), respectively. Unfortunately, the domains of definition of these operators are too small. Indeed, the nodal functionals N_i^1 and N_i^2, $i = 0,\ldots,2$, require point values of functions which are not defined for $H_0^1(\Omega)^2$ and $H_0^1(\Omega)$, respectively.

One way to overcome this is the Scott–Zhang interpolation of less smooth functions developed in Section 4.6. Let S_h denote the Scott–Zhang operator defined on $H_0^1(\Omega)$ into the space of continuous, piecewise linear functions. The application of S on a vector-valued function is understood in a componentwise manner. Further let the linear operators

$$\pi_h^1 : H_0^1(\Omega)^2 \to \{\theta \in H_0^1(\Omega)^2 : \theta|_K \in (\operatorname{span} \lambda_1\lambda_2\lambda_3)^2\}$$
$$\pi_h^2 : H_0^1(\Omega) \to \{w \in H_0^1(\Omega) : w|_K \in \operatorname{span}(\lambda_1\lambda_2, \lambda_2\lambda_3, \lambda_3\lambda_1)\}$$

locally defined by

$$\int_K (\pi_h^1\theta - \theta)\,\mathrm{d}x = 0, \qquad \int_E (\pi_h^2 w - w)\,\mathrm{d}\gamma = 0 \quad \text{for all edges } E \subset \partial K.$$

Then, the operators

$$\Pi_h^1\theta = S_h\theta + \pi_h^1(\theta - S_h\theta) \quad \text{and} \quad \Pi_h^2\theta = S_h\theta + \pi_h^2(\theta - S_h\theta)$$

satisfy all assumptions of Lemma 7.3.

Theorem 7.3. *Let the solution of the continuous problem* (7.9), (7.10) *belong to* $H^2(\Omega)^2 \times H^2(\Omega) \times H^1(\Omega)^2$ *and let* (θ_h, w_h, ξ_h) *be the solution of the discrete problem* (7.11), (7.12) *based on the finite element spaces* (7.15)–(7.17). *Then, the error estimate*

$$\|\theta - \theta_h\|_1 + \|w - w_h\|_1 + |||\xi - \xi_h|||_t \le Ch(\|\theta\|_2 + \|w\|_2 + \|\xi\|_1)$$

holds true uniformly in t.

Proof. Use the quasi-optimality stated in Lemma 7.3 and the standard interpolation error estimates. ∎

CHAPTER 8

SYSTEMS IN FLUID MECHANICS

In this chapter, we discuss finite element methods for the incompressible Stokes and Navier–Stokes equations. We mainly treat the Stokes equations, in particular, the concept of mixed finite element methods for the Stokes equations is presented. Nevertheless, it can easily be extended to Navier–Stokes equations. In addition, the inclusion of different boundary conditions and the use of equal order stabilized finite elements are discussed. A comprehensive list of inf–sup stable finite elements are also presented.

8.1. Conservation of Mass and Momentum

One fundamental principle in classical physics says that mass can neither be destroyed nor be created. Thus, the mass of a fluid in a small control volume ω can change over time t in a given time interval $[0,\mathrm{I}]$ only by flow in and/or out of the boundary $\partial\omega$, that is,

$$\frac{\mathrm{d}}{\mathrm{d}t}\int_{\omega}\rho\,\mathrm{d}x+\int_{\partial\omega}\rho\,\mathbf{u}\cdot\mathbf{n}\,\mathrm{d}s=0.$$

Here, ρ denotes the density of the fluid, $\mathbf{u}=(u_1,\ldots,u_d)$ the fluid velocity and $\mathbf{n}=(n_1,\ldots,n_d)$ the outer normal to ω. The first term represents the rate of change of mass in ω and the second term denotes the rate of mass loss through the domain boundary $\partial\omega$. Using Gauss theorem to replace the surface integral we get for all control volumes ω

$$\int_{\omega}\left(\frac{\partial\rho}{\partial t}+\mathrm{div}\,(\rho\mathbf{u})\right)\mathrm{d}x=0.$$

As a consequence we have

$$\frac{\partial\rho}{\partial t}+\mathrm{div}\,(\rho\mathbf{u})=0. \tag{8.1}$$

Incompressible fluids are characterized by a constant density, thus for incompressible fluids mass conservation can be expressed by

$$\operatorname{div}\mathbf{u} = 0. \tag{8.2}$$

Besides the mass conservation, Newton's second law states the conservation of momentum. The rate of change of momentum equals the net force $\mathbf{F}$ acting on it. Taking into consideration that the momentum can be transported in and out of the boundary $\partial\omega$ of the control volume ω, we obtain

$$\frac{\mathrm{d}}{\mathrm{d}t}\int_\omega \rho\mathbf{u}\,\mathrm{d}x + \int_{\partial\omega} \rho\mathbf{u}\mathbf{u}\cdot\mathbf{n}\,\mathrm{d}s = \mathbf{F}.$$

Again by Gauss theorem we can replace the surface integral in the i-th equation

$$\int_{\partial\omega} \rho u_i \mathbf{u}\cdot\mathbf{n}\,\mathrm{d}s = \int_\omega \operatorname{div}(\rho u_i \mathbf{u})\,\mathrm{d}x = \int_\omega (u_i \operatorname{div}(\rho\mathbf{u}) + \rho(\mathbf{u}\cdot\nabla)u_i)\,\mathrm{d}x.$$

Using the chain rule we can replace the other term in the i-th equation by

$$\frac{\mathrm{d}}{\mathrm{d}t}\int_\omega \rho u_i\,\mathrm{d}x = \int_\omega \left(u_i\frac{\partial\rho}{\partial t} + \rho\frac{\partial u_i}{\partial t}\right)\mathrm{d}x.$$

Summing up the two equations and taking the conservation of mass property (8.1) into consideration, we end up with the following equation written in vector form

$$\int_\omega \rho\left(\frac{\partial\mathbf{u}}{\partial t} + (\mathbf{u}\cdot\nabla)\mathbf{u}\right)\mathrm{d}x = \mathbf{F}.$$

Next, we write the net force as the sum of internal and external forces

$$\mathbf{F} = \int_\omega (\operatorname{div}\mathbb{S}(\mathbf{u},p) + \mathbf{f})\,\mathrm{d}x$$

with the stress tensor $\mathbb{S}$ and a given body load $\mathbf{f}$, for example, like gravity. Viscous effects, like friction of the fluid particles, and the pressure can be considered as internal forces. The stress tensor for the Newtonian fluid is defined by

$$\mathbb{S}(\mathbf{u},p) = 2\eta\mathbb{D}(u) - p\mathbb{I}, \qquad \mathbb{D}(\mathbf{u})_{ij} = \frac{1}{2}\left(\frac{\partial u_i}{\partial x_j} + \frac{\partial u_j}{\partial x_i}\right), \quad i,j = 1,\dots,d,$$

where η is the dynamic viscosity of the fluid, $\mathbb{D}(\mathbf{u})$ the velocity deformation tensor, p the pressure and $\mathbb{I}$ the identity tensor. Fluids obeying such a constitutive law are called Newtonian.

Finally, we obtain for all control volumes

$$\int_\omega \rho\left(\frac{\partial\mathbf{u}}{\partial t} + (\mathbf{u}\cdot\nabla)\mathbf{u}\right)\mathrm{d}x = \int_\omega (\operatorname{div}\mathbb{S}(\mathbf{u},p) + \mathbf{f})\,\mathrm{d}x$$

and consequently

$$\rho\left(\frac{\partial\mathbf{u}}{\partial t} + (\mathbf{u}\cdot\nabla)\mathbf{u}\right) - \operatorname{div}\mathbb{S}(\mathbf{u},p) = \mathbf{f}. \tag{8.3}$$

For an incompressible fluid we have

$$\operatorname{div}\mathbb{S}(\mathbf{u},p) = \eta\Delta\mathbf{u} + \eta\nabla(\operatorname{div}\mathbf{u}) - \nabla p = \eta\Delta\mathbf{u} - \nabla p.$$

Thus, the conservation of mass and momentum for an incompressible Newtonian fluid in a bounded domain Ω can be expressed by

$$\rho\left(\frac{\partial\mathbf{u}}{\partial t} + (\mathbf{u}\cdot\nabla)\mathbf{u}\right) - \eta\Delta\mathbf{u} + \nabla p = \mathbf{f} \quad \text{in } \Omega\times(0,\mathrm{I}] \tag{8.4}$$

$$\operatorname{div}\mathbf{u} = 0 \quad \text{in } \Omega\times(0,\mathrm{I}]. \tag{8.5}$$

The set of PDEs (8.4) and (8.5) that describe the fluid velocity $\mathbf{u}$ and the pressure p in $\Omega\times(0,\mathrm{I}]$ are called Navier–Stokes equations. To complete the problem we have to set initial and boundary conditions. The initial condition prescribes a given velocity field $\mathbf{u}_0$ at time $t=0$

$$\mathbf{u}(x,0) = \mathbf{u}_0(x) \quad x\in\Omega. \tag{8.6}$$

In case that the flow does not depend on time, the time derivative of $\mathbf{u}$ vanishes in Equation (8.4), the initial condition (8.6) is omitted and consequently we obtain the steady Navier–Stokes equations.

We have different types of boundary conditions that can be imposed on the whole or a part of the boundary. At fixed walls of the fluid domain the no slip condition $\mathbf{u}=\mathbf{0}$ is used in general. At the inlet (and the outlet) of the fluid domain a known velocity field can be prescribed. Both cases are classified as Dirichlet boundary conditions

$$\mathbf{u} = \mathbf{g} \quad \text{on } \Gamma_{\mathrm{D}}, \tag{8.7}$$

where $\mathbf{g}$ is some given velocity field on the Dirichlet part, Γ_{D}, of the boundary. Alternatively, for allowing slip at fixed walls of the fluid domain, as it can be assumed for flows with moving contact lines, for instance, the Navier-Slip boundary condition

$$\mathbf{u}\cdot\mathbf{n} = 0, \qquad \mathbf{u}\cdot\tau_i + \beta\,\tau_i\cdot\mathbb{S}(\mathbf{u},p)\cdot\mathbf{n} = 0, \quad i=1,\ldots,d-1 \text{ on } \Gamma_{\mathrm{S}} \tag{8.8}$$

is often used. Here, the first expression models the no penetration of fluid into the solid wall Γ_{S}, β is the slip coefficient and τ_i, $i=1,\ldots,d-1$, denote a set of linearly independent unit vectors tangential to Γ_{S}. In the limit $\beta\to 0$ we obtain formally the no slip condition. If β tends to infinity, we obtain the free slip condition

$$\mathbf{u}\cdot\mathbf{n} = 0, \qquad \tau_i\cdot\mathbb{S}(\mathbf{u},p)\cdot\mathbf{n} = 0, \quad i=1,\ldots,d-1 \text{ on } \Gamma_{\mathrm{S}}.$$

In free surface flows, the boundary condition along the free surface Γ_{F} becomes

$$\mathbb{S}(\mathbf{u},p)\cdot\mathbf{n} = \sigma\mathcal{K}\mathbf{n} + \nabla_\Gamma\sigma \quad \text{on } \Gamma_{\mathrm{F}} \tag{8.9}$$

where σ denotes the surface tension, $\mathcal{K}$ the sum of the principal curvatures and $\nabla_\Gamma = (\mathbb{I}-\mathbf{n}\otimes\mathbf{n})\nabla$ the surface gradient. The second term, $\nabla_\Gamma\sigma$, called Marangoni force, drives tangential motion of the fluid when the surface tension is not constant over Γ_{F}. The first term, $\sigma\mathcal{K}\mathbf{n}$ is the surface tension force, and is the only effect of surface energy if σ is constant and Γ_{F} is a closed hypersurface.

Navier–Stokes problem

The Navier–Stokes problem consists in finding a velocity field $\mathbf{u}$ and a pressure p which satisfy Equations (8.4) and (8.5), the initial condition (8.6) and one of the boundary conditions (8.7) and (8.8) on disjoint parts Γ_D, Γ_S and Γ_F of the boundary $\partial\Omega = \Gamma_D \cup \Gamma_S \cup \Gamma_F$.

Oseen problem

We now consider some special cases of the Navier–Stokes equations. Let $t = t_0 = 0 < t_1 < \cdots < t_n$ be discrete time instances. Using an implicit time discretization and/or a fixed point iteration to linearize Equation (8.4), we get besides Equation (8.5) for $t = t_n$

$$\rho\left(\frac{\mathbf{u}^n - \mathbf{u}^{n-1}}{\Delta t} + (\mathbf{u}^{n-1} \cdot \nabla)\mathbf{u}^n\right) - \eta \Delta \mathbf{u}^n + \nabla p^n = \mathbf{f}^n \quad \text{in } \Omega,$$

where $\mathbf{f}^n = \mathbf{f}(\cdot, t_n)$ and $\mathbf{u}^n$, p^n denote approximations of $\mathbf{u}(\cdot, t_n)$ and $p(\cdot, t_n)$, respectively. Then, in each time/iteration step one has to solve a problem like

$$-\nu \Delta \mathbf{u} + (\mathbf{b} \cdot \nabla)\mathbf{u} + \sigma \mathbf{u} + \nabla p = \tilde{\mathbf{f}} \quad \text{in } \Omega,$$
$$\operatorname{div} \mathbf{u} = 0 \quad \text{in } \Omega,$$

completed with the associated boundary conditions. Here, the first equation has been divided by ρ which gives the kinematic viscosity $\nu = \eta/\rho$, the convection field $\mathbf{b} = \mathbf{u}^{n-1}$, $\sigma = 1/\Delta t$ or $\sigma = 0$ in the stationary case. This set of equations with appropriate boundary conditions is called Oseen problem. It is a linear boundary value problem. Although the Oseen problem with Dirichlet boundary conditions admits a unique solution for all viscosities ν and data $\tilde{\mathbf{f}}$, which is not the case for the incompressible Navier–Stokes equations, it still has the two basic challenges of handling the effect of dominate advection, $\nu \ll \|\mathbf{b}\|$ and handling the incompressibility constraint.

Stokes problem

Suppose the viscous effects dominate the advection, that is, $\|\mathbf{b}\| \ll \nu$, and the flow is time-independent, we obtain the steady Stokes equation

$$-\nu \Delta \mathbf{u} + \nabla p = \mathbf{f} \quad \text{in } \Omega, \qquad \operatorname{div} \mathbf{u} = 0 \quad \text{in } \Omega,$$

which is the simplest model to study incompressible flows.

8.2. Weak formulation of the Stokes problem

In the following, we consider the Stokes problem in a bounded domain $\Omega \subset \mathbb{R}^d$, $d = 2, 3$

$$-\nu \Delta \mathbf{u} + \nabla p = \mathbf{f} \quad \text{in } \Omega, \qquad \operatorname{div} \mathbf{u} = 0 \quad \text{in } \Omega, \qquad \mathbf{u} = \mathbf{g} \quad \text{on } \Gamma = \partial\Omega, \tag{8.10}$$

where $\mathbf{u} = (u_1, \ldots, u_d)$ denotes the fluid velocity, p the pressure, $\mathbf{f} = (f_1, \ldots, f_d)$ a body force and $\mathbf{g} = (g_1, \ldots, g_d)$ the given velocity field on the boundary. The Gauss theorem shows that the velocity at the boundary $\mathbf{g}$ cannot be prescribed arbitrarily, in fact, the Dirichlet data needs to satisfy

$$\int_\Gamma \mathbf{g} \cdot \mathbf{n} \, ds = \int_\Gamma \mathbf{u} \cdot \mathbf{n} \, ds = \int_\Omega \operatorname{div} \mathbf{u} \, dx = 0.$$

In order to derive the weak formulation of the Stokes problem we multiply the first d equations by a test function $\mathbf{v} = (v_1, \ldots, v_d) \in C_0^\infty(\Omega)^d$, integrate over Ω, and apply integration by parts to obtain

$$\begin{aligned}(-\nu \Delta \mathbf{u} + \nabla p, \mathbf{v}) &= -\nu \int_\Omega \sum_{i,j=1}^{d} \frac{\partial^2 u_i}{\partial x_j^2} v_i \, dx + \int_\Omega \sum_{i=1}^{d} \frac{\partial p}{\partial x_i} v_i \, dx \\ &= \nu \int_\Omega \sum_{i,j=1}^{d} \frac{\partial u_i}{\partial x_j} \frac{\partial v_i}{\partial x_j} \, dx - \int_\Omega p \sum_{i=1}^{d} \frac{\partial v_i}{\partial x_i} \, dx.\end{aligned}$$

Thus, using the notation $(\cdot,\cdot)$ for the inner product in $L^2(\Omega)$ and its vector-valued and matrix-valued versions, we get for all $\mathbf{v} \in C_0^\infty(\Omega)^d$

$$\nu(\nabla \mathbf{u}, \nabla \mathbf{v}) - (p, \operatorname{div} \mathbf{v}) = (\mathbf{f}, \mathbf{v}). \tag{8.11}$$

The second equation follows by multiplying the incompressibility constraint with a test function $q \in L^2(\Omega)$

$$(q, \operatorname{div} \mathbf{u}) = 0. \tag{8.12}$$

The expressions in Equations (8.11) and (8.12) are welldefined for $\mathbf{u}, \mathbf{v} \in H^1(\Omega)^d$ and $p, q \in L^2(\Omega)$, $\mathbf{f} \in L^2(\Omega)^d$. Note further, that if Equation (8.11) holds for all $\mathbf{v} \in C_0^\infty(\Omega)^d$ it also holds for all $\mathbf{v} \in H_0^1(\Omega)^d$ since $C_0^\infty(\Omega)^d$ is dense in $H_0^1(\Omega)^d$. Let $(\mathbf{u}, p)$ be a solution of the problem (8.11), (8.12). Then, $(\mathbf{u}, p + \text{const})$ is a solution too, due to

$$(1, \operatorname{div} \mathbf{v}) = \int_\Gamma \mathbf{v} \cdot \mathbf{n} \, ds = 0 \quad \text{for all } \mathbf{v} \in H_0^1(\Omega)^d.$$

It remains to impose the boundary condition, the trace of $\mathbf{u} \in H^1(\Omega)^d$ onto $\Gamma = \partial\Omega$ should be equal to $\mathbf{g}$. Since the image of $H^1(\Omega)^d$ with respect to the trace operator is $H^{1/2}(\Gamma)^d$, we assume that the Dirichlet data $\mathbf{g}$ belongs to $H^{1/2}(\Gamma)^d$ satisfying $\int_\Gamma \mathbf{g} \cdot \mathbf{n} \, ds = 0$. Note that, for the test function $q = 1$, Equation (8.12) is automatically satisfied because of

$$(1, \operatorname{div} \mathbf{u}) = \int_\Gamma \mathbf{u} \cdot \mathbf{n} \, ds = \int_\Gamma \mathbf{g} \cdot \mathbf{n} \, ds = 0.$$

In order to remove the constants from the test space and to fix the additive constant in the pressure, we restrict the solution space for the pressure to

$$L_0^2(\Omega) := \{q \in L^2(\Omega) : (q, 1) = 0\}.$$

Now the weak formulation of problem (8.10) reads

Given $\mathbf{f} \in L^2(\Omega)^d$ and $\mathbf{g} \in H^{1/2}(\Gamma)^d$ satisfying $\int_\Gamma \mathbf{g} \cdot \mathbf{n}\, ds = 0$,
find $(\mathbf{u}, p) \in H^1(\Omega)^d \times L_0^2(\Omega)$ with $\mathbf{u} = \mathbf{g}$ on Γ such that

$$\nu(\nabla \mathbf{u}, \nabla \mathbf{v}) - (p, \operatorname{div} \mathbf{v}) = (\mathbf{f}, \mathbf{v}),$$
$$(q, \operatorname{div} \mathbf{u}) = 0$$

for all $(\mathbf{v}, q) \in H_0^1(\Omega)^d \times L_0^2(\Omega)$.

It is convenient to work with the same ansatz and test space for the velocity. This can be achieved by homogenization of the boundary conditions. Let $\tilde{\mathbf{g}} \in H^1(\Omega)^d$ be an extension of the boundary data $\mathbf{g} \in H^{1/2}(\Gamma)^d$ into Ω. Then, instead of looking for $\mathbf{u} \in H^1(\Omega)^d$ with $\mathbf{u} = \mathbf{g}$ on Γ, we look for $\tilde{\mathbf{u}} \in H_0^1(\Omega)^d$ in $\mathbf{u} = \tilde{\mathbf{u}} + \tilde{\mathbf{g}}$. Eliminating $\mathbf{u}$ in the above equations, we end up with the problem

Given $\mathbf{f} \in L^2(\Omega)^d$ and $\mathbf{g} \in H^{1/2}(\Gamma)^d$ satisfying $\int_\Gamma \mathbf{g} \cdot \mathbf{n}\, ds = 0$,
find $(\tilde{\mathbf{u}}, p) \in H_0^1(\Omega)^d \times L_0^2(\Omega)$ such that

$$\nu(\nabla \tilde{\mathbf{u}}, \nabla \mathbf{v}) - (p, \operatorname{div} \mathbf{v}) = (\mathbf{f}, \mathbf{v}) - (\nabla \tilde{\mathbf{g}}, \nabla \mathbf{v}) =: \langle \tilde{\mathbf{f}}, \mathbf{v} \rangle,$$
$$(q, \operatorname{div} \tilde{\mathbf{u}}) = -(q, \operatorname{div} \tilde{\mathbf{g}}) =: \langle g, q \rangle$$

for all $(\mathbf{v}, q) \in H_0^1(\Omega)^d \times L_0^2(\Omega)$.

Here, we used the notation $\langle \cdot, \cdot \rangle$ for the duality paring between the spaces $H_0^1(\Omega)^d$, $L_0^2(\Omega)$ and their dual spaces. Note that, $\tilde{\mathbf{f}} \in H^{-1}(\Omega)^d$ by definition and g belong to the dual space of $L_0^2(\Omega)$ since

$$\langle g, 1 \rangle = -(1, \operatorname{div} \tilde{\mathbf{g}}) = -\int_\Gamma \tilde{\mathbf{g}} \cdot \mathbf{n}\, ds = -\int_\Gamma \mathbf{g} \cdot \mathbf{n}\, ds = 0.$$

Let $V = H_0^1(\Omega)^d$ be equipped with the norm $|\cdot|_1$ and $Q = L_0^2(\Omega)$ be equipped with the norm $\|\cdot\|_0$. For the existence and uniqueness of solutions of the Stokes problem we skip the tilde and asked for the solution of the following problem:

Find $(\mathbf{u}, p) \in V \times Q$ such that for all $(\mathbf{v}, q) \in V \times Q$

$$\nu(\nabla \mathbf{u}, \nabla \mathbf{v}) - (p, \operatorname{div} \mathbf{v}) = \langle \mathbf{f}, \mathbf{v} \rangle, \tag{8.13}$$
$$(q, \operatorname{div} \mathbf{u}) = \langle g, q \rangle. \tag{8.14}$$

We see that the left hand side is a continuous bilinear form and the right hand side a continuous linear form on the product space $V \times Q$ equipped with the norm $(\mathbf{v}, q) \mapsto \sqrt{|\mathbf{v}|_1^2 + \|q\|_0^2}$, therefore, we can write

Find $(\mathbf{u}, p) \in V \times Q$ such that for all $(\mathbf{v}, q) \in V \times Q$

$$A\big((\mathbf{u}, p); (\mathbf{v}, q)\big) = \ell\big((\mathbf{v}, q)\big), \tag{8.15}$$

where

$$A\big((\mathbf{u}, p); (\mathbf{v}, q)\big) := \nu(\nabla \mathbf{u}, \nabla \mathbf{v}) - (p, \operatorname{div} \mathbf{v}) + (q, \operatorname{div} \mathbf{u}),$$
$$\ell\big((\mathbf{v}, q)\big) := \langle \mathbf{f}, \mathbf{v} \rangle + \langle g, q \rangle.$$

Equations (8.13) and (8.14) are equivalent to Equation (8.15), Equation (8.15) follows from Equations (8.13) and (8.14) by summation, and Equations (8.13) and (8.14) from (8.15) by testing with $(\mathbf{v},q)=(\mathbf{v},0)$ and $(\mathbf{v},q)=(\mathbf{0},q)$, respectively.

The continuity of $A(\cdot;\cdot)$ follows from Cauchy–Schwarz inequality for sums

$$\begin{aligned}\left|A\left((\mathbf{u},p);(\mathbf{v},q)\right)\right| &\le \nu|\mathbf{u}|_1|\mathbf{v}|_1+\sqrt{d}\|p\|_0|\mathbf{v}|_1+\sqrt{d}\|q\|_0|\mathbf{u}|_1\\ &\le C\sqrt{|\mathbf{u}|_1^2+\|p\|_0^2}\sqrt{|\mathbf{v}|_1^2+\|q\|_0^2}.\end{aligned}$$

Similarly, we can prove the continuity of $\ell(\,\cdot\,)$

$$\left|\ell\left((\mathbf{v},q)\right)\right|\le\|\mathbf{f}\|_{V^*}|\mathbf{v}|_1+\|g\|_{Q^*}\|q\|_0\le C\sqrt{|\mathbf{v}|_1^2+\|q\|_0^2}.$$

Since V and Q are Hilbert spaces, the product space $V\times Q$ is a Hilbert space. In order to apply the Lax–Milgram theorem, we have to show the coercivity of $A(\cdot;\cdot)$ on the product space. Unfortunately, the bilinear form is not coercive on $V\times Q$. Indeed, for any $q\in Q$ and $\mathbf{v}=\mathbf{0}\in V$ we get

$$A\left((\mathbf{v},q);(\mathbf{v},q)\right)=\nu|\mathbf{v}|_1^2=0.$$

As a consequence, we need a different concept to show existence and uniqueness of a solution of the problem (8.13) and (8.14).

The new concept that we have to develop is the mixed finite element method. It directly applies to Equations (8.13) and (8.14) with two Hilbert spaces V and Q, two continuous bilinear forms $a\colon V\times V\to\mathbb{R}$ and $b\colon V\times Q\to\mathbb{R}$, and two continuous linear forms $\langle\mathbf{f},\cdot\rangle : V\to\mathbb{R}$ and $\langle g,\cdot\rangle : Q\to\mathbb{R}$. In our case the bilinear forms are given as

$$a(\mathbf{u},\mathbf{v}):=\nu(\nabla\mathbf{u},\nabla\mathbf{v}),\qquad b(q,\mathbf{v}):=-(q,\operatorname{div}\mathbf{v})\quad\text{for all }\mathbf{u},\mathbf{v}\in V, q\in Q.$$

The generalized formulation of Equations (8.13) and (8.14) reads as

Find $(\mathbf{u},p)\in V\times Q$ such that for all $(\mathbf{v},q)\in V\times Q$

$$a(\mathbf{u},\mathbf{v})+b(\mathbf{v},p)=\langle\mathbf{f},\mathbf{v}\rangle,\tag{8.16}$$

$$-b(\mathbf{u},q)=\langle g,q\rangle.\tag{8.17}$$

We transform Equations (8.16) and (8.17) to an equivalent operator equation. The mapping $\mathbf{v}\mapsto a(\mathbf{u},\mathbf{v})$ is a continuous linear form on V for any $\mathbf{u}\in V$. Theorem 1.2 (Riesz representation theorem) tells us that for any $\mathbf{u}\in V$ there is an element $A\mathbf{u}\in V^*$ with

$$\langle A\mathbf{u},\mathbf{v}\rangle=a(\mathbf{u},\mathbf{v})\quad\text{for all }\mathbf{u},\mathbf{v}\in V.$$

Moreover, the linearity and continuity of the mapping $\mathbf{u}\mapsto a(\mathbf{u},\mathbf{v})$ for any fixed $\mathbf{v}\in V$ implies that the operator $A\colon V\to V^*$ is linear and continuous. Similar arguments prove that there are linear continuous operators $B\colon V\to Q^*$ and $B^*: Q\to V^*$ satisfying

$$\langle B\mathbf{v},q\rangle=b(\mathbf{v},q),\qquad\langle B^*q,\mathbf{v}\rangle=b(\mathbf{v},q)\quad\text{for all }q\in Q,\mathbf{v}\in V.$$

Problem (8.16), (8.17) is equivalent to

Find $(\mathbf{u}, p) \in V \times Q$ such that

$$A\mathbf{u} + B^* p = f \quad \text{in } V^*, \tag{8.18}$$

$$-B\mathbf{u} = g \quad \text{in } Q^*. \tag{8.19}$$

We introduce the closed subspace

$$W = \{\mathbf{v} \in V : b(\mathbf{v}, q) = 0 \text{ for all } q \in Q\}$$

and the affine manifold

$$W_g = \{\mathbf{v} \in V : b(\mathbf{v}, q) = \langle g, q\rangle \text{ for all } q \in Q\}.$$

Now Equations (8.14) and (8.19) state that the velocity $\mathbf{u}$ belongs to W_g. Therefore, taking $\mathbf{v} \in W$ in Equation (8.13) we see that the velocity part $\mathbf{u}$ of the solution $(\mathbf{u}, p)$ has to satisfy

$$\text{Find } \mathbf{u} \in W_g \text{ such that for all } \mathbf{v} \in W: \quad a(\mathbf{u}, \mathbf{v}) = \langle \mathbf{f}, \mathbf{v}\rangle. \tag{8.20}$$

Suppose for a moment that this problem is uniquely solvable. Then, Equation (8.17) is satisfied and Equation (8.16) has been required, up to now, only on the subspace $W \subset V$. That means, for the pressure $p \in Q$ we are left with Equation (8.16) on the orthogonal complement $W^\perp$ of W in $V = W \oplus W^\perp$. The problem to determine the pressure reads

Given $\mathbf{u} \in W_g$ as solution of Equation (8.20). Find $p \in Q$ such that

$$b(\mathbf{v}, p) = \langle \mathbf{f}, \mathbf{v}\rangle - a(\mathbf{u}, \mathbf{v}) \quad \text{for all } \mathbf{v} \in W^\perp. \tag{8.21}$$

The existence and uniqueness of solutions of the problem (8.16), (8.17) is split into two subproblems to determine the velocity $\mathbf{u} \in W_g$ first as a solution of Equation (8.20), and afterwards $p \in Q$ as solution of the problem (8.21).

In case of the Stokes problem, the bilinear form a is coercive on V and thus also on the subspace W. Therefore, the unique solvability of the velocity problem (8.20) follows from Theorem 2.1 (Lax–Milgram) provided that W_g is not empty. The latter means that $g \in Q^*$ should belong to the image of the operator $B : V \to Q^*$. Indeed, if $W_g \neq \emptyset$ there is a $\mathbf{u}_0 \in W_g$ and problem (8.20) turns out to be equivalent to

$$\text{Find } \mathbf{u} - \mathbf{u}_0 \in W \text{ such that for all } \mathbf{v} \in W : a(\mathbf{u} - \mathbf{u}_0, \mathbf{v}) = \langle \mathbf{f}, \mathbf{v}\rangle - a(\mathbf{u}_0, \mathbf{v}).$$

Since the bilinear form a is continuous and coercive, the Lax–Milgram theorem can be applied to establish the existence and uniqueness of a solution $\mathbf{u} - \mathbf{u}_0 \in W$, therefore $\mathbf{u} \in W_g$.

Now we are looking for the solvability of problem (8.21). For this we introduce the polar set

$$W^0 := \{\mathbf{f} \in V^* : \langle \mathbf{f}, \mathbf{v}\rangle = 0 \text{ for all } \mathbf{v} \in W\}.$$

The following lemma is fundamental in the theory of mixed finite element methods.

Lemma 8.1. *The following three properties are equivalent*

(i) *there is a constant $\beta > 0$ such that*

$$\inf_{q \in Q} \sup_{\mathbf{v} \in V} \frac{b(\mathbf{v}, q)}{\|\mathbf{v}\|_V \, \|q\|_Q} \geq \beta, \tag{8.22}$$

(ii) *the operator* B^* *is an isomorphism from* Q *onto* W^0 *and*

$$\|B^* q\|_{V^*} \geq \beta \|q\|_Q \quad \textit{for all } q \in Q, \tag{8.23}$$

(iii) *the operator* B *is an isomorphism from* $W^\perp$ *onto* Q^* *and*

$$\|B\mathbf{v}\|_{Q^*} \geq \beta \|\mathbf{v}\|_V \quad \textit{for all } \mathbf{v} \in W^\perp. \tag{8.24}$$

Proof. See Lemma I.4.1 of Girault and Raviart 1986. ∎

Let us assume that the inf–sup condition (8.22) is satisfied. Then, B is an isomorphism from $W^\perp$ onto Q^*, that means, the velocity problem (8.20) has a unique solution $\mathbf{u} \in W_g$. Now consider the pressure Equation (8.21) written in the form

$$\langle B^* p, \mathbf{v}\rangle = \langle \mathbf{f}, \mathbf{v}\rangle - a(\mathbf{u}, \mathbf{v}) \quad \text{for all } \mathbf{v} \in W^\perp.$$

The right hand side represents a continuous linear functional on V that vanishes on W because $\mathbf{u}$ solves Equation (8.20). This shows that the right hand side belongs to the polar set W^0, shortly

$$\langle \mathbf{f}, \cdot\rangle - a(\mathbf{u}, \cdot) \in W^0.$$

According to (8.23) $B^* : Q \to W^0$ is injective and surjective, consequently, there is a unique solution of the pressure Equation (8.21).

Lemma 8.2. *There exists a unique solution* $(\mathbf{u}, p) \in V \times Q$ *of the Stokes problem* (8.13) *and* (8.14).

Proof. One can show that the image of the div-operator applied to $V = H_0^1(\Omega)^d$ is just the space $Q = L_0^2(\Omega)$ (see Corollary I.2.4 of Girault and Raviart 1986). For $q \in Q$ there is a function $\mathbf{v} \in V$ with $\operatorname{div}\mathbf{v} = q$ and $\|\mathbf{v}\|_V \leq C\|q\|_Q$. This gives for $q \in Q$ the lower bound

$$\sup_{\mathbf{v}\in V} \frac{b(\mathbf{v}, q)}{\|\mathbf{v}\|_V} = \sup_{\mathbf{v}\in V} \frac{(q, \operatorname{div}\mathbf{v})}{\|\mathbf{v}\|_V} \geq \frac{1}{C} \frac{(q, q)}{\|q\|_Q} = \frac{1}{C}\|q\|_Q$$

from which the inf–sup condition follows immediately. ∎

The above well-posedness condition (8.22) is also known as Ladyzhenskaya–Babuška–Brezzi (LBB) condition.

8.3. Conforming discretizations of the Stokes problem

Let $V_h \subset V$ and $Q_h \subset Q$ be finite element spaces for approximating the velocity space V and the pressure space Q, respectively. Then, the standard Galerkin discretization of the Stokes equations reads

Find $(\mathbf{u}_h, p_h) \in V_h \times Q_h$ such that for all $(\mathbf{v}_h, q_h) \in V_h \times Q_h$

$$\nu(\nabla \mathbf{u}_h, \nabla \mathbf{v}_h) - (p_h, \operatorname{div} \mathbf{v}_h) = \langle \mathbf{f}, \mathbf{v}_h \rangle, \tag{8.25}$$

$$(q_h, \operatorname{div} \mathbf{u}_h) = \langle g, q_h \rangle. \tag{8.26}$$

In order to show the well-posedness of the discrete problem (8.25), (8.26), we mimic the continuous case in the previous section. First, we introduce the sets

$$W_h := \{\mathbf{v}_h \in V_h : (q_h, \operatorname{div} \mathbf{v}_h) = 0 \text{ for all } q_h \in Q_h\},$$
$$W_{gh} := \{\mathbf{v}_h \in V_h : (q_h, \operatorname{div} \mathbf{v}_h) = \langle g, q_h \rangle \text{ for all } q_h \in Q_h\}.$$

The velocity part of the solution has to satisfy the problem

$$\text{Find } \mathbf{u}_h \in W_{gh} \text{ such that for all } \mathbf{v}_h \in W_h: \quad \nu(\nabla \mathbf{u}_h, \nabla \mathbf{v}_h) = \langle \mathbf{f}, \mathbf{v}_h \rangle. \tag{8.27}$$

Now the unique solvability of this problem can be shown by applying the Lax–Milgram theorem provided W_{gh} is not empty. Indeed, take an arbitrary $\mathbf{u}_{0h} \in W_{gh}$ and solve

$$\text{Find } \mathbf{u}_h - \mathbf{u}_{0h} \in W_h \text{ such that for all } \mathbf{v}_h \in W_h$$
$$\nu(\nabla(\mathbf{u}_h - \mathbf{u}_{0h}), \nabla \mathbf{v}_h) = \langle \mathbf{f}, \mathbf{v}_h \rangle - \nu(\nabla \mathbf{u}_{0h}, \nabla \mathbf{v}_h).$$

As in the continuous case, the discrete pressure can be obtained from $\mathbf{u}_h \in W_{gh}$ and restricting Equation (8.25) on the orthogonal complement $W_h^{\perp}$ of the space of discretely divergence free functions. It reads

Find $p_h \in Q_h$ such that for all $\mathbf{v}_h \in W_h^{\perp}$

$$(p_h, \operatorname{div} \mathbf{v}_h) = -\langle \mathbf{f}, \mathbf{v}_h \rangle + \nu(\nabla \mathbf{u}_h, \nabla \mathbf{v}_h). \tag{8.28}$$

Lemma 8.1 holds analogous in the discrete setting. In particular, if the discrete version of the inf–sup condition (8.22),

$$\inf_{q_h \in Q_h} \sup_{\mathbf{v}_h \in V_h} \frac{(q_h, \operatorname{div} \mathbf{v}_h)}{|\mathbf{v}_h|_1 \, \|q_h\|_0} \geq \beta > 0 \tag{8.29}$$

holds, we conclude that W_{gh} is not empty and the pressure problem (8.28) is uniquely solvable.

Lemma 8.3. *Let the discrete version of the inf–sup condition* (8.29) *be satisfied. Then, there is a unique solution* $(\mathbf{u}_h, p_h) \in V_h \times Q_h$ *of the discrete Stokes problem* (8.25) *and* (8.26).

Remark 8.1. Note that linearity, continuity and coercivity of a bilinear form transform from a continuous space V, $V \times V$, Q, $V \times Q$ to the finite dimensional subspaces V_h, $V_h \times V_h$, Q_h, $V_h \times Q_h$. The continuous inf–sup condition (8.22), however, does not imply the discrete inf–sup condition (8.29) for all possible pairs of finite elements spaces $V_h \subset V$ and $Q_h \subset Q$. Therefore, we have to verify (8.29) separately for each finite element pair (V_h, Q_h).

Remark 8.2. For existence and uniqueness of a solution of the discrete problem on a fixed mesh, it is sufficient if the special choice of finite element spaces satisfy the inf–sup condition on that mesh. Of course, the left hand side of (8.29) depends on the mesh. Nevertheless, on a mesh in which it is positive, we have existence and uniqueness for the solution of the discrete problem. However, we are often interested in using a sequence of meshes with mesh size $h \to 0$. Thus, in order to guarantee optimal error estimates one has to assume a fixed mesh-independent lower bound $\beta > 0$, that is (8.29) has to hold uniformly with respect to h.

Remark 8.3. One could think to solve problem (8.27) for the velocity $\mathbf{u}_h$ separately, since its dimension is smaller than the dimension of the coupled problem. However, this requires the knowledge of a basis in the space W_h of discretely divergence free functions, which is not a simple task. It has been shown that in the case of multiconnected domains Ω, one looses the locality of the support of the basis functions, see Turek (1994).

Next we show that the discrete problem (8.25), (8.26) is equivalent to a linear algebraic system of equations with a certain structure. Let $\{\mathbf{\Phi}_i\}$, $i = 1,\dots,N$, and $\{\Psi_k\}$, $k = 1,\dots,M$, be a basis in the spaces V_h and Q_h, respectively. Setting $\mathbf{v}_h = \mathbf{\Phi}_i$ in Equation (8.25) and $q_h = \Psi_k$ in Equation (8.26), we get

$$\begin{aligned} \nu(\nabla \mathbf{u}_h, \nabla \mathbf{\Phi}_i) - (p_h, \operatorname{div} \mathbf{\Phi}_i) &= \langle \mathbf{f}, \mathbf{\Phi}_i \rangle, \quad i = 1,\dots,N, \\ (\Psi_k, \operatorname{div} \mathbf{u}_h) &= \langle g, \Psi_k \rangle, \ k = 1,\dots,M. \end{aligned} \tag{8.30}$$

We write the discrete solution $\mathbf{u}_h \in V_h$, its gradient $\nabla \mathbf{u}_h$, its divergence $\operatorname{div} \mathbf{u}_h$, and $p_h \in Q_h$ in terms of the basis functions of V_h and Q_h as

$$\begin{aligned} \mathbf{u}_h(x) &= \sum_{j=1}^{N} U_j \mathbf{\Phi}_j(x), & \nabla \mathbf{u}_h(x) &= \sum_{j=1}^{N} U_j \nabla \mathbf{\Phi}_j(x), \\ \operatorname{div} \mathbf{u}_h(x) &= \sum_{j=1}^{N} U_j \operatorname{div} \mathbf{\Phi}_j(x), & p_h(x) &= \sum_{l=1}^{M} P_l \Psi_l(x), \end{aligned} \tag{8.31}$$

where U_j, $j = 1,\dots,N$ and P_l, $l = 1,\dots,M$, are the unknown coefficients to be determined. Substituting (8.31) in (8.30) leads to

$$\begin{aligned} \sum_{j=1}^{N} \nu(\nabla \mathbf{\Phi}_j, \nabla \mathbf{\Phi}_i) U_j - \sum_{l=1}^{M} (\Psi_l, \operatorname{div} \mathbf{\Phi}_i) P_l &= \langle \mathbf{f}, \mathbf{\Phi}_i \rangle, \quad i = 1,\dots,N, \\ \sum_{j=1}^{N} U_j (\Psi_k, \operatorname{div} \mathbf{\Phi}_j) &= \langle g, \Psi_k \rangle, \quad k = 1,\dots,M. \end{aligned}$$

Setting the entries $a_{ij} := \nu(\nabla \mathbf{\Phi}_j, \nabla \mathbf{\Phi}_i)$, $b_{kj} := -(\Psi_k, \operatorname{div} \mathbf{\Phi}_j)$, $f_i := \langle \mathbf{f}, \mathbf{\Phi}_i \rangle$ and $g_k := -\langle g, \Psi_k \rangle$ for $i,j = 1,\dots,N$ and $k = 1,\dots,M$, we get the system of algebraic equations

$$\begin{bmatrix} A & B^T \\ B & 0 \end{bmatrix} \begin{bmatrix} U \\ P \end{bmatrix} = \begin{bmatrix} F \\ G \end{bmatrix}, \tag{8.32}$$

where the blocks are given by

$$A=\begin{bmatrix} a_{11} & a_{12} & \cdots & a_{1N} \\ a_{21} & a_{22} & \cdots & a_{2N} \\ \vdots & \vdots & \ddots & \vdots \\ a_{N1} & a_{N2} & \cdots & a_{NN} \end{bmatrix}, \qquad B=\begin{bmatrix} b_{11} & b_{12} & \cdots & b_{1N} \\ \vdots & \vdots & & \vdots \\ b_{M1} & b_{M2} & \cdots & b_{MN} \end{bmatrix},$$

$$U=\begin{bmatrix} U_1 \\ U_2 \\ \vdots \\ U_N \end{bmatrix}, \qquad F=\begin{bmatrix} f_1 \\ f_2 \\ \vdots \\ f_N \end{bmatrix}, \qquad P=\begin{bmatrix} P_1 \\ \vdots \\ P_M \end{bmatrix}, \qquad G=\begin{bmatrix} g_1 \\ \vdots \\ g_M \end{bmatrix}.$$

We have seen that when $(\mathbf{u}_h, p_h)$ is a solution of the discrete problem (8.25), (8.26), the associated coefficient vectors U and P satisfy the algebraic system of Equations (8.32). For the equivalence of Equation (8.32) with the problem (8.25), (8.26) we still have to show that if the problem (8.25), (8.26) is satisfied for a basis $\mathbf{v}_h = \mathbf{\Phi}_i$ and $q_h = \Psi_k$, respectively, they hold for all $(\mathbf{v}_h, q_h) \in V_h \times Q_h$. Due to the linearity of (8.25) with respect to $\mathbf{v}_h$ and (8.26) with respect to q_h, these equations hold for any linear combination of the basis in V_h and Q_h, that is, for all $(\mathbf{v}_h, q_h) \in V_h \times Q_h$.

The system (8.32) has the typical structure of a saddle point problem. Since $A^T = A$, the coefficient matrix is symmetric, however, it is indefinite. Nevertheless, there are many algorithms for an efficient solution. For an overview we refer to Benzi et al. (2005).

In the derivation above $\mathbf{\Phi}_i$, $i = 1, \ldots, N$, denoted a basis in the vector-valued space V_h. In most cases, each component of the velocity is approximated by the same scalar finite element space S_h. Now let $\{\Phi^i\}$, $i = 1, \ldots, N$, be a basis of S_h and $V_h = S_h^d$ such that $N = \dim V_h$ has been replaced by Nd. Simplifying the notation we restrict the presentation on the case $d = 2$. The extension to an arbitrary $d \in \mathbb{N}$ is straightforward. Let us generate a basis in the vector-valued space V_h from the basis in the scalar space S_h by setting

$$\mathbf{\Phi}_i := \Phi^i \begin{bmatrix} 1 \\ 0 \end{bmatrix}, \qquad \mathbf{\Phi}_{i+N} := \Phi^i \begin{bmatrix} 0 \\ 1 \end{bmatrix}, \qquad i = 1, \ldots, N.$$

Then, by direct computation we observe for $i, j = 1, \ldots, N$

$$\begin{aligned} a_{ij} &= \nu(\nabla \mathbf{\Phi}_j, \nabla \mathbf{\Phi}_i) = \nu(\nabla \Phi^j, \nabla \Phi^i), \\ a_{i,j+N} &= a_{j+N,i} = \nu(\nabla \mathbf{\Phi}_{j+N}, \nabla \mathbf{\Phi}_i) = 0, \\ a_{i+N,j+N} &= \nu(\nabla \mathbf{\Phi}_{j+N}, \nabla \mathbf{\Phi}_{i+N}) = \nu(\nabla \Phi^j, \nabla \Phi^i). \end{aligned}$$

Thus the matrix A, the solution vector U and the right hand side vector F in (8.32) are decomposed into a block diagonal matrix and splited solution vectors such that

$$\begin{bmatrix} A & B^T \\ B & 0 \end{bmatrix} = \begin{bmatrix} A_{11} & 0 & B_1^T \\ 0 & A_{22} & B_2^T \\ B_1 & B_2 & 0 \end{bmatrix}, \qquad \begin{bmatrix} U \\ P \end{bmatrix} = \begin{bmatrix} U^I \\ U^{II} \\ P \end{bmatrix}, \qquad \begin{bmatrix} F \\ G \end{bmatrix} = \begin{bmatrix} F^I \\ F^{II} \\ G \end{bmatrix}. \tag{8.33}$$

Note that $A_{11} = A_{22}$ is the ν-fold of the stiffness matrix associated with the Laplace operator. For further details on the structure we refer to Chapter 9.4.

Finally, we provide an *a priori* error estimate for the Stokes equations.

Theorem 8.1. *Suppose that W_{gh} is not empty. Then, problem (8.27) has a unique solution $u_h \in W_{gh}$ and there is a constant C_1 such that the error bound*

$$|\mathbf{u}-\mathbf{u}_h|_1 \leq C_1 \left(\inf_{\mathbf{v}_h \in W_{gh}} |\mathbf{u}-\mathbf{v}_h|_1 + \inf_{q_h \in Q_h} \|p-q_h\|_0 \right) \tag{8.34}$$

holds. Let the discrete version of the inf–sup condition (8.29) be satisfied. Then, W_{gh} is not empty and there is a constant C_2 such that for the unique solution $(\mathbf{u}_h, p_h) \in V_h \times Q_h$ of (8.25), (8.26), it holds

$$|\mathbf{u}-\mathbf{u}_h|_1 + \|p-p_h\|_0 \leq C_2 \left(\inf_{\mathbf{v}_h \in V_h} |\mathbf{u}-\mathbf{v}_h|_1 + \inf_{q_h \in Q_h} \|p-q_h\|_0 \right). \tag{8.35}$$

Proof. We start with the first part in which W_{gh} is not empty. The solvability of (8.27) has been already established. We show the error estimate. Choose an arbitrary element $\mathbf{v}_h \in W_{gh}$. Then, $\mathbf{w}_h := \mathbf{u}_h - \mathbf{v}_h \in W_h$. Taking into consideration that $\mathbf{u}$ and $\mathbf{u}_h$ are the solutions of the continuous and discrete problem, respectively, we obtain

$$\begin{aligned}
\nu|\mathbf{u}_h-\mathbf{v}_h|_1^2 &= \nu(\nabla(\mathbf{u}_h-\mathbf{v}_h), \nabla\mathbf{w}_h) \\
&= \nu(\nabla(\mathbf{u}-\mathbf{v}_h), \nabla\mathbf{w}_h) + \langle \mathbf{f}, \mathbf{w}_h\rangle - \nu(\nabla\mathbf{u}, \nabla\mathbf{w}_h) \\
&= \nu(\nabla(\mathbf{u}-\mathbf{v}_h), \nabla\mathbf{w}_h) - (p, \operatorname{div}\mathbf{w}_h).
\end{aligned}$$

Note that, the last term on the right hand side does not vanish, since $\mathbf{w}_h$ belongs to the space of discretly divergence free functions $W_h \not\subset W$ but not in the space of divergence free functions W. But we have $(q_h, \operatorname{div}\mathbf{w}_h) = 0$ for all $q_h \in Q_h$. Introducing this in the last term on the right hand side and using Cauchy–Schwarz inequalities we obtain with a constant C independent of ν

$$\nu|\mathbf{u}_h-\mathbf{v}_h|_1^2 \leq \left(\nu|\mathbf{u}-\mathbf{v}_h|_1 + C\|p-q_0\|_0\right) |\mathbf{w}_h|_1.$$

Applying the triangle inequality and using the derived estimate for $\mathbf{u}_h - \mathbf{v}_h$

$$\begin{aligned}
\nu|\mathbf{u}-\mathbf{u}_h|_1 &\leq \nu|\mathbf{u}-\mathbf{v}_h|_1 + \nu|\mathbf{v}_h-\mathbf{u}_h|_1 \\
&\leq 2\nu|\mathbf{u}-\mathbf{v}_h|_1 + C\|p-q_0\|_0.
\end{aligned}$$

This is the stated estimate with $C_1 = \max(2, C/\nu)$.

Now let the discrete inf–sup condition be satisfied. Then, W_{gh} is not empty and there is a unique solution of (8.25) and (8.26) follows from Lemma 8.1 applied in the discrete case. It remains to show the error estimate. We show first that the error of the best approximation of $\mathbf{u} \in W_g$ in W_{gh} can be bounded (modulo a multiplicative constant) by the best approximation of a function $\mathbf{u} \in V$ in V_h, which means that there is a constant C_3 such that

$$\inf_{\mathbf{w}_h \in W_{gh}} |\mathbf{u}-\mathbf{w}_h|_1 \leq C_3 \inf_{\mathbf{v}_h \in V_h} |\mathbf{u}-\mathbf{v}_h|_1. \tag{8.36}$$

Let $\mathbf{v}_h$ be an arbitrary element of V_h. We define $\mathbf{z}_h \in W_h^{\perp}$ as solution of

$$\langle B_h \mathbf{z}_h, q_h \rangle = (q_h, \operatorname{div} \mathbf{z}_h) = (q_h, \operatorname{div}(\mathbf{u} - \mathbf{v}_h)) \quad \text{for all } q_h = Q_h$$

and obtain

$$|\mathbf{z}_h|_1 \leq \frac{1}{\beta} \|B_h \mathbf{z}_h\| = \frac{1}{\beta} \sup_{q_h \in Q_h} \frac{(q_h, \operatorname{div}(\mathbf{u} - \mathbf{v}_h))}{\|q_h\|_0} \leq C |\mathbf{u} - \mathbf{v}_h|_1.$$

Now we choose $\mathbf{w}_h = \mathbf{z}_h + \mathbf{v}_h$ and get $\mathbf{w}_h \in W_{gh}$ due to

$$\begin{aligned} (q_h, \operatorname{div} \mathbf{w}_h) &= (q_h, \operatorname{div} \mathbf{z}_h) + (q_h, \operatorname{div} \mathbf{v}_h) \\ &= (q_h, \operatorname{div}(\mathbf{u} - \mathbf{v}_h)) + (q_h, \operatorname{div} \mathbf{v}_h) \\ &= \langle g, q_h \rangle \quad \text{for all } q_h \in Q_h. \end{aligned}$$

We have shown that for an arbitrary $\mathbf{v}_h \in V_h$ there is a $\mathbf{w}_h \in W_{gh}$ such that

$$\inf_{\mathbf{w}_h \in W_{gh}} |\mathbf{u} - \mathbf{w}_h|_1 \leq |\mathbf{u} - \mathbf{w}_h|_1 \leq |\mathbf{u} - \mathbf{v}_h|_1 + |\mathbf{z}_h|_1 \leq (1 + C)\, |\mathbf{u} - \mathbf{v}_h|_1.$$

Taking the infimum over all $\mathbf{v}_h \in V_h$ we get (8.36).

The pressure estimate starts with the discrete inf–sup condition. For an arbitrary $q_h \in Q_h$ we have

$$\|p_h - q_h\|_0 \leq \frac{1}{\beta} \sup_{\mathbf{v}_h \in V_h} \frac{(p_h - q_h, \operatorname{div} \mathbf{v}_h)}{|\mathbf{v}_h|_1}.$$

The numerator can be rewritten using the fact that p and p_h are solutions of the continuous and discrete problem

$$\begin{aligned} (p_h - q_h, \operatorname{div} \mathbf{v}_h) &= (p - q_h, \operatorname{div} \mathbf{v}_h) - (p - p_h, \operatorname{div} \mathbf{v}_h) \\ &= (p - q_h, \operatorname{div} \mathbf{v}_h) - \nu (\nabla(\mathbf{u} - \mathbf{u}_h), \nabla \mathbf{v}_h), \\ |(p_h - q_h, \operatorname{div} \mathbf{v}_h)| &\leq \left(\|p - q_h\|_0 + \nu |\mathbf{u} - \mathbf{u}_h|_1 \right) |\mathbf{v}_h|_1. \end{aligned}$$

Using the triangle inequality, the velocity estimate and taking the infimum over $q_h \in Q_h$ proves (8.35). ∎

Remark 8.4. The error estimate (8.35) establishes the *quasi-optimality* of the Galerkin method applied to the Stokes problem. The error with respect to the natural norm in $H_0^1(\Omega)^d \times L_0^2(\Omega)$ is modulo a multiplicative factor bounded by the error of the best approximation in $H_0^1(\Omega)^d \times L_0^2(\Omega)$. In this sense, Theorem 8.1 can be considered as the generalization of Cea's lemma for the Poisson problem applied to the Stokes problem.

Remark 8.5. The error bound (8.35) suggests to choose finite element spaces V_h and Q_h for which the best approximation in the H^1 norm and the L^2 norm, respectively, are of equal order. Thus, the pair

$$\begin{aligned} V_h &:= \{\mathbf{v}_h \in H_0^1(\Omega)^d : \mathbf{v}_h|_K \in P_1(K)^d \text{ for all } K \in \mathcal{T}_h\}, \\ Q_h &:= \{q_h \in L_0^2(\Omega) : q_h|_K \in P_0(K) \text{ for all } K \in \mathcal{T}_h\} \end{aligned}$$

of continuous, piecewise linear velocities and discontinuous, piecewise constant pressures would be a suitable choice. Indeed, the errors of best approximations are bounded by the interpolation error as follows

$$\inf_{\mathbf{v}_h\in V_h} |\mathbf{u}-\mathbf{v}_h|_1 + \inf_{q_h\in Q_h} \|p-q_h\|_0 \le |\mathbf{u}-i_h\mathbf{u}|_1 + \|p-j_hp\|_0 \le Ch(|\mathbf{u}|_2+|p|_1).$$

Unfortunately, this finite element pair does not satisfy the inf–sup stability condition (8.29) as Exercise 8.1 shows.

Remark 8.6. As for the Poisson problem in Section 4.5, one can ask whether the velocity converges in the L^2 norm one order better than in the H^1 norm. For sufficient regularity this is the case. The essential assumption is the $H^2(\Omega)^d \times H^1(\Omega)$ regularity of the adjoint problem.

Remark 8.7. It should be highlighted that in Theorem 8.1 the velocity error and the pressure error depend on the error of best approximation of both the velocity and the pressure. Therefore, it can (it does!) happen that the velocity error is influenced by the best approximation error of the pressure. This type of nonrobustness and ways on how to minimize this influence have been recently discussed in Linke (2014) and Linke et al. (2016).

Exercise 8.1. Consider the decomposition $\mathcal{T}_h$ of $\Omega = (-1,+1)^2$ into four triangles K_i, $i=1,\ldots,4$, given by

$$K_1 := \{(x_1,x_2)\in\mathbb{R}^2 : -x_1\le x_2\le x_1,\ 0\le x_1\le +1\},$$
$$K_2 := \{(x_1,x_2)\in\mathbb{R}^2 : -x_2\le x_1\le x_2,\ 0\le x_2\le +1\},$$
$$K_3 := \{(x_1,x_2)\in\mathbb{R}^2 : x_1\le x_2\le -x_1,\ -1\le x_1\le 0\},$$
$$K_4 := \{(x_1,x_2)\in\mathbb{R}^2 : x_2\le x_1\le -x_2,\ -1\le x_2\le 0\}.$$

We introduce the finite element spaces

$$V_h := \{\mathbf{v}_h \in H_0^1(\Omega)^2 : \mathbf{v}_h|_K \in P_1(K)^2 \text{ for all } K\in\mathcal{T}_h\},$$
$$Q_h := \{q_h \in L_0^2(\Omega) : q_h|_K \in P_0(K) \text{ for all } K\in\mathcal{T}_h\}$$

and define the functions

$$\psi_1 := \begin{cases} +1 & \text{in } K_1\cup K_2 \\ -1 & \text{in } K_3\cup K_4 \end{cases}, \qquad \psi_2 := \begin{cases} +1 & \text{in } K_2\cup K_3 \\ -1 & \text{in } K_4\cup K_1 \end{cases},$$

$$\psi_3 := \begin{cases} +1 & \text{in } K_1\cup K_3 \\ -1 & \text{in } K_2\cup K_4 \end{cases}.$$

Prove that $\dim V_h = 2$, $\dim Q_h = 3$ and $Q_h = \operatorname{span}(\psi_1,\psi_2,\psi_3)$. Further, we define for $i=1$, 2, 3,

$$W_{i,h} := \{\mathbf{v}_h \in V_h : (q_h, \operatorname{div}\mathbf{v}_h) = \langle \psi_i, q_h\rangle \text{ for all } q_h\in Q_h\}.$$

Show that $W_{1,h} = \{0\}$ and $W_{3,h} = \emptyset$.

Exercise 8.2. Consider the decomposition $\mathcal{T}_h$ of $\Omega = (-1,+1)^2$ into four squares $K_1 = (0,1)^2$, $K_2 = (-1,0) \times (0,1)$, $K_3 = (-1,0)^2$ and $K_4 = (0,1) \times (-1,0)$. We introduce the finite element spaces

$$V_h := \{\mathbf{v}_h \in H_0^1(\Omega)^2 : \mathbf{v}_h|_K \in Q_1(K)^2 \text{ for all } K \in \mathcal{T}_h\},$$
$$Q_h := \{q_h \in L_0^2(\Omega) : q_h|_K \in Q_0(K) \text{ for all } K \in \mathcal{T}_h\}$$

and define the functions

$$\psi_1 := \begin{cases} +1 & \text{in } K_1 \cup K_2 \\ -1 & \text{in } K_3 \cup K_4 \end{cases}, \qquad \psi_2 := \begin{cases} +1 & \text{in } K_2 \cup K_3 \\ -1 & \text{in } K_4 \cup K_1 \end{cases},$$
$$\psi_3 := \begin{cases} +1 & \text{in } K_1 \cup K_3 \\ -1 & \text{in } K_2 \cup K_4 \end{cases}.$$

Show that there is a $q_h^* \in Q_h$ satisfying

$$(q_h^*, \operatorname{div} \mathbf{v}_h) = 0 \qquad \text{for all } \mathbf{v}_h \in V_h.$$

Does it hold the inf–sup condition of Equation (8.29)?

8.4. Nonconforming discretizations of the Stokes problem

In the previous section, we used conforming discretizations for approximating velocity and pressure, that is, the finite element spaces V_h and Q_h satisfy $V_h \subset V$ and $Q_h \subset Q$. For discretizing the biharmonic equation in Section 5.4, we used the option to relax the assumption $V_h \subset V$ to avoid the construction of C^1 finite elements. Although in case of the Stokes problem only C^0 finite elements for the velocity components and discontinuous finite elements for the pressure are needed for a conforming discretization, the relaxed assumption that $V_h \not\subset V$ can help to satisfy the inf–sup condition between V_h and Q_h more easily. This is demonstrated in the next section. Here, we study the solvability and *a priori* error estimates for nonconforming methods on an abstract level. As in the conforming case the starting point is the continuous formulation (8.16) and (8.17)

Find $(\mathbf{u}, p) \in V \times Q$ such that for all $(\mathbf{v}, q) \in V \times Q$

$$a(\mathbf{u}, \mathbf{v}) + b(\mathbf{v}, p) = \langle \mathbf{f}, \mathbf{v} \rangle,$$
$$-b(\mathbf{u}, q) = \langle g, q \rangle.$$

If $V_h \not\subset V$, the bilinear forms $a : V \times V \to \mathbb{R}$ and $b : V \times Q \to \mathbb{R}$ are, in general, not defined on the finite element spaces $V_h \times V_h$ and $V_h \times Q_h$, respectively. Therefore, in the first step, we extend the forms onto $(V + V_h) \times (V + V_h)$ and $(V + V_h) \times (Q + Q_h)$. A natural choice consists in computing the bilinear forms cellwise

$$a_h(\mathbf{u},\mathbf{v}) := \nu \sum_{K\in\mathcal{T}_h} \int_K \nabla\mathbf{u} : \nabla\mathbf{v}\, dx, \qquad b_h(\mathbf{v},\mathbf{q}) := -\sum_{K\in\mathcal{T}_h} \int_K q \operatorname{div}\mathbf{v}\, dx$$

for all $\mathbf{u},\mathbf{v} \in V + V_h$, $q \in Q$. Here, we assumed that the restriction $v_h|_K$ of a function $v_h \in V_h$ belongs to $H^1(K)^d$ for all $K \in \mathcal{T}_h$ and $Q_h \subset Q$. Further, we introduce the broken H^1 seminorm

$$|\mathbf{v}|_{1,h} := \left(\sum_{K\in\mathcal{T}_h} \int_K |\nabla\mathbf{v}|^2\, dx \right)^{1/2} \qquad \text{for all } \mathbf{v} \in V + V_h. \tag{8.37}$$

The nonconforming finite element discretization of the Stokes problem reads as

Find $(\mathbf{u}_h, p_h) \in V_h \times Q_h$ such that for all $(\mathbf{v}_h, q_h) \in V_h \times Q_h$

$$a_h(\mathbf{u}_h,\mathbf{v}_h) + b_h(\mathbf{v}_h,p_h) = \langle \mathbf{f},\mathbf{v}_h\rangle, \tag{8.38}$$

$$-b_h(\mathbf{u}_h,q_h) = \langle g,q_h\rangle. \tag{8.39}$$

Remark 8.8. The structure of the algebraic system of the nonconforming finite element discretization is identical to that of the conforming one given in (8.32). Note that the assembling of the matrices and the right hand side vectors follow the same ideas as in the conforming case; compute the local contributions of each cell K and add them to get the global matrix entries. For more details, see Chapter 9.4.

Theorem 8.2. *Let the broken H^1 seminorm in (8.37) be a norm on $V + V_h$. Then, the bilinear form a_h is continuous and coercive on $V + V_h$; the bilinear form b_h is continuous on $(V + V_h) \times (Q + Q_h)$, in particular, there is a positive constant C such that*

$$a_h(\mathbf{v},\mathbf{v}) \ge \nu\, |\mathbf{v}|_{1,h}^2 \quad \text{for all } \mathbf{v} \in V + V_h,$$

$$|a_h(\mathbf{u},\mathbf{v})| \le \nu\, |\mathbf{u}|_{1,h} |\mathbf{v}|_{1,h} \quad \text{for all } \mathbf{u},\mathbf{v} \in V + V_h,$$

$$|b_h(\mathbf{v},q)| \le C |\mathbf{v}|_{1,h} \|q\|_0 \quad \text{for all } \mathbf{v} \in V + V_h, q \in Q + Q_h.$$

Proof. The definition of the broken H^1 seminorm implies for all $\mathbf{v} \in V + V_h$

$$a_h(\mathbf{v},\mathbf{v}) = \nu \sum_{K\in\mathcal{T}_h} \int_K |\nabla\mathbf{v}|^2\, dx = \nu\, |\mathbf{v}|_{1,h}^2.$$

Applying Cauchy–Schwarz inequality for integrals and sums we get

$$|a_h(\mathbf{u},\mathbf{v})| = \left| \nu \sum_{K\in\mathcal{T}_h} \int_K \nabla\mathbf{u} : \nabla\mathbf{v}\, dx \right| \le \nu \sum_{K\in\mathcal{T}_h} \left(\int_K |\nabla\mathbf{u}|^2\, dx \right)^{1/2} \left(\int_K |\nabla\mathbf{v}|^2\, dx \right)^{1/2}$$

$$\le \nu \left(\sum_{K\in\mathcal{T}_h} \int_K |\nabla\mathbf{u}|^2\, dx \right)^{1/2} \left(\sum_{K\in\mathcal{T}_h} \int_K |\nabla\mathbf{v}|^2\, dx \right)^{1/2} = \nu\, |\mathbf{u}|_{1,h} |\mathbf{v}|_{1,h}$$

for all $\mathbf{u}, \mathbf{v} \in V + V_h$. Finally, Cauchy–Schwarz inequality yields

$$|b_h(\mathbf{v}, q)| = \left| \sum_{K \in \mathcal{T}_h} \int_K q \operatorname{div} \mathbf{v} \, dx \right| \leq \sum_{K \in \mathcal{T}_h} \left(\int_K q^2 \, dx \right)^{1/2} \left(\int_K |\operatorname{div} \mathbf{v}|^2 \, dx \right)^{1/2}$$

$$\leq C \left(\sum_{K \in \mathcal{T}_h} \int_K q^2 \, dx \right)^{1/2} \left(\sum_{K \in \mathcal{T}_h} \int_K |\nabla \mathbf{v}|^2 \, dx \right)^{1/2} \leq C \|q\|_0 \, |\mathbf{v}|_{1,h}$$

for all $\mathbf{v} \in V + V_h$ and $q \in Q + Q_h$, respectively. ∎

The solvability of the discrete problem (8.38), (8.39) can be studied again by introducing the subspaces

$$W_{gh} := \{\mathbf{v}_h \in V_h : b_h(\mathbf{v}_h, q_h) = -\langle g, q_h \rangle \text{ for all } q_h \in Q_h\},$$
$$W_h := \{\mathbf{v}_h \in V_h : b_h(\mathbf{v}_h, q_h) = 0 \text{ for all } q_h \in Q_h\}.$$

If $(\mathbf{u}_h, p_h)$ is a solution of problem (8.38), (8.39), then $\mathbf{u}_h$ solves

Find $\mathbf{u}_h \in W_{gh}$ such that for all $\mathbf{v}_h \in W_h$

$$a_h(\mathbf{u}_h, \mathbf{v}_h) = \langle \mathbf{f}, \mathbf{v}_h \rangle. \tag{8.40}$$

If a solution $\mathbf{u}_h \in W_{gh}$ of problem (8.40) exists, the discrete pressure $p_h \in Q_h$ can be obtained by restricting (8.38) on the orthogonal complement $W_h^\perp$ of the space of discretely divergence free functions

Find $p_h \in Q_h$ such that for all $\mathbf{v}_h \in W_h^\perp$

$$b_h(\mathbf{v}_h, p_h) = \langle \mathbf{f}, \mathbf{v}_h \rangle - a_h(\mathbf{u}_h, \mathbf{v}_h). \tag{8.41}$$

If W_{gh} is not empty and $\mathbf{u}_{0h} \in W_{gh}$, (8.40) is equivalent to

Find $\mathbf{u}_h - \mathbf{u}_{0h} \in W_h$ such that for all $\mathbf{v}_h \in W_h$

$$a_h(\mathbf{u}_h - \mathbf{u}_{0h}, \mathbf{v}_h) = \langle \mathbf{f}, \mathbf{v}_h \rangle - a_h(\mathbf{u}_{0h}, \mathbf{v}_h).$$

This problem is uniquely solvable according to Theorem 2.1 (Lax–Milgram). As in the conforming case we cannot use arbitrary pairs of finite elements to discretize velocity and pressure. It can be shown that the inf–sup stability condition

$$\inf_{q_h \in Q_h} \sup_{\mathbf{v}_h \in V_h} \frac{b_h(\mathbf{v}_h, q_h)}{|\mathbf{v}_h|_{1,h} \|q_h\|_0} \geq \beta > 0 \tag{8.42}$$

guarantees that W_{gh} is not empty and the pressure problem (8.41) is uniquely solvable (compare Lemma 8.1).

Theorem 8.3. *Let the broken H^1 seminorm (8.37) be a norm on $V + V_h$ and assume that W_{gh} is not empty. Then, problem (8.40) has a unique solution $\mathbf{u}_h \in W_{gh}$ and there is a constant C_1 such* *that the error bound*

$$|\mathbf{u}-\mathbf{u}_h|_{1,h} \le C_1\Big(\inf_{\mathbf{v}_h\in W_{gh}} |\mathbf{u}-\mathbf{v}_h|_{1,h} + \inf_{q_h\in Q_h} \|p-q_h\|_0 + \sup_{\mathbf{w}_h\in V_h} \frac{a_h(\mathbf{u},\mathbf{w}_h)+b_h(\mathbf{w}_h,p)-\langle \mathbf{f},\mathbf{w}_h\rangle}{|\mathbf{w}_h|_{1,h}}\Big)$$

holds. Let in addition, the inf–sup condition (8.42) *be satisfied. Then,* W_{gh} *is not empty and there is a constant* C_2 *such that for the unique solution* $(\mathbf{u}_h, p_h) \in V_h \times Q_h$ *of problem* (8.38), (8.39), *it holds*

$$|\mathbf{u}-\mathbf{u}_h|_{1,h} + \|p-p_h\|_0 \le C_2\Big(\inf_{\mathbf{v}_h\in V_h} |\mathbf{u}-\mathbf{v}_h|_{1,h} + \inf_{q_h\in Q_h} \|p-q_h\|_0 + \sup_{\mathbf{w}_h\in V_h} \frac{a_h(\mathbf{u},\mathbf{w}_h)+b_h(\mathbf{w}_h,p)-\langle \mathbf{f},\mathbf{w}_h\rangle}{|\mathbf{w}_h|_{1,h}}\Big).$$

Proof. The unique solvability of the discrete problem has been already discussed above. It remains to prove the error bounds. Choose an arbitrary element $\mathbf{v}_h \in W_{gh}$. Then, $\mathbf{w}_h := \mathbf{u}_h - \mathbf{v}_h \in W_h$. Taking into consideration that $\mathbf{u}_h$ is the solution of the discrete problem and $\mathbf{w}_h \in W_h$ is discretely divergence free, we get

$$\begin{aligned}\nu|\mathbf{u}_h-\mathbf{v}_h|^2_{1,h} &= a_h(\mathbf{u}_h-\mathbf{v}_h,\mathbf{w}_h)\\ &= a_h(\mathbf{u}-\mathbf{v}_h,\mathbf{w}_h)+b_h(\mathbf{w}_h,p)+\langle \mathbf{f},\mathbf{w}_h\rangle - a_h(\mathbf{u},\mathbf{w}_h)-b_h(\mathbf{w}_h,p)\\ &= a_h(\mathbf{u}-\mathbf{v}_h,\mathbf{w}_h)+b_h(\mathbf{w}_h,p-q_h)+\langle \mathbf{f},\mathbf{w}_h\rangle - a_h(\mathbf{u},\mathbf{w}_h)-b_h(\mathbf{w}_h,p).\end{aligned}$$

The continuity of the bilinear forms a_h and b_h yield

$$|\mathbf{u}_h-\mathbf{v}_h|_{1,h} \le |\mathbf{u}-\mathbf{v}_h|_{1,h} + \frac{d}{\nu}\|p-q_h\|_0 + \sup_{\mathbf{w}_h\in V_h} \frac{\langle \mathbf{f},\mathbf{w}_h\rangle - a_h(\mathbf{u},\mathbf{w}_h)-b_h(\mathbf{w}_h,p)}{\nu\,|\mathbf{w}|_{1,h}},$$

where d is the dimension of the problem. Apply the triangle inequality

$$|\mathbf{u}-\mathbf{u}_h|_{1,h} \le |\mathbf{u}-\mathbf{v}_h|_{1,h} + |\mathbf{v}_h-\mathbf{u}_h|_{1,h}$$

and take the infimum over all $(\mathbf{v}_h, q_h) \in W_{gh} \times Q_h$ to get the first inequality of Theorem 8.3 with $C_1 = \max(2, d/\nu, 1/\nu)$.

Now let the discrete inf–sup condition (8.42) be satisfied. Then, Lemma 8.1 shows that W_{gh} is not empty and that there is a unique solution of problem (8.38), (8.39). As in the proof of Theorem 8.1 one can show that the error of the best approximation of $\mathbf{u} \in W_g$ in W_{gh} can be bounded (modulo a multiplicative constant) by the best approximation of a function $\mathbf{u} \in V$ in V_h, that is, there is a constant C_3 such that

$$\inf_{\mathbf{w}_h\in W_{gh}} |\mathbf{u}-\mathbf{w}_h|_{1,h} \le C_3 \inf_{\mathbf{v}_h\in V_h} |\mathbf{u}-\mathbf{v}_h|_{1,h}.$$

Finally, we establish the pressure estimate. Let $q_h \in Q_h$ be arbitrary. We apply the discrete inf–sup condition

$$\|p_h - q_h\|_0 \le \frac{1}{\beta} \sup_{\mathbf{v}_h \in V_h} \frac{b_h(\mathbf{v}_h, p_h - q_h)}{|\mathbf{v}_h|_{1,h}}$$

and rewrite the numerator using the fact that p_h is a solution of the discrete problem

$$\begin{aligned} b_h(\mathbf{v}_h, p_h - q_h) &= b_h(\mathbf{v}_h, p - q_h) + \langle \mathbf{f}, \mathbf{v}_h \rangle - a_h(\mathbf{u}_h, \mathbf{v}_h) - b_h(\mathbf{v}_h, p) \\ &= b_h(\mathbf{v}_h, p - q_h) + a_h(\mathbf{u} - \mathbf{u}_h, \mathbf{v}_h) + \langle \mathbf{f}, \mathbf{v}_h \rangle - a_h(\mathbf{u}, \mathbf{v}_h) - b_h(\mathbf{v}_h, p). \end{aligned}$$

Use the continuity of the bilinear forms a_h and b_h to obtain

$$\begin{aligned} \frac{|b_h(\mathbf{v}_h, p_h - q_h)|}{|\mathbf{v}_h|_{1,h}} \le C\Big(&\|p - q_h\|_0 + |\mathbf{u} - \mathbf{u}_h|_{1,h} \\ &+ \sup_{\mathbf{w}_h \in V_h} \frac{\langle \mathbf{f}, \mathbf{w}_h \rangle - a_h(\mathbf{u}, \mathbf{w}_h) - b_h(\mathbf{w}_h, p)}{\nu\, |\mathbf{w}|_{1,h}} \Big). \end{aligned}$$

Apply the triangle inequality

$$\|p - p_h\|_0 \le \|p - q_h\|_0 + \|q_h - p_h\|_0,$$

use the already proven estimate for $|\mathbf{u} - \mathbf{u}_h|_{1,h}$ and take the infimum over all $q_h \in Q_h$ to get the stated estimate for $\|p - p_h\|_0$. ■

Remark 8.9. Comparing Theorem 8.3 with Theorem 8.1 we see that (as in Section 5.4) an additional consistency error has to be estimated in the nonconforming case. Upto the multiplicative constant C_2, the three terms on the right hand side of the *a priori* error estimates correspond to the best approximation of the velocity in V_h, the best approximation of the pressure in Q_h, and the consistency error due to the nonconformity of the method.

8.5. The nonconforming Crouzeix–Raviart element

In this section, we consider the simplest first order method proposed in Crouzeix and Raviart (1973). Let $\Omega \subset \mathbb{R}^d$ be decomposed into simplices $K \in \mathcal{T}_h$. The vertices of K are denoted by $a_0, \dots, a_d$, the barycenter of K by a_K, the face opposite to a_i by f_i, $i = 0, \dots, d$, and the centroid of the face f_i by b_i, $i = 0, \dots, d$. The local approximation spaces for velocity and pressure are

$$\mathcal{P}_K^u = P_1(K)^d, \qquad \mathcal{P}_K^p = P_0(K).$$

The dimensions are given by $\dim \mathcal{P}_K^u = d(d+1)$ and $\dim \mathcal{P}_K^p = 1$. The sets of dof Σ_K^u and Σ_K^p are given by the (vector- and scalar-valued) nodal functionals

$$N_i^u(\mathbf{v}) = \frac{1}{|f_i|} \int_{f_i} \mathbf{v}\, d\gamma, \quad i = 0, \dots, d, \qquad N(q) = \frac{1}{|K|} \int_K q\, dx.$$

Lemma 8.4. *The sets of dof Σ_K^u and Σ_K^p in the nonconforming Crouzeix–Raviart elements are $\mathcal{P}_K^u$ and $\mathcal{P}_K^p$ unisolvent, respectively.*

Exercise 8.3. Prove Lemma 8.4.

Since the midpoint rule integrates polynomials of degree less than or equal to one, we conclude

$$N_i^u(\mathbf{v}) = \mathbf{v}(b_i) \text{ for all } \mathbf{v} \in \mathcal{P}_K^u,\ i = 0,\dots,d, \qquad N(q) = q(a_K) \text{ for all } q \in \mathcal{P}_K^p.$$

This allows us to define the finite element spaces by means of integral mean values or by point values. Note that, the finite element spaces will be independent of this choice, however, the canonical interpolation operators are different. In the first case, the nodal functional N is defined on the space $L^1(\Omega)$ and N_i^u by the trace theorem at least on the space $H^1(\Omega)^d$, that is, on much larger spaces than continuous function spaces needed for point values.

Now, the Crouzeix–Raviart element generates the finite element spaces

$$\begin{aligned} V_h &:= \{\mathbf{v}_h \in L^2(\Omega)^d : \mathbf{v}_h|_K \in \mathcal{P}_K^u, \mathbf{v}_h \text{ continuous at the centroids of inner faces,} \\ &\qquad\qquad \mathbf{v}_h(b_i) = \mathbf{0} \text{ for } b_i \in \Gamma\}, \\ Q_h &:= \{q_h \in L_0^2(\Omega) : q_h|_K \in \mathcal{P}_K^p \text{ for all } K \in \mathcal{T}_h\} \end{aligned}$$

for approximating the solution of the Stokes problem belonging to $V = H_0^1(\Omega)^d$ and $Q = L_0^2(\Omega)$, respectively.

Lemma 8.5. *On the space V_h of nonconforming Crouzeix–Raviart elements $\mathbf{v}_h \mapsto |\mathbf{v}_h|_{1,h}$ is a norm.*

Proof. Since $\mathbf{v}_h \mapsto |\mathbf{v}_h|_{1,h}$ is a seminorm, we have to only show that $|\mathbf{v}_h|_{1,h} = 0$ implies $\mathbf{v}_h = \mathbf{0}$. Let $\mathbf{v}_h \in V_h$ with $|\mathbf{v}_h|_{1,h} = 0$. Then, $\mathbf{v}_h$ is constant on each cell $K \in \mathcal{T}_h$. Since $\mathbf{v}_h$ is continuous at the centroids of inner faces we conclude that $\mathbf{v}_h$ is constant on Ω. Taking into consideration that $\mathbf{v}_h$ vanishes at $b_i \in \Gamma$, we see that $\mathbf{v}_h = \mathbf{0}$. Thus, $|\cdot|_{1,h}$ is a norm on V_h. ■

Note that the families of elements are affine equivalent. Indeed, according to Lemma 4.1 we have for the affine mapping $F_K : \widehat{K} \to K$ with $F_K(\hat{x}) = B_K\hat{x} + b_K$

$$N(q) = \frac{1}{|K|}\int_K q \,\mathrm{d}x = \frac{1}{|K|}\int_{\widehat{K}} \hat{q} \det B_K \,\mathrm{d}\hat{x} = \frac{1}{|\widehat{K}|}\int_{\widehat{K}} \hat{q} \,\mathrm{d}\hat{x} = \widehat{N}(\hat{q}).$$

Analogously, it holds for $i = 0, 1, \dots, d$

$$N_i^u(\mathbf{v}) = \frac{1}{|f_i|}\int_{f_i} \mathbf{v} \,\mathrm{d}\gamma = \frac{1}{|\hat{f}_i|}\int_{\hat{f}_i} \hat{\mathbf{v}} \,\mathrm{d}\hat{\gamma} = \widehat{N}_i^u(\hat{\mathbf{v}}).$$

As a consequence of Theorem 4.3, we get for the canonical interpolation operators $\Pi_K^u : H^1(K)^d \to \mathcal{P}_K^u$ and $\Pi_K^p : L^1(K) \to \mathcal{P}_K^p$ the estimates

$$\begin{aligned} \|\mathbf{v} - \Pi_K^u\mathbf{v}\|_{0,K} + h_K|\mathbf{v} - \Pi_K^u\mathbf{v}|_{1,K} &\le C h_K^m |\mathbf{v}|_{m,K} \quad \text{for all } \mathbf{v} \in H^m(K),\ m = 1,2, \\ \|q - \Pi_K^p q\|_{0,K} &\le C h_K |q|_{1,K} \quad \text{for all } q \in H^1(K). \end{aligned}$$

For establishing the inf–sup condition of Equation (8.42) the following lemma will be useful.

Lemma 8.6. *Suppose there is a linear, continuous operator $\Pi_h^u : H_0^1(\Omega)^d \to V_h$ satisfying*

$$b_h(\Pi_h^u \mathbf{v}, q_h) = b_h(\mathbf{v}, q_h) \quad \textit{for all } q_h \in Q_h,\ \mathbf{v} \in H_0^1(\Omega)^d, \tag{8.43}$$

$$|\Pi_h^u \mathbf{v}|_{1,h} \le C|\mathbf{v}|_1 \quad \textit{for all } \mathbf{v} \in H_0^1(\Omega)^d \tag{8.44}$$

with a constant $C > 0$ independent of h. Then, the inf–sup condition (8.42) *holds with a constant $\beta > 0$ independent of h.*

Proof. We use the two properties of the operator Π_h^u to get

$$\sup_{\mathbf{v}_h \in V_h} \frac{b_h(\mathbf{v}_h, q_h)}{|\mathbf{v}_h|_{1,h}} \ge \frac{b_h(\Pi_h^u \mathbf{v}, q_h)}{|\Pi_h^u \mathbf{v}|_{1,h}} \ge \frac{1}{C} \frac{b_h(\mathbf{v}, q_h)}{|\mathbf{v}|_1}$$

for all $\mathbf{v} \in H_0^1(\Omega)^d$ and $q_h \in Q_h$. Taking the supremum over all $\mathbf{v} \in H_0^1(\Omega)^d$, we get from the continuous inf–sup condition with a constant $\beta^* > 0$ the estimate

$$\sup_{\mathbf{v}_h \in V_h} \frac{b_h(\mathbf{v}_h, q_h)}{|\mathbf{v}_h|_{1,h}} \ge \frac{1}{C} \sup_{\mathbf{v} \in H_0^1(\Omega)^d} \frac{b_h(\mathbf{v}, q_h)}{|\mathbf{v}|_1} \ge \frac{\beta^*}{C} \|q_h\|_0.$$

Thus, the discrete inf–sup condition (8.42) holds with $\beta = \beta^*/C$. ■

Remark 8.10. The operator satisfying (8.43) and (8.44) is called Fortin operator, see Fortin (1977). It can be shown that the existence of such a Fortin operator is also necessary for the validity of the uniform inf–sup condition, see Fortin (1977) and Girault and Raviart (1986).

Let us show that the canonical interpolation of the Crouzeix–Raviart element satisfies the assumptions of the Fortin operator. Before that, we define the global interpolation $\Pi_h^u : H_0^1(\Omega)^d \to V_h$ in the piecewise manner

$$(\Pi_h^u \mathbf{v})|_K = \Pi_K^u(\mathbf{v}|_K) \quad \text{for all } \mathbf{v} \in H_0^1(\Omega)^d,\ K \in \mathcal{T}_h.$$

Lemma 8.7. *The canonical interpolation $\Pi_h^u : H_0^1(\Omega)^d \to V_h$ satisfies* (8.43) *and* (8.44) *with $C = 1$. Thus, the Crouzeix–Raviart element satisfies the discrete inf–sup condition* (8.42).

Proof. We start with (8.43). Apply Theorem 1.4 (Gauss theorem) and use the local definition $N_i^u(\mathbf{v} - \Pi_h^u \mathbf{v}) = 0$ for all i to get

$$\begin{aligned}
b_h(\Pi_h^u \mathbf{v}, q_h) &= \sum_{K \in \mathcal{T}_h} q_h|_K \int_K \operatorname{div} \Pi_h^u \mathbf{v} \, dx = \sum_{K \in \mathcal{T}_h} q_h|_K \sum_{f \in \partial K} \int_f \Pi_h^u \mathbf{v} \, d\gamma \cdot \mathbf{n}_K \\
&= \sum_{K \in \mathcal{T}_h} q_h|_K \sum_{f \in \partial K} \int_f \mathbf{v} \, d\gamma \cdot \mathbf{n}_K = \sum_{K \in \mathcal{T}_h} q_h|_K \int_K \operatorname{div} \mathbf{v} \, dx \\
&= b_h(\mathbf{v}, q_h).
\end{aligned}$$

The first derivative of $\Pi_K^u \mathbf{v}$ is constant on K, therefore, we obtain by the Gauss theorem and by the definition of the local interpolation Π_K^u

$$\frac{\partial \Pi_K^u \mathbf{v}}{\partial x_j} = \frac{1}{|K|}\int_K \frac{\partial \Pi_K^u \mathbf{v}}{\partial x_j}\,\mathrm{d}x = \frac{1}{|K|}\int_{\partial K} \Pi_K^u \mathbf{v} n_j\,\mathrm{d}\gamma$$
$$= \frac{1}{|K|}\int_{\partial K} \mathbf{v} n_j\,\mathrm{d}\gamma = \frac{1}{|K|}\int_K \frac{\partial \mathbf{v}}{\partial x_j}\,\mathrm{d}x.$$

From the Cauchy–Schwarz inequality we get

$$\left(\frac{\partial \Pi_K^u \mathbf{v}}{\partial x_j}\right)^2 \le \frac{1}{|K|^2}\int_K \mathrm{d}x \int_K \left(\frac{\partial \mathbf{v}}{\partial x_j}\right)^2 \mathrm{d}x = \frac{1}{|K|}\int_K \left(\frac{\partial \mathbf{v}}{\partial x_j}\right)^2 \mathrm{d}x.$$

Summation over $j = 1, \ldots, d$ and integration over K leads to

$$|\Pi_K^u \mathbf{v}|_{1,K}^2 \le |\mathbf{v}|_{1,K}^2.$$

We sum up over $K \in \mathcal{T}_h$ and obtain (8.44). Finally, apply Lemma 8.6 to conclude the discrete inf–sup condition (8.42). ■

Based on Theorem 8.3 it remains to estimate the consistency error

$$\sup_{\mathbf{w}_h \in V_h} \frac{a_h(\mathbf{u}, \mathbf{w}_h) + b_h(\mathbf{w}_h, p) - \langle \mathbf{f}, \mathbf{w}_h \rangle}{|\mathbf{w}_h|_{1,h}}.$$

The key idea in the consistency estimate is an appropriate representation of the numerator. Assuming that the solution of the Stokes problem $(\mathbf{u}, p)$ belongs to $(H_0^1(\Omega)^d \cap H^2(\Omega)^d) \times (L_0^2(\Omega \cap H^1(\Omega))$, it holds

$$-\nu\Delta\mathbf{u} + \nabla p = \mathbf{f} \quad \text{in } L^2(\Omega)^d, \text{ for all } K \in \mathcal{T}_h.$$

Multiplying with a function $\mathbf{w}_h \in V_h$, integrating over K, elementwise integrating by parts and sum up over all $K \in \mathcal{T}_h$, we obtain

$$E(\mathbf{u}, p; \mathbf{w}_h) := a_h(\mathbf{u}, \mathbf{w}_h) + b_h(\mathbf{w}_h, p) - \langle \mathbf{f}, \mathbf{w}_h \rangle$$
$$= \sum_{K \in \mathcal{T}_h} \int_{\partial K} \left(\nu \frac{\partial \mathbf{u}}{\partial \mathbf{n}_K} \cdot \mathbf{w}_h - p\mathbf{w}_h \cdot \mathbf{n}_K\right) \mathrm{d}\gamma.$$

Next, we derive profit from the property

$$\int_f [\mathbf{w}_h]_f\,\mathrm{d}\gamma = 0 \text{ for all inner faces } f \quad \text{and} \quad \int_f \mathbf{w}_h\,\mathrm{d}\gamma = 0 \text{ for all boundary faces } f \subset \Gamma$$

of the Crouzeix–Raviart finite element space V_h. Here, $[\mathbf{w}_h]_f$ stands for the jump of $\mathbf{w}_h \in V_h$ across an inner face. It allows us to represent $E(\mathbf{u}, p; \mathbf{w}_h)$ as

$$E(\mathbf{u}, p; \mathbf{w}_h) = \sum_{K \in \mathcal{T}_h} \int_{\partial K} \left[\nu \frac{\partial \mathbf{u}}{\partial \mathbf{n}_K} - p\mathbf{n}_K - \Pi_0\left(\nu \frac{\partial \mathbf{u}}{\partial \mathbf{n}_K} - p\mathbf{n}_K\right)\right] \cdot \mathbf{w}_h\,\mathrm{d}\gamma,$$

where $\Pi_0 : L^2(f) \to P_0(f)$ denotes the L^2 projection into the space of piecewise constant functions. Now applying a generalization of the Bramble–Hilbert lemma (similar to that used for the rectangular Adini element in Section 5.4), we get

$$|E(\mathbf{u},p;\mathbf{w}_h)| \le Ch(|\mathbf{u}|_2 + |p|_1)\,|\mathbf{w}_h|_{1,h}.$$

For details, see Crouzeix and Raviart (1973).

Theorem 8.4. *Let the solution* $(\mathbf{u},p)$ *of the Stokes problem* (8.16), (8.17) *belong to* $(H_0^1(\Omega)^d \cap H^2(\Omega)^d) \times (L_0^2(\Omega \cap H^1(\Omega)))$ *and* $(\mathbf{u}_h,p_h)$ *denote the solution of the discrete problem* (8.38), (8.39) *with the Crouzeix–Raviart element. Then, the error estimate*

$$|\mathbf{u}-\mathbf{u}_h|_{1,h} + \|p-p_h\|_0 \le Ch(|\mathbf{u}|_2 + |p|_1)$$

holds true.

Proof. Collect the estimates for the approximation and consistency error, respectively, and apply Theorem 8.3. ■

Exercise 8.4. Consider the decomposition $\mathcal{T}_h$ of $\Omega = (-1,+1)^2$ into two triangles K_1, K_2 given by

$$K_1 := \{(x_1,x_2) \in \mathbb{R}^2 : -1 \le x_2 \le x_1, -1 \le x_1 \le +1\},$$
$$K_2 := \{(x_1,x_2) \in \mathbb{R}^2 : x_1 \le x_2 \le +1, -1 \le x_1 \le +1\}.$$

We introduce the Crouzeix–Raviart finite element spaces

$$V_h := \{\mathbf{v}_h \in L^2(\Omega)^2 : \mathbf{v}_h|_K \in P_1(K)^2 \text{ for all } K \in \mathcal{T}_h, \mathbf{v}_h \text{ continuous at } (0,0),$$
$$\mathbf{v}_h(1,0) = \mathbf{v}_h(0,1) = \mathbf{v}_h(-1,0) = \mathbf{v}_h(0,-1) = 0\},$$
$$Q_h := \{q_h \in L_0^2(\Omega) : q_h|_K \in P_0(K) \text{ for all } K \in \mathcal{T}_h\}.$$

Prove that $\dim V_h = 2$ and $\dim Q_h = 1$. Find a basis for the discretely divergence free subspace

$$W_h := \{\mathbf{v}_h \in V_h : (q_h, \operatorname{div} \mathbf{v}_h) = 0 \text{ for all } q_h \in Q_h\}$$

and show that

$$W_{gh} := \{\mathbf{v}_h \in V_h : (q_h, \operatorname{div} \mathbf{v}_h) = \langle g, q_h \rangle \text{ for all } q_h \in Q_h\} \ne \emptyset$$

for all $g \in L_0^2(\Omega)$. Compare the results with Exercise 8.1.

8.6. Further inf–sup stable finite element pairs

In Sections 8.3 and 8.4, we derived *a priori* error estimates for velocity and pressure provided that a discrete version of the inf–sup condition is satisfied. We also saw that arbitrary finite element pairs for approximating velocity and pressure do not satisfy this stability condition.

Lemma 8.8. *Suppose that the pair* (V_h, Q_h) *satisfies the inf–sup condition* (8.42) *with the constant* $\beta > 0$. *Then, for any pair* $(\widetilde{V}_h, \widetilde{Q}_h)$ *with* $V_h \subset \widetilde{V}_h$ *and* $\widetilde{Q}_h \subset Q_h$ *we have*

$$\inf_{q_h \in \widetilde{Q}_h} \sup_{\mathbf{v}_h \in \widetilde{V}_h} \frac{b_h(\mathbf{v}_h, q_h)}{|\mathbf{v}_h|_{1,h} \, \|q_h\|_0} \geq \beta.$$

Proof. Any $q_h \in \widetilde{Q}_h$ belongs to Q_h and (V_h, Q_h) satisfies (8.42). Thus, for all $q_h \in \widetilde{Q}_h$ we get

$$\beta \|q_h\|_0 \leq \sup_{\mathbf{v}_h \in V_h} \frac{b_h(\mathbf{v}_h, q_h)}{|\mathbf{v}_h|_{1,h}} \leq \sup_{\mathbf{v}_h \in \widetilde{V}_h} \frac{b_h(\mathbf{v}_h, q_h)}{|\mathbf{v}_h|_{1,h}}$$

from which the statement immediately follows. ∎

Remark 8.11. Lemma 8.8 tells us, roughly speaking, that the space of velocity approximations should be rich enough compared to the space of pressure approximation. Note, however, that the requirement $\dim V_h > \dim Q_h$ does not guarantee a mesh independent inf–sup constant β. See Boland and Nicolaides (1984) for a counter example and more details.

In the following section we give a short overview on finite element pairs that satisfy the discrete inf–sup condition.

Conforming triangular finite elements

The *a priori* error bound (8.35) motivates continuous, piecewise linear velocity approximations and discontinuous, piecewise constant pressure approximations. However, as we already mentioned in Remark 8.5 for this pair the discrete inf–sup condition does not hold. In view of Lemma 8.8, the first idea is to enrich the velocity space to continuous, piecewise quadratic functions, see the first element in Table 8.1. The good news is that the inf–sup condition holds for this (P_2, P_0) element, however, the bad news is that the approximation is not balanced ($\mathcal{O}(h^2)$ in $H^1(\Omega)^d$ and only $\mathcal{O}(h)$ in $L_0^2(\Omega)$, respectively), see Fortin (1972). Later (Fortin (1981); Bernardi and Raugel (1985)), it became clear that it is enough to enrich the normal component of the velocity over the edges by quadratic functions as the second element in Table 8.1. Let $\mathbf{n}_i$ denote the outer normal to the edge $\lambda_i = 0$ opposite to the vertex a_i. Then, the enrichment of the $P_1(K)^2$ velocity space is defined by

$$\mathcal{P}_K^u := P_1(K)^2 \oplus \operatorname{span}(\mathbf{w}_1, \mathbf{w}_2, \mathbf{w}_3)$$

with the vector-valued functions $\mathbf{w}_i = \lambda_j \lambda_k \mathbf{n}_i$, $j, k \neq i$, $j \neq k$. Note that $\dim \mathcal{P}_K = 9$ and the dof are the values at the vertices and the mean values of the normal component over the edges.

The third element in Table 8.1 stands for the lowest order case $k = 1$, the MINI element, of a family of elements

$$\mathcal{P}_K^u := \left(P_k(K) + \operatorname{span} \lambda_1 \lambda_2 \lambda_3 P_{k-1}(K)\right)^2, \qquad \mathcal{P}_K^p := P_k(K), \quad k \geq 1.$$

It belongs to the class of continuous pressure approximations and has been proposed and analyzed by Arnold et al. (1984). Note that for $k \geq 3$ the sum in the definition of $\mathcal{P}_K^u$ is not a direct sum, since for example, the bubble function $\lambda_1 \lambda_2 \lambda_3$ belongs to $P_3(K)$ and to $\lambda_1 \lambda_2 \lambda_3 P_2(K)$.

Table 8.1 Conforming inf–sup stable triangular finite elements

Velocity and pressure	P_K^u	P_K^p	Order
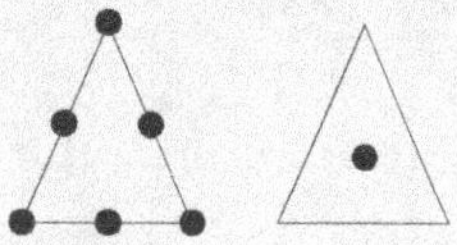	P_2	P_0	$\mathcal{O}(h)$
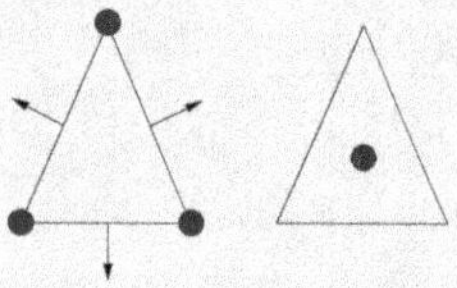	$P_1 \oplus \text{span}\,(\mathbf{w}_1, \mathbf{w}_2, \mathbf{w}_3)^2$	P_0	$\mathcal{O}(h)$
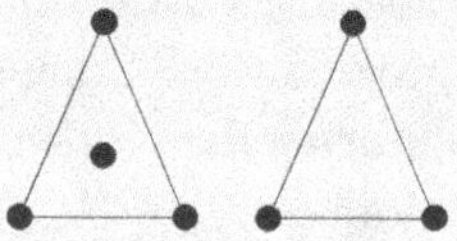	$P_1 \oplus \text{span}\,(\lambda_1\lambda_2\lambda_3)$	P_1	$\mathcal{O}(h)$
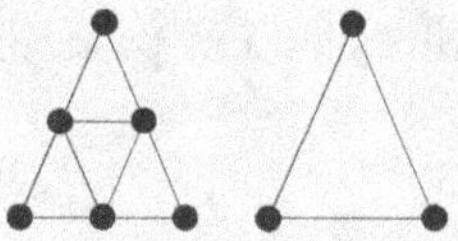	$4xP_1$	P_1	$\mathcal{O}(h)$
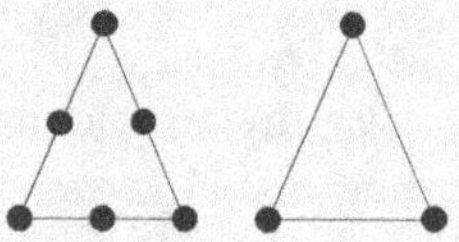	P_2	P_1	$\mathcal{O}(h^2)$
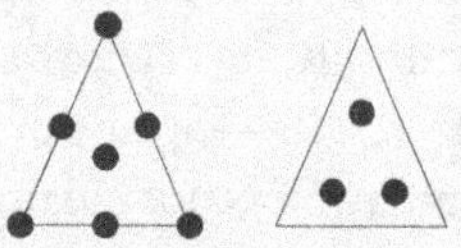	$P_2 \oplus \text{span}\,(\lambda_1\lambda_2\lambda_3)$	P_1^{disc}	$\mathcal{O}(h^2)$

[1] $\mathbf{w}_i = \lambda_j\lambda_k\mathbf{n}_i$.

The fourth and fifth element in Table 8.1 have the same dof but from the viewpoint of balanced approximation, the fifth element should be used. It also belongs to the class of continuous pressure approximations and is the first element of the Taylor–Hood family of elements

$$\mathcal{P}_K^u := P_k(K)^2 \qquad \mathcal{P}_K^p = P_{k-1}(K), \quad k \geq 2.$$

The Taylor–Hood elements satisfy the discrete inf–sup condition, see Bercovier and Pironneau (1979); Brezzi and Falk (1991); Girault and Raviart (1986) and Boffi et al. (2013) and is well-balanced from approximation point of view. The lowest order case P_2, P_1 element belongs to the most commonly used elements due to its properties : inf–sup stable, second order accurate on triangular (and tetrahedral) meshes.

Nevertheless, discontinuous pressure approximations are preferred in cases when mass conservation is crucial. The reason is that a discontinuous pressure approximation in P_{k-1}^{disc} satisfies

$$\int_K q_h \operatorname{div} \mathbf{u}_h \, dx = 0 \quad \text{for all } q_h \in P_{k-1}(K), K \in \mathcal{T}_h,$$

which means that on each cell $K \in \mathcal{T}_h$ the first moments of $\operatorname{div} \mathbf{u}_h$ vanish, whereas continuous pressure approximations guarantee mass conservation only over a patch of cells. In addition, discontinuous pressure approximations are intrinsic parallel since no communication is needed between the pressure dof of different cells in the Stokes solver. Unfortunately, the $(P_k, P_{k-1}^{\text{disc}})$ element is not inf–sup stable on general meshes. Therefore, the velocity space P_k has to be enriched by certain edge/face/cell bubble ($k+1$ order of polynomial) functions in order to guarantee the inf–sup condition, see Bernardi and Raugel (1985) and Girault and Raviart (1986) for $k = 1$, Crouzeix and Raviart (1973) for $k = 2, 3$, and Mansfield (1982) for a general k. For introducing the family of elements we denote the space of homogeneous polynomials by $\widetilde{P}_k$ of degree k with $\dim \widetilde{P}_k = k + 1$. Then, the family of discontinuous pressure approximations is given by

$$\mathcal{P}_K^u := \left[P_k(K) \oplus \operatorname{span} \lambda_1 \lambda_2 \lambda_3 \widetilde{P}_{k-2}(K)\right]^2, \qquad \mathcal{P}_K^p := P_{k-1}^{\text{disc}}(K), \quad k \geq 2.$$

Note that $\dim \mathcal{P}_K^u = k(k+5)$. The set of dof for the velocity consists of the $2 \times 3k$ function values of the standard P_k element on the edges of K and the $(k-1)k$ moments

$$\mathbf{v}_h \mapsto \frac{1}{|K|} \int_K \mathbf{v}_h \cdot \mathbf{w}_h \, dx \quad \text{for all } \mathbf{w}_h \in P_{k-2}^2.$$

Also within this family the second order element, shortly denoted by $(P_2^+, P_1^{\text{disc}})$ element, is the most commonly used one.

Exercise 8.5. Prove that the set of dof for the P_2^+ element is $\mathcal{P}_K^u$ unisolvent.

Conforming quadrilateral finite elements

We now consider conforming finite element approximations on quadrilaterals. As in the case of a triangular mesh, the simplest and approximation balanced choice $\mathcal{P}_K^u := Q_1(K)^2$ and $\mathcal{P}_K^p := Q_0(K)$ is not inf–sup stable. Nevertheless, enriching the velocity space to mapped biquadratic functions we get the first order inf–sup stable pair Q_2, Q_0, as indicated in Table 8.2. This enrichment increases the dimension of the local velocity space from 2×4 to 2×9. However, Fortin (1981) has shown that an enrichment of the $Q_1(K)^2$ space by four vector-valued functions in the normal direction $\mathbf{n}_K$ is enough to guarantee the inf–sup condition, see Theorem II.3.1 in Girault and Raviart (1986).

The second element in Table 8.2 is the lowest order element of the Taylor–Hood element family on quadrilaterals given by

$$\mathcal{P}_K^u := Q_k(K)^2, \qquad \mathcal{P}_K^p := Q_{k-1}(K), \quad k \geq 2.$$

It is well-balanced and satisfies the inf–sup condition, see Bercovier and Pironneau (1979); Girault and Raviart (1986); Verfürth (1984) and Boffi et al. (2013).

Due to its good mass conservation properties, the family of discontinuous pressure approximations

$$\mathcal{P}_K^u := Q_k(K)^2, \qquad \mathcal{P}_K^p := P_{k-1}^{\text{disc}}(K), \quad k \geq 2$$

Table 8.2 Conforming inf–sup stable quadrilateral finite elements

Velocity and pressure	P_K^u	P_K^p	Order
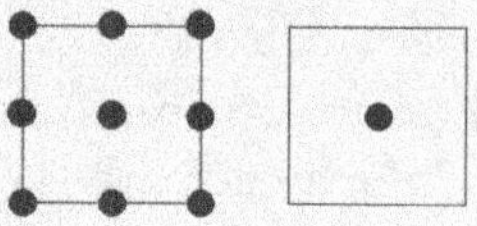	Q_2	Q_0	$\mathcal{O}(h)$
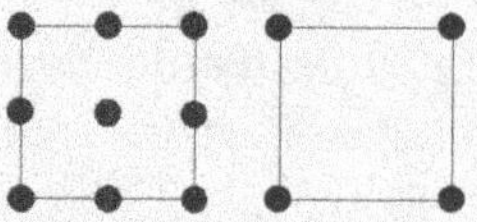	Q_2	Q_1	$\mathcal{O}(h^2)$
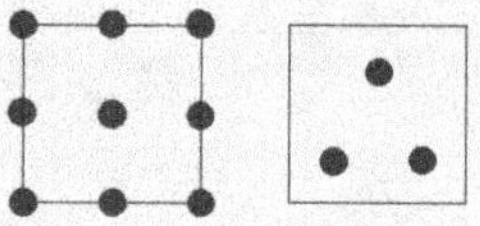	Q_2	P_1 mapped	$\mathcal{O}(h^2)$[1]
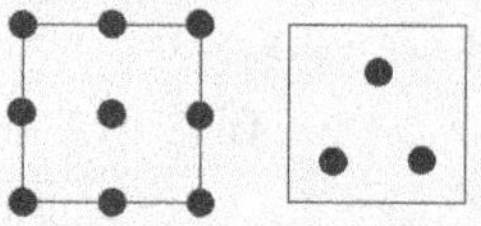	Q_2	P_1 unmapped	$\mathcal{O}(h^2)$

[1]On uniformly refined meshes.

is quite popular. The choice of the pressure space needs some discussion. Using a P_m space on a quadrilateral is not a standard procedure. Indeed, due to the, in general, nonaffine mapping there are two different versions for the pressure approximation. One can take discontinuous, piecewise polynomials of degree less than or equal to m on $K \in \mathcal{T}_h$ or one takes polynomials of degree less than or equal to m on the reference square $\widehat{K}$ and maps them to the general element $K \in \mathcal{T}_h$ by the bilinear (bijective) mapping $F_K : \widehat{K} \to K$. We call the first way as the unmapped approach and the second as the mapped approach. If F_K is a bilinear, nonaffine mapping we have

$$P_m(K) \neq \{q : K \to \mathbb{R} : q = \hat{q} \circ F_K^{-1}, \hat{q} \in P_m(\widehat{K})\}.$$

Exercise 8.6. Find the bilinear mapping $F_K : \widehat{K} \to K$ of the reference square $\widehat{K} = (-1,+1)^2$ onto the quadrangle with the vertices $(1,1)$, $(-1,1)$, $(-2,-1)$, $(2,-1)$. Show that the mapped P_1 space

$$\{q : K \to \mathbb{R} : q = \hat{q} \circ F_K^{-1}, \hat{q} \in P_1(\widehat{K})\}$$

is different from the unmapped space $P_1(K)$.

Both finite element pairs, the unmapped version shown in Stenberg (1984); Girault and Raviart (1986) and Boffi et al. (2013) as well as the mapped version in Boffi and Gastaldi (2002) and Boffi et al. (2013) for $k = 2$ and Matthies and Tobiska (2002) for any $k \geq 2$ and any space dimensions $d \geq 2$ satisfy the inf–sup condition. However, it can happen as shown in Matthies

(2001) and Boffi and Gastaldi (2002) that the error of best approximation of the pressure is reduced on certain sequences of meshes, which influences the rate of convergence of the velocity and pressure according to (8.35). It should be mentioned that the optimal order of convergence can be retained on uniformly refined meshes (the next finer mesh level is generated by connecting the midpoints of opposite edges of the quadrilateral), see Matthies (2001) and Matthies and Tobiska (2002). Such meshes are often used in the framework of multigrid solvers.

Nonconforming triangular finite elements

The first element in Table 8.3 is the Crouzeix–Raviart element studied in detail in Section 8.5. Nonconforming finite element approximations are more attractive for parallel solvers, as the velocity approximation needs less communication between the processors since each velocity dof is shared by two cells at the most. In addition, discontinuous pressure approximations are embarrassingly parallel, and hence the nonconforming elements in Table 8.3 are the first choice for efficient parallel algorithms. The dof of the velocity components for the first, second and

Table 8.3 Nonconforming inf–sup stable triangular finite elements

Velocity and pressure	P_K^u	P_K^p	Order
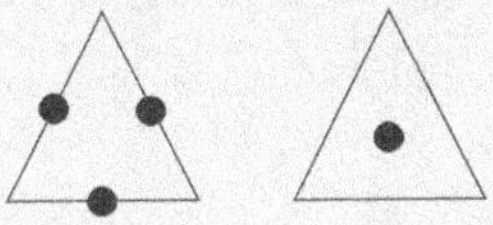	P_1^{nc}	P_0	$\mathcal{O}(h)$
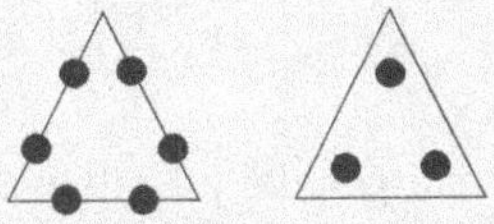	P_2^{nc}	P_1^{disc}	$\mathcal{O}(h^2)$
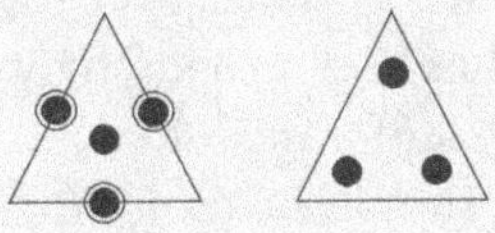	$P_2 \oplus \operatorname{span} \tilde{b}^1$	P_1^{disc}	$\mathcal{O}(h^2)$
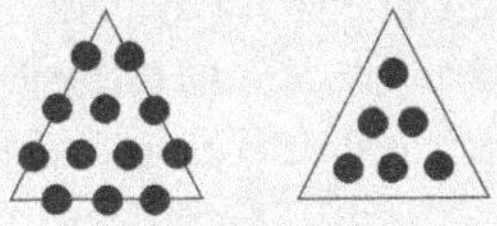	$P_3 \oplus \operatorname{span}(\lambda_1 b, \lambda_2 b)^2$	P_2^{disc}	$\mathcal{O}(h^3)$
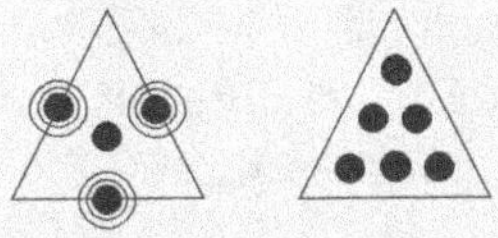	$P_3 \oplus \operatorname{span}(\lambda_1 \tilde{b}, \lambda_2 \tilde{b})^1$	P_2^{disc}	$\mathcal{O}(h^3)$

[1] $\tilde{b} := (\lambda_1 - \lambda_2)(\lambda_2 - \lambda_3)(\lambda_3 - \lambda_1)$.
[2] $b := \lambda_1 \lambda_2 \lambda_3$.

fourth pair are function values at the corresponding points. For the third and fifth pair, the dof are mean values over the edges and over the cell, respectively.

Consider the third and fifth element as examples of a whole family in more detail. Let f_i denote the edge opposite to the vertex a_i, $i = 1,2,3$ (modulo 3). We set $\tilde{b} := (\lambda_1 - \lambda_2)(\lambda_2 - \lambda_3)(\lambda_3 - \lambda_1)$ and for $k \geq 2$

$$\mathcal{P}_K^u := \left[P_k(K) \oplus \mathrm{span}_{0\leq j\leq k-2}\, (\tilde{b}\lambda_1^{k-2-j}\lambda_2^j)\right]^2, \qquad \mathcal{P}_K^p := P_{k-1}^{\mathrm{disc}}(K)$$

and chose the following nodal functionals

$$N_i(\mathbf{v}) = \frac{1}{|f_i|}\int_{f_i} q_{ij}\,\mathbf{v}\,d\gamma\,, \quad j = 0,\ldots,k-1, \;\; i = 1,2,3,$$

$$N^m(\mathbf{v}) = \frac{1}{|K|}\int_K w_m\,\mathbf{v}\,dx, \quad m = 1,\ldots,(k-1)k/2.$$

where span $(q_{i0},\ldots,q_{i,k-1}) = P_{k-1}(f_i)$ and span $(w_1,\ldots,w_{(k-1)k/2}) = P_{k-2}(K)$. This family of elements satisfies the inf–sup condition, see Matthies and Tobiska (2005). The reader who is interested in a proof of the inf–sup stability of the other elements listed in Table 8.3 has to refer to Section 8.5 (first element), Fortin and Soulie (1983) (second element) and Crouzeix and Raviart (1973) (fourth element). In view of Lemma 8.8 we finally note that the discontinuous pressure approximations P_{k-1}^{disc}, $k \geq 2$, can be replaced by continuous approximations P_{k-1} without losing the inf–sup stability.

Exercise 8.7. Show that for one velocity component of the third element in Table 8.3 the set of dof Σ_K is $(P_2 \oplus \mathrm{span}\,\tilde{b})$ unisolvent.

Nonconforming quadrilateral finite elements

One of the first nonconforming inf–sup stable finite element on quadrilaterals is the rectangular element proposed by Han (1984). The dof are the four mean values over the edges and the mean value over the cell. The second element in Table 8.4 has been proposed by Rannacher and Turek (1992). The case of $\mathcal{O}(h^k)$ convergent families, $k \geq 2$, of nonconforming quadrilateral elements has been investigated by Matthies (2007). One should be aware that in case of nonaffine mappings from the reference square to convex quadrilaterals two versions of these elements, a mapped and an unmapped version, exist. As for the mapped $(Q_k, P_{k-1}^{\mathrm{disc}})$ element the optimal approximation property can be missed on arbitrary meshes (but is still retained on uniformly refined meshes). This applies, in particular, to the P_1^{disc} discretizations of the third, fourth and fifth element in Table 8.4. Note, however, that the local finite element spaces are in general not invariant with respect to bilinear transformations, therefore, also mapped and unmapped versions of the local velocity spaces can exist. For more details, refer to Rannacher and Turek (1992) and Matthies (2007).

As we discussed earlier, nonconforming finite element approximations for the velocity on quadrilateral meshes combined with discontinuous pressure approximation are a good choice for efficient parallel algorithms. On the one hand, the nonconforming velocity dof avoid

Table 8.4 Nonconforming inf–sup stable quadrilateral finite elements

Velocity and pressure	P_K^u	P_K^p	Order
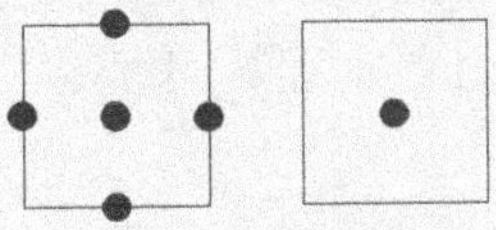	$P_1 \oplus \mathrm{span}\,(\varphi(x_1), \varphi(x_2))$ [1]	Q_0	$\mathcal{O}(h)$
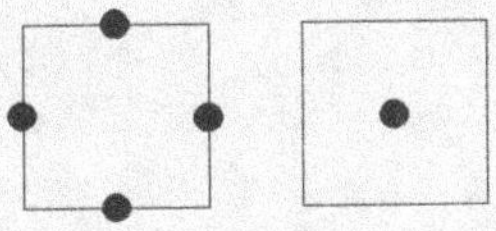	$Q_1^{\mathrm{rot}} = \mathrm{span}\,(1, x_1, x_2, x_1^2 - x_2^2)$	Q_0	$\mathcal{O}(h)$
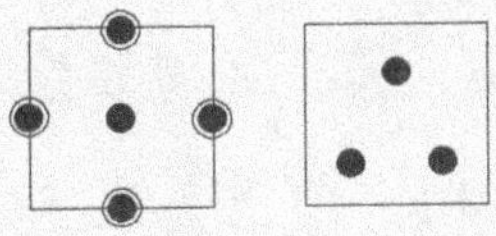	$P_2 \oplus B_{\mathrm{NC}}^1$ [2]	P_1^{disc}	$\mathcal{O}(h^2)$
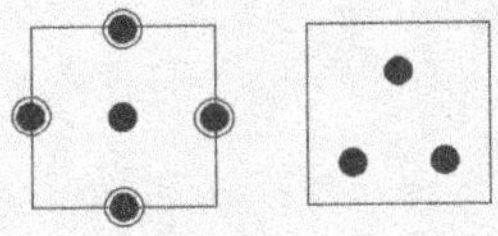	$P_2 \oplus B_{\mathrm{NC}}^2$ [3]	P_1^{disc}	$\mathcal{O}(h^2)$
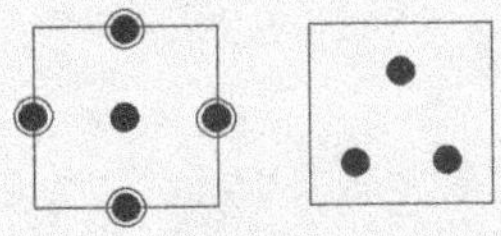	$P_2 \oplus B_{\mathrm{NC}}^3$ [4]	P_1^{disc}	$\mathcal{O}(h^2)$

[1] $\varphi(t) := (5t^4 - 3t^2)/2$.
[2] $B_{\mathrm{NC}}^1 := \mathrm{span}\,(x_1^2 x_2, x_1 x_2^2, x_1^3 x_2 - x_1 x_2^3)$.
[3] $B_{\mathrm{NC}}^2 := \mathrm{span}\,(x_1(x_1^2 - x_2^2), x_2(x_1^2 - x_2^2), x_1^3 x_2 - x_1 x_2^3)$.
[4] $B_{\mathrm{NC}}^3 := \mathrm{span}\,(x_1^3, x_2^3, x_1^3 x_2 - x_1 x_2^3)$.

cross-vertex communications. On the other hand, discontinuous pressure approximations are embarrassingly parallel.

Inf–sup stable finite elements in 3d

Many of the inf–sup stable 2d finite elements have extensions to the 3d case. Nevertheless, there are some essential differences resulting from the fact that the pair (P_2, P_0) is not inf–sup stable in 3d. Furthermore, the standard bubble function $\lambda_1\lambda_2\lambda_3\lambda_4$ belongs to P_4. Some of the inf–sup stable 3d finite element pairs are listed in Table 8.5. More details can be found in Girault and Raviart (1986) and Boffi et al. (2013). In particular, $\mathcal{O}(h^k)$ convergent families of nonconforming finite elements on brick meshes with the discontinuous pressure approximation P_{k-1}^{disc} are constructed in Matthies (2007).

The first element in Table 8.5 is the Crouzeix–Raviart element studied in Section 8.5 on simplices of any space dimension. The second element has been proposed by Bernardi and

Table 8.5 Some inf–sup stable finite elements in 3d

Velocity and pressure	P_K^u	P_K^p	Order
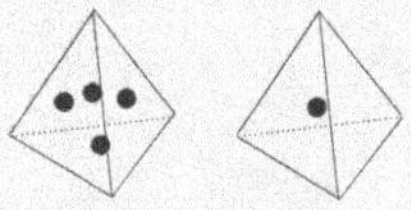	P_1^{nc}	P_0	$\mathcal{O}(h)$
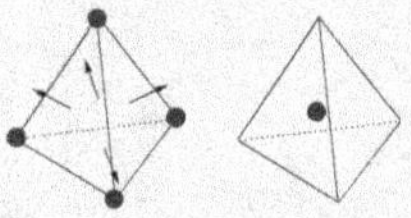	$P_1 \oplus \mathrm{span}\,(\mathbf{w}_1,\ldots,\mathbf{w}_4)$[1]	P_0	$\mathcal{O}(h)$
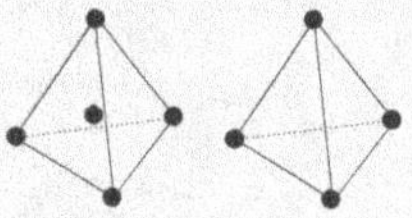	$P_1 \oplus \mathrm{span}\,(\lambda_1,\ldots,\lambda_4)$	P_1	$\mathcal{O}(h)$
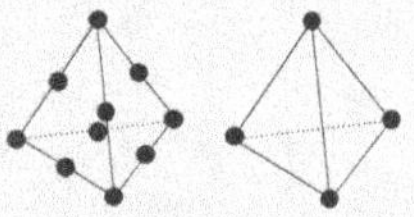	P_2	P_1	$\mathcal{O}(h^2)$
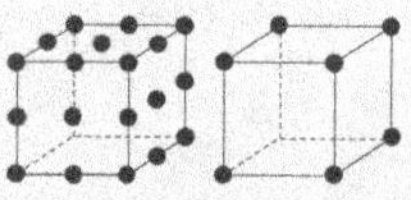	Q_2	Q_1	$\mathcal{O}(h^2)$
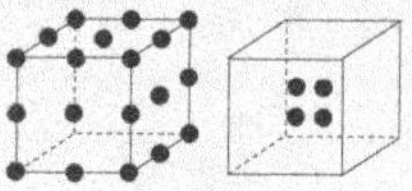	Q_2	P_1^{disc}	$\mathcal{O}(h^2)$[2]

[1] $\mathbf{w}_i := \lambda_j\lambda_k\lambda_l\mathbf{n}_i$.
[2] On uniformly refined meshes for mapped pressure spaces.

Raugel (1985) and is the extension of the associated triangular element (second element in Table 8.1). Since the face bubbles are in 3d cubic functions, we now have

$$\mathcal{P}_K^u := P_1(K)^3 \oplus \mathrm{span}\,(\mathbf{w}_1,\mathbf{w}_2,\mathbf{w}_3,\mathbf{w}_4) \subset P_3(K)^3$$

with the vector-valued functions $\mathbf{w}_i = \lambda_j\lambda_k\lambda_l\mathbf{n}_i$, $j,k,l \neq i$, $j,k \neq l$, $j \neq k$. Again we used the notation $\mathbf{n}_i$ for the outer normal to the face $\lambda_i = 0$ opposite to the vertex a_i, $i = 0,1,2,3$.

The third element in Table 8.5 stands for the lowest order case $k = 1$ of the MINI element family

$$\mathcal{P}_K^u := (P_k(K) + \lambda_1\lambda_2\lambda_3\lambda_4 P_{k-1}(K))^3, \qquad \mathcal{P}_K^p := P_k(K), \quad k \geq 1.$$

The fourth and fifth element in Table 8.5 are the lowest order elements of the Taylor–Hood family on tetrahedra and hexahedra, respectively, given by

$$\mathcal{P}_K^u := P_k(K)^3, \qquad \mathcal{P}_K^p := P_{k-1}(K), \quad k \geq 2,$$
$$\mathcal{P}_K^u := Q_k(K)^3, \qquad \mathcal{P}_K^p := Q_{k-1}(K), \quad k \geq 2.$$

The MINI element and the Taylor–Hood families use continuous pressure approximations. For improving the mass conservation we can also use the popular family of discontinuous pressure approximations given by

$$\mathcal{P}_K^u := Q_k(K)^3, \qquad \mathcal{P}_K^p := P_{k-1}^{\text{disc}}(K), \quad k \geq 2.$$

We recall that there are two versions of the pressure space in case of nonaffine mappings, both of them are inf–sup stable, see Girault and Raviart (1986); Boffi et al. (2013) and Matthies and Tobiska (2002). The optimal approximation property of the unmapped version can be retained on uniformly refined meshes, which are often used in practise.

Note that, nonconforming velocity approximations combined with discontinuous pressure approximations allow high parallel efficiency, since no cross-vertex and cross-edge communication is needed. A family of nonconforming velocity and discontinuous pressure approximations on brick meshes has been proposed in Matthies (2007).

Some special cases of inf–sup stable elements

In order to improve the mass conservation of the Taylor–Hood family, the pressure space Q_h has been enriched by piecewise constant functions, see Thatcher (1990). It has been shown (Thatcher (1990); Jinsheng and Xiaoliang (1997); Boffi et al. (2012) and Arndt (2013)), that the families

$$(P_k, (P_{k-1} + P_0)), \qquad (Q_k, (Q_{k-1} + Q_0))$$

satisfy the inf–sup condition.

On general meshes, the family $(P_k, P_{k-1}^{\text{disc}})$ does not satisfy the inf–sup condition. However, if the mesh results from a triangular or a tetrahedral mesh by connecting the barycenter of each cell with its vertices, then the inf–sup condition holds provided $k \geq d$, $d = 2, 3$, see Scott and Vogelius (1985a;b) and Zhang (2005). The element is called Scott–Vogelius element and can be also considered as a composite element. It has the favorite property that discretely divergence free functions are divergence free.

The velocity space of the nonconforming triangular element $(P_3^{\text{nc}}, P_2^{\text{disc}})$ had to be enriched by two quartic bubble functions to become inf–sup stable, see the fourth element in Table 8.3. With an additional assumption on the mesh the quartic bubbles are not needed for the inf–sup stability, for details see Crouzeix and Raviart (1973) and Crouzeix and Falk (1989).

An unmapped version of the nonconforming Han element on the quadrilateral meshes has been recently proposed in Zhang (2016). This finite element pair is based on a nonconforming enriched P_1 velocity space

$$\mathcal{P}_K^u := (P_1(K) \oplus \text{span}\,(\xi^2, \eta^2))^2$$

combined with the discontinuous pressure space $P_1^{\text{disc}}(K)$. As the nonconforming Crouzeix–Raviart element, it possesses the property of being piecewise divergence free.

8.7. Equal order stabilized finite elements

As we have seen from the above discussion, the finite element space V_h for approximating the velocity has to be rich enough in comparison with the finite element space Q_h approximating the pressure. Under this condition, there is a unique solution of the Stokes problem and the error in the $H_0^1(\Omega)^d \times L_0^2(\Omega)$ norm is quasioptimal (that is, proportional to the error of best approximation).

Unfortunately, equal order finite element approximations for both the velocity and pressure do not satisfy the discrete inf–sup condition, and induce oscillations in the solution, especially in the pressure. Due to its simplicity of using equal order approximations, for example, continuous, piecewise linear functions for velocity and pressure, ways to circumvent the inf–sup condition were studied in the eighties. The idea of a residual-based stabilization method consists in adding weighted residuals of the strong form of the differential equation to its weak formulation. The stabilization modifies the discrete problem and the corresponding bilinear form becomes coercive in the product space $V_h \times Q_h$. This facilitated to use the Lax–Milgram theorem to show the well-posedness of the stabilized Stokes problem with equal order interpolation.

The Petrov–Galerkin (PSPG) formulation has been proposed by Hughes et al. (1986) to circumvent the discrete inf–sup condition and to use an equal order interpolation for the Stokes problem. Let $\mathcal{T}_h$ be an admissible triangulation of Ω and $Y_h \subset H^1(\Omega)$ be a continuous, scalar-valued finite element space. We set $V_h := (Y_h \cap H_0^1(\Omega))^d$ and $Q_h := Y_h \cap L_0^2(\Omega)$ for the spaces approximating velocity and pressure, respectively. We assume that the solution of the Stokes problem (8.13), (8.14) belongs to $(H_0^1(\Omega) \cap H^2(\Omega))^d \times (L_0^2(\Omega) \cap H^1(\Omega))$. Then, it holds

$$-\nu \Delta \mathbf{u} + \nabla p = \mathbf{f} \quad \text{in } L^2(K) \quad \text{for all } K \in \mathcal{T}_h.$$

We add weighted residuals

$$(-\nu \Delta \mathbf{u}_h + \nabla p_h - \mathbf{f}, \delta_K \nabla q_h)_K$$

to the discretized weak formulation. Recall that $(\cdot,\cdot)_K$ denotes the inner product in $L^2(K)^d$. Now the PSPG form of the Stokes equations reads

Find $(\mathbf{u}_h, p_h) \in V_h \times Q_h$ such that for all $(\mathbf{v}_h, q_h) \in V_h \times Q_h$

$$\begin{aligned} \nu(\nabla \mathbf{u}_h, \nabla \mathbf{v}_h) - (p_h, \operatorname{div} \mathbf{v}_h) &= \langle \mathbf{f}, \mathbf{v}_h \rangle, \\ (q_h, \operatorname{div} \mathbf{u}_h) + \sum_{K \in \mathcal{T}_h} (-\nu \Delta \mathbf{u}_h + \nabla \mathbf{p}_h - \mathbf{f}, \delta_K \nabla q_h)_K &= \langle g, q_h \rangle, \end{aligned}$$

where $\delta_K > 0$ is the local stabilization parameter. The solvability of the stabilized PSPG method depends on the properties of the mesh-dependent bilinear form $A_h : (V_h \times Q_h) \times (V_h \times Q_h) \to \mathbb{R}$

and the linear form $\ell_h : V_h \times Q_h \to \mathbb{R}$ given by

$$A_h((\mathbf{u}_h, p_h); (\mathbf{v}_h, q_h)) := \nu(\nabla \mathbf{u}_h, \nabla \mathbf{v}_h) - (p_h, \operatorname{div} \mathbf{v}_h) + (q_h, \operatorname{div} \mathbf{u}_h) + \sum_{K \in \mathcal{T}_h} (-\nu \Delta \mathbf{u}_h + \nabla p_h, \delta_K \nabla q_h)_K,$$

$$\ell_h((\mathbf{v}_h, q_h)) := \langle \mathbf{f}, \mathbf{v}_h \rangle + \langle g, q_h \rangle + \sum_{K \in \mathcal{T}_h} (\mathbf{f}, \delta_K \nabla q_h)_K.$$

The continuity of the pressure space Q_h guarantees that

$$(\mathbf{v}_h, q_h) \mapsto \left(\nu |\mathbf{v}_h|_1^2 + \sum_{K \in \mathcal{T}_h} \delta_K \|\nabla q_h\|_{0,K}^2 \right)^{1/2}$$

is a (mesh-dependent) norm on the product space $V_h \times Q_h$. A detailed study shows that the bilinear form A_h is coercive with respect to that norm, provided that $\delta_K = \mathcal{O}(h_K^2/\nu)$. The application of Theorem 2.1 (Lax–Milgram) yields the unique solvability of the PSPG method. Since the residual of the strong form of the Stokes equations is used to modify the discrete bilinear form, we still have the Galerkin orthogonality

$$A_h((\mathbf{u} - \mathbf{u}_h, p - p_h); (\mathbf{v}_h, q_h)) = 0 \quad \text{for all } (\mathbf{v}_h, q_h) \in V_h \times Q_h.$$

Thus the PSPG discretisation enhances the stability without altering the consistency of the problem. Even though the PSPG formulation allows to use equal order finite elements for the Stokes problem, an optimal choice of the stabilization parameter δ_K is an open problem.

The PSPG formulation contains many stabilization terms, and the second derivatives of the basis functions are needed to assemble the stabilization terms. Alternatively, the local projection stabilization (LPS) method has been proposed in Becker and Braack (2001) for the Stokes problem with equal order finite elements (see also Matthies et al. (2007) and Ganesan et al. (2008)). LPS is based on a projection $\pi_h : V_h \to D_h$ of the finite element space V_h into a discontinuous space D_h. In LPS, stabilization is achieved by adding terms which give a weighted L^2 control over the fluctuations $\kappa_h := id - \pi_h$ of the gradients of the quantity of interest. Define $D_h(K) := \{q_h|_K : q_h \in D_h\}$, and let $\pi_K : L^2(K) \to D_h(K)$ be a local projection that defines the global projection $\pi_h : L^2(\Omega) \to D_h$ by $(\pi_h w)|_K := \pi_K(w|_K)$. After applying the stabilization, the LPS formulation for the Stokes problem reads

Find $(\mathbf{u}_h, p_h) \in V_h \times Q_h$ such that for all $(\mathbf{v}_h, q_h) \in V_h \times Q_h$

$$\nu(\nabla \mathbf{u}_h, \nabla \mathbf{v}_h) - (p_h, \operatorname{div} \mathbf{v}_h) = \langle \mathbf{f}, \mathbf{v}_h \rangle,$$

$$(q_h, \operatorname{div} \mathbf{u}_h) + \sum_{K \in \mathcal{T}_h} \delta_K (\kappa_h \nabla p_h, \kappa_h \nabla q_h) = \langle g, q_h \rangle,$$

where $\delta_K > 0$ is the LPS stabilization parameter. Though the LPS method is not a consistence method, optimal order of convergence can be shown, if the projection space D_h is chosen appropriately. On the one hand, it should be rich enough to guarantee a certain order of consistency and on the other hand, small enough that the interpolation error is orthogonal to the projection space. A careful error analysis shows that one should choose $\delta_K \sim h_K^2/\nu$ but in contrast to the PSPG approach there is no upper bound for δ_K. For more details and several variants of the LPS method, see Matthies et al. (2007); Ganesan et al. (2008) and the references therein.

8.8. Navier–Stokes problem with mixed boundary conditions

In the previous sections, the Stokes equations with Dirichlet boundary conditions have been considered. This type of boundary conditions is taken into consideration in defining the velocity space V as a subspace of $H_0^1(\Omega)^d$. Now, we show how other type of boundary conditions can be incorporated in the variational form. Our formulation is related to the nonsteady Navier–Stokes problem. Modifications that are needed to handle the steady Stokes and steady Navier–Stokes problem, respectively, are straightforward.

Assume that the boundary consists of disjoint parts Γ_{D}, Γ_{F} and Γ_{S}, that is, $\partial\Omega = \Gamma_{\mathrm{D}} \cup \Gamma_{\mathrm{F}} \cup \Gamma_{\mathrm{S}}$ with $\operatorname{meas}(\Gamma_{\mathrm{D}} \cup \Gamma_{\mathrm{S}}) > 0$. Further, impose the following mixed boundary conditions

$$\begin{aligned} &\mathbf{u} = \mathbf{g} \quad \text{on } \Gamma_{\mathrm{D}},\\ \mathbf{u}\cdot\mathbf{n} = 0, \qquad &\beta\,\tau_i \cdot \mathbb{S}(\mathbf{u},p)\cdot\mathbf{n} = -\tau_i\cdot\mathbf{u}, \quad i = 1,\dots,d-1 \ \text{ on } \Gamma_{\mathrm{S}},\\ &\mathbb{S}(\mathbf{u},p)\cdot\mathbf{n} = \sigma\,\mathcal{K}\mathbf{n} + \nabla_\Gamma \sigma \quad \text{on } \Gamma_{\mathrm{F}}. \end{aligned}$$

Here, $\mathbf{n}$ is the outward normal on the associated part of the boundary, $\mathcal{K}$ the sum of the principal curvatures, ∇_Γ the tangential gradient, σ the surface tension of the fluid and β the slip parameter.

Instead of using Equations (8.4) and (8.5) to describe the fluid motion and the pressure in a given time interval $(0,\mathrm{I}]$, we use its equivalent form (8.3) and (8.5)

$$\rho\left(\frac{\partial \mathbf{u}}{\partial t} + (\mathbf{u}\cdot\nabla)\mathbf{u}\right) - \operatorname{div}\mathbb{S}(\mathbf{u},p) = \mathbf{f} \quad \text{in } \Omega\times(0,\mathrm{I}], \tag{8.45}$$

$$\operatorname{div}\mathbf{u} = 0 \quad \text{in } \Omega\times(0,\mathrm{I}]. \tag{8.46}$$

We cannot expect uniqueness of the pressure when the free surface part of the boundary Γ_{F} is empty. Indeed, we have for $\tilde{p} = p + \text{const}$

$$\mathbb{S}(\mathbf{u},\tilde{p}) = 2\eta\mathbb{D}(\mathbf{u}) - \tilde{p}\mathbb{I}$$

from which

$$\operatorname{div}\mathbb{S}(\mathbf{u},\tilde{p}) = \operatorname{div}\mathbb{S}(\mathbf{u},p), \qquad \tau_i\cdot\mathbb{S}(\mathbf{u},\tilde{p})\cdot\mathbf{n} = \tau_i\cdot\mathbb{S}(\mathbf{u},p)\cdot\mathbf{n}$$

follow. This nonuniqueness of the pressure upto an additive constant is handled as in the Stokes problem by seeking the pressure in $L_0^2(\Omega)$. In addition, we cannot impose arbitrary values for

Dirichlet type boundary conditions when $\text{meas}(\Gamma_\text{F}) = 0$ since

$$0 = \int_\Omega \operatorname{div} \mathbf{u} \, dx = \int_{\Gamma_\text{D}} \mathbf{u} \cdot \mathbf{n} \, ds + \int_{\Gamma_\text{S}} \mathbf{u} \cdot \mathbf{n} \, ds = \int_{\Gamma_\text{D}} \mathbf{g} \cdot \mathbf{n} \, ds \tag{8.47}$$

has to be satisfied.

Contrarily, the situation is different when the free surface part on the boundary exists, that is, $\text{meas}(\Gamma_\text{F}) > 0$. Since no condition for $\mathbf{u} \cdot \mathbf{n}$ on Γ_F is imposed, the Dirichlet value on other parts of the boundary can be arbitrary as $\mathbf{u} \cdot \mathbf{n}$ accommodates appropriate values in order to satisfy the divergence condition (8.47). Further, we have for the normal component of the normal stress on Γ_F

$$\mathbf{n} \cdot \mathbb{S}(\mathbf{u}, \tilde{p}) \cdot \mathbf{n} - \mathbf{n} \cdot \mathbb{S}(\mathbf{u}, p) \cdot \mathbf{n} = p - \tilde{p} \neq 0.$$

Hence, if $(\mathbf{u}, p)$ satisfies the boundary condition on Γ_F, $(\mathbf{u}, p + \text{const})$ does not satisfy this boundary condition. Therefore, we seek the pressure p in $L^2(\Omega)$ and choose the test functions q for the incompressibility constraint from $L^2(\Omega)$.

We are now ready to derive the weak formulation of Navier–Stokes equations with the set of boundary conditions given above. Defining the test space for the momentum equation as

$$V := \{\mathbf{v} \in H^1(\Omega)^d : \mathbf{v} = \mathbf{0} \text{ on } \Gamma_D, \mathbf{v} \cdot \mathbf{n} = 0 \text{ on } \Gamma_S\},$$

we seek the solution $\mathbf{u}$ in the space $\tilde{\mathbf{g}} + V$, where $\tilde{\mathbf{g}}$ stands for an extension of the Dirichtlet data $\mathbf{g}$ into Ω. The choice of the pressure space Q and the necessity of additional conditions on the Dirichtlet data $\mathbf{g}$ depend on Γ_F:

	Choice of Q	$(1, \mathbf{g} \cdot \mathbf{n})_{\Gamma_\text{D}} = 0$ needed?
$\text{meas}(\Gamma_\text{F}) > 0$	$L^2(\Omega)$	no
$\text{meas}(\Gamma_\text{F}) = 0$	$L^2_0(\Omega)$	yes

The first vector equation (8.45) is multiplied by a test function $\mathbf{v} \in V$ and integrated over Ω. Integration by parts is then applied to the term involving the stress tensor to get

$$-\left(\operatorname{div} \mathbb{S}(\mathbf{u}, p), \mathbf{v}\right) = \left(\mathbb{S}(\mathbf{u}, p), \nabla \mathbf{v}\right) - \int_{\partial\Omega} \mathbf{v} \cdot \mathbb{S}(\mathbf{u}, p) \cdot \mathbf{n} \, ds.$$

Since the stress tensor is symmetric, we can rewrite the first term on the right

$$\left(\mathbb{S}(\mathbf{u}, p), \nabla \mathbf{v}\right) = \left(\mathbb{S}(\mathbf{u}, p), \frac{1}{2}\left(\nabla \mathbf{v} + \nabla \mathbf{v}^T\right)\right) = 2\eta\left(\mathbb{D}(\mathbf{u}), \mathbb{D}(\mathbf{v})\right) - (p, \operatorname{div} \mathbf{v}).$$

Further, the integral over the boundary is split into its different parts

$$\int_{\partial\Omega} \mathbf{v} \cdot \mathbb{S}(\mathbf{u}, p) \cdot \mathbf{n} \, ds = \int_{\Gamma_\text{D}} \ldots \, ds + \int_{\Gamma_\text{F}} \ldots \, ds + \int_{\Gamma_\text{S}} \ldots \, ds.$$

Here, the integral over the Dirichlet part of the boundary Γ_{D} vanishes due to $\mathbf{v} = \mathbf{0}$ on Γ_{D}. On Γ_{F}, the normal stress is given by the boundary condition, that is,

$$\int_{\Gamma_{\mathrm{F}}} \mathbf{v} \cdot \mathbb{S}(\mathbf{u}, p) \cdot \mathbf{n} \, ds = \int_{\Gamma_{\mathrm{F}}} (\sigma \mathcal{K} \mathbf{v} \cdot \mathbf{n} + \nabla_\Gamma \sigma \cdot \mathbf{v}) \, ds.$$

Note that, the sum of the principal curvature $\mathcal{K}$ can be computed as the negative of the surface divergence of the normal, that is, $\mathcal{K} = -\operatorname{div}_\Gamma \mathbf{n}$. Thus, in order to evaluate the surface tension and Marangoni forces, second order derivatives of the parametric representation of the free surface and first order derivatives of the surface tension coefficient have to be computed. This can be avoided by deriving profit from the representation

$$\operatorname{div}_\Gamma (\sigma \mathbb{P}) = \sigma \mathcal{K} \mathbf{n} + \nabla_\Gamma \sigma, \quad \mathbb{P} := \mathbb{I} - \mathbf{n} \otimes \mathbf{n}$$

and an integration by parts on surface integral. In case of a closed hypersurface Γ_{F}, the boundary $\partial \Gamma_{\mathrm{F}}$ is empty and

$$\int_{\Gamma_{\mathrm{F}}} \operatorname{div}_\Gamma (\sigma \mathbb{P}) \cdot \mathbf{v} \, ds = -\int_{\Gamma_{\mathrm{F}}} \sigma \nabla_\Gamma \mathrm{id} : \nabla_\Gamma \mathbf{v} \, ds,$$

where $\mathrm{id} : \Gamma_{\mathrm{F}} \to \Gamma_{\mathrm{F}}$ is on Γ_{F} the identity. The formula also holds true in the case when $\partial \Gamma_{\mathrm{F}}$ is part of Γ_{D} since $\mathbf{v} = \mathbf{0}$ on Γ_{D}. If $\partial \Gamma_{\mathrm{F}}$ is part of Γ_{S} an additional term depending on the contact angle appears, see Ganesan (2015) and Ganesan et al. (2016). For details of integration by parts on surfaces we refer to Dziuk and Elliott (2013).

Further, we assumed that the normal component $\mathbf{v} \cdot \mathbf{n}$ of the test function vanishes on Γ_{S}. Thus, in the integral over Γ_{S} the vector $\mathbf{v}$ can be represented as

$$\mathbf{v} = \sum_{i=1}^{d-1} (\mathbf{v} \cdot \tau_i) \tau_i$$

which gives

$$\int_{\Gamma_{\mathrm{S}}} \mathbf{v} \cdot \mathbb{S}(\mathbf{u}, p) \cdot \mathbf{n} \, ds = -\frac{1}{\beta} \int_{\Gamma_{\mathrm{S}}} \sum_{i=1}^{d-1} \mathbf{u} \cdot \tau_i \mathbf{v} \cdot \tau_i \, ds.$$

Collecting all terms and multiplying the second scalar equation (8.46) by a test function $q \in Q$, we have the following weak formulation for the Navier–Stokes equations with mixed boundary conditions

Find $(\mathbf{u}, p) \in (\tilde{\mathbf{g}} + V) \times Q$ such that for all $(\mathbf{v}, q) \in V \times Q$

$$\rho \left(\frac{\partial \mathbf{u}}{\partial t} + (\mathbf{u} \cdot \nabla) \mathbf{u}, \mathbf{v} \right) + 2\eta (\mathbb{D}(\mathbf{u}), \mathbb{D}(\mathbf{v})) + \frac{1}{\beta} \int_{\Gamma_{\mathrm{S}}} \left(\sum_{i=1}^{d-1} \mathbf{u} \cdot \tau_i \mathbf{v} \cdot \tau_i \right) ds$$

$$- (p, \operatorname{div} \mathbf{v}) + (q, \operatorname{div} \mathbf{u}) = (\mathbf{f}, \mathbf{v}) + \int_{\Gamma_{\mathrm{F}}} \operatorname{div}_\Gamma (\sigma \mathbb{P}) \cdot \mathbf{v} \, ds.$$

The initial condition is imposed as the $L^2(\Omega)$ projection

$$(\mathbf{u}(0), \mathbf{w}) = (\mathbf{u}_0, \mathbf{w}) \quad \text{for all } \mathbf{w} \in L^2(\Omega)^d.$$

Remark 8.12. Note that boundary conditions of nonDirichlet type lead to the bilinear form $(\mathbf{u},\mathbf{v}) \mapsto 2\eta(\mathbb{D}(\mathbf{u}),\mathbb{D}(\mathbf{v}))$ instead of the form $(\mathbf{u},\mathbf{v}) \mapsto \nu(\nabla\mathbf{u},\nabla\mathbf{v})$. Due to the Poincare inequality and Korn's inequality both bilinear forms are coercive on the velocity space V. In case of a nonconforming finite element approximation $V_h \not\subset V$, however, the cellwise defined discrete bilinear form

$$a_h(\mathbf{u},\mathbf{v}) := 2\eta \sum_{K\in\mathcal{T}_h} (\mathbb{D}(\mathbf{u}),\mathbb{D}(\mathbf{v}))_K$$

is not automatically coercive on $(V+V_h)$. For the Crouzeix–Raviart element $a_h : V_h \times V_h \to \mathbb{R}$ is not coercive as shown in Arnold (1993).

8.9. Time discretization and linearization of the Navier–Stokes problem

In the following, we present a simple strategy for a full discretization of the nonsteady Navier–Stokes equation. For advanced methods including turbulent flows, we refer to John (2016). Choosing an inf–sup stable finite element discretization we obtain the semidiscrete problem, which is a nonlinear system of ODEs. Use a time discretization and a linearization strategy to get a sequence of Oseen type problems. For simplicity of presentation, we consider here the implicit backward Euler Scheme. The application of any other time stepping schemes discussed in Chapter 6 is carried out in the same manner.

Applying the backward method, one time step at $t = t_n$ in the fully discrete form of the Navier–Stokes equations read as

Given $\mathbf{u}_h^{n-1}$, find $(\mathbf{u}_h^n, p_h^n) \in (\tilde{\mathbf{g}}_h + V_h) \times Q_h$ such that for all $(\mathbf{v}_h, q_h) \in V_h \times Q_h$

$$\rho\left(\frac{\mathbf{u}_h^n - \mathbf{u}_h^{n-1}}{\Delta t} + (\mathbf{u}_h^n\cdot\nabla)\mathbf{u}_h^n, \mathbf{v}_h\right) + 2\eta\left(\mathbb{D}(\mathbf{u}_h^n),\mathbb{D}(\mathbf{v}_h)\right) + \frac{1}{\beta}\int_{\Gamma_S}\left(\sum_{i=1}^{d-1}\mathbf{u}_h^n\cdot\tau_i\mathbf{v}_h\cdot\tau_i\right)ds$$

$$-(p_h^n,\operatorname{div}\mathbf{v}) + (q_h,\operatorname{div}\mathbf{u}_h^n) = (\mathbf{f}^n,\mathbf{v}_h) + \int_{\Gamma_F}\operatorname{div}_\Gamma(\sigma\mathbb{P})\cdot\mathbf{v}_h\, ds.$$

The nonlinear convection term in the above equation can be handled in different ways. The most simple linearization is obtained by replacing the convective term with the term at the previous time step value, that is,

$$(\mathbf{u}_h^n\cdot\nabla)\mathbf{u}_h^n \approx (\mathbf{u}_h^{n-1}\cdot\nabla)\mathbf{u}_h^n.$$

Instead of this simple linearization, an iteration of fixed point type can be used, when the nonlinearity has to be considered implicitly. Let $\mathbf{u}^{n,0} = \mathbf{u}^{n-1}$. Then, solve the discrete equation for $(\mathbf{u}_h^{n,k}, p^{n,k}) \in (\tilde{\mathbf{g}}_h + V_h) \times Q_h$ with the linearized convection term

$$(\mathbf{u}_h^n\cdot\nabla)\mathbf{u}_h^n \approx (\mathbf{u}_h^{n,k-1}\cdot\nabla)\mathbf{u}_h^{n,k}, \quad k = 1,2,3,\ldots.$$

Alternatively, the Newton type variant is to use

$$(\mathbf{u}_h^n \cdot \nabla)\mathbf{u}_h^n \approx (\mathbf{u}_h^{n,k-1} \cdot \nabla)\mathbf{u}_h^{n,k} + (\mathbf{u}_h^{n,k} \cdot \nabla)\mathbf{u}_h^{n,k-1} - (\mathbf{u}_h^{n,k-1} \cdot \nabla)\mathbf{u}_h^{n,k-1}, \quad k = 1,2,\ldots.$$

For second order temporal schemes an extrapolation of second order

$$(\mathbf{u}_h^n \cdot \nabla)\mathbf{u}_h^n \approx \left(\frac{3}{2}\mathbf{u}_h^{n-1} - \frac{1}{2}\mathbf{u}_h^{n-2} \cdot \nabla\right)\mathbf{u}_h^n$$

can be used.

Finally, we discuss a semiimplicit time discretization for the surface tension term in free surface problems with the arbitrary Lagrangian–Eulerian approach, see Ganesan and Tobiska (2009; 2012). The free surface is moved with the velocity field according to

$$\frac{X^n - X^{n-1}}{\Delta t} = \mathbf{u}_h^n \quad \text{approximating} \quad \frac{\mathrm{d}X}{\mathrm{d}t} = \mathbf{u}.$$

We assume that the free surface Γ_F is closed without boundary. Since at time $t = t^n$ the velocity field $\mathbf{u}_h^n$ is computed and the position of the free surface X^n is unknown, we cannot evaluate the surface tension force fully implicitly. Instead, we approximate it semiimplicitly in the following manner

$$\begin{aligned}
\int_{\Gamma_\mathrm{F}} \mathrm{div}_\Gamma(\sigma\mathbb{P}) \cdot v\,\mathrm{d}s\Big|_{t=t^n} &\approx -\int_{\Gamma_\mathrm{F}^{n-1}} \sigma \nabla_{\Gamma_\mathrm{F}^{n-1}} X^n : \nabla_{\Gamma_\mathrm{F}^{n-1}} \mathbf{v}\,\mathrm{d}s \\
&\approx -\int_{\Gamma_\mathrm{F}^{n-1}} \Delta t \sigma \nabla_{\Gamma_\mathrm{F}^{n-1}} \mathbf{u}_h^n : \nabla_{\Gamma_\mathrm{F}^{n-1}} \mathbf{v}\,\mathrm{d}s + \int_{\Gamma_\mathrm{F}^{n-1}} \sigma \nabla_{\Gamma_\mathrm{F}^{n-1}} X^{n-1} : \nabla_{\Gamma_\mathrm{F}^{n-1}} \mathbf{v}\,\mathrm{d}s.
\end{aligned}$$

Note that the first term on the right hand side without the sign is a semipositive bilinear form which improves the stability of the algebraic system to compute the new velocity and pressure field $(\mathbf{u}_h^n, p_h^n) \in V_h \times Q_h$.

CHAPTER 9

IMPLEMENTATION OF THE FINITE ELEMENT METHOD

This chapter deals with the development of finite element algorithms. The key steps involved in finite element computations are the mesh generation, construction of the finite element space, generating sparse matrix stencils for the used finite elements, matrix assembling and the solution of the systemof algebraic equations. In this chapter, we will discuss the practical implementation of these steps in an object-oriented programming finite element package. The finite element algorithms presented in this chapter are based on the knowledge and experience gained through the development of our in-house finite element package, MooNMD, see John and Matthies (2004) and its parallel version ParMooN, see Ganesan et al. (2016b) and Wilbrandt et al. (2017).

9.1. Mesh handling and data structure

In this section, we discuss the terminologies and a data structure that is used to store finite element meshes. In addition to the commercial mesh generators, there are many open source mesh generators for academic research. *Triangle* is one of the open source mesh generator that generates high quality triangular meshes, see Shewchuk (1996). It also provides a lot of options to control the properties of the mesh. *TetGen* is analogous to the Triangle mesh generator but for three-dimensional polyhedral domains, see Si (2015). TetGen generates exact constrained Delaunay tetrahedralizations, boundary conforming Delaunay meshes, and Voronoi partitions. *Gmsh* is another popular public mesh generator that can generate 1d, 2d and 3d meshes, see Geuzaine and Remacle (2009) and Toulorge et al. (2013). An interface can be written to call these mesh generators directly from a finite element code, and it allows to control the mesh quality and avoid the storage of meshes separately.

Even though one can write their own code for generating a structured mesh on simple geometries, it is recommended to use an existing mesh generator for finite element computations. Mesh generation itself is an active field of research, and therefore there are several advantages of using an existing (commercial/open source) mesh generator. Apart from the possibilities of controlling the quality of the mesh, handling complex geometries, pre-adaptive meshing, etc., new methods and algorithms for meshing might be available through updates. Therefore, the focus of this section is on the mesh storage and its data structure rather than the mesh generation.

In order to assemble finite element matrices efficiently, the generated mesh needs to be stored in a structured way. The mesh data structure in a finite element package should be general enough to handle different meshes, irrespective of the dimension and the type of the mesh. Most commonly used mesh types are depicted in Figure 9.1. The choice of triangular/quadrilateral mesh and tetrahedral/hexahedral mesh in 2d and 3d, respectively, depends on the complexity of the geometry. A simplex (triangular/tetrahedral) mesh is preferred for finite element models, when the geometry has curved/complex boundaries as a simplex mesh approximates the curved/complex geometries better. Nevertheless, a quadrilateral/hexahedral mesh is preferred whenever possible due to it less computational complexity.

We now discuss a data structure that stores these meshes. In addition to the geometrical data such as vertices, edges, cells, etc., a collection of joints and cells are stored in a finite element mesh data structure. The joint collection contains a set of vertices, edges, triangles/quadrilaterals in 1d, 2d and 3d, respectively. Similarly, a collection of cells contains a set of edges, triangles/quadrilaterals and tetrahedra/hexahedra in 1d, 2d and 3d, respectively, see Table 9.1 for details. These information need to be stored with appropriate details. For example, a joint in a finite element mesh data structure should have the information of all cells that contain this joint, that is, the neighbouring cells of the joint. Further, the joint should contain a marker to classify it as an inner or boundary joint with a boundary type. The boundary joint can further

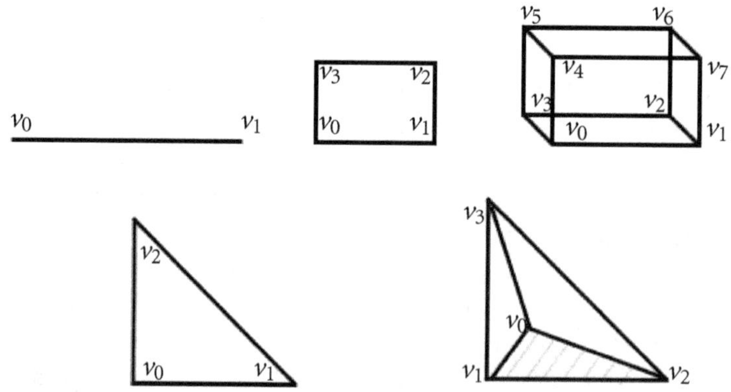

FIGURE 9.1 A description of finite element cells in one, two and three dimensions. A joint and a cell in 1d interval are defined as $J_0 = \{v_0\}$ and $K = \{v_0, v_1\}$, in 2d quadrilateral $J_0 = \{v_0, v_1\}$ and $K = \{v_0, v_1, v_2, v_3\}$, in 2d triangle $J_0 = \{v_0, v_1\}$ and $K = \{v_0, v_1, v_2\}$, in 3d tetrahedron $J_0 = \{v_0, v_1, v_2\}$ and $K = \{v_0, v_1, v_2, v_3\}$, and in 3d hexahedron $J_0 = \{v_0, v_1, v_2, v_3\}$ and $K = \{v_0, v_1, v_2, v_3, v_4, v_5, v_6, v_7\}$.

Table 9.1 Joints and cells in a finite element mesh.

Dimension	Joints	Cells
1	Vertices	Intervals
2	Intervals	Triangles/quadrilaterals
3	Triangles/quadrilaterals	Tetrahedra/hexahedra

be classified into straight line (plane in 3d) and curved boundary joint in order to use isoparametric mapping on the cell that contains this curved joint. Similarly, the data structure of a cell should contain the information of all its vertices and joints. In addition, an array of clipboard variables to mark the cell, joints and vertices are defined.

In object-oriented finite element codes, the joint class and the cell class are defined first, and then each joint in the collection of joints and each cell in the collection of cells are derived as instances of their respective classes. For the definition of class and object, and examples of the classes in the context of finite elements, see Section 9.7.

Construction of a finite element space is one of the key steps in a finite element package. This module is also known as a dof manager, and it is the core of any finite element package. The main purpose of the dof manager is to map the local dofs in a cell to the global dofs. The mapping from the local dofs in a cell to the global dofs of the mesh needs to be constructed only once at the beginning of the computations even for computations on deforming meshes. Nevertheless, the mapping has to be reconstructed when the domain is adaptively refined or remeshed.

In object-oriented finite element packages, a finite element class with a set of variables to store the information of the finite element type (conforming/nonconforming/discontinuous), number of dofs, number of interior dofs, number of dofs on each joint, etc. is constructed. After that each finite element is stored as an instance of the finite element class. For example, the finite element data for continuous piecewise bilinear and biquadratic elements are given in Table 9.2. Using these data, Q_1 and Q_2 finite elements are derived as instances of the finite element class. Similarly, all other finite elements, P_k, $P_k^{\rm nc}$ $P_k^{\rm disc}$, Q_k, $Q_k^{\rm nc}$, $k = 0, 1, 2, \ldots$, can be implemented as instances of the finite element class.

We next discuss the construction of a finite element space using finite elements. Let us consider the continuous piecewise bilinear element Q_1 on a quadrilateral mesh containing four cells. Construction of a finite element space is nothing but allocating a finite element on each

Table 9.2 Finite element data for the finite element class in 2d.

Finite element	Type	No. of dof	No. of inner dof	No. of joint dof
Q_1	Conforming	4	0	2
Q_2	Conforming	9	1	3

cell and numbering the global dofs. In the example considered, first allocate Q_1 finite element on each cell, see Figure 9.2 (left), and then use the information stored in the Q_1 finite element to number global dofs. Often, the inner dofs are numbered first, and then the dofs on the boundary are numbered. In particular, the Dirichlet dofs are numbered at the end to set the Dirichlet values efficiently. To identify the total number of dofs in the mesh for the used finite element, first loop over all cells, and reset all clipboard variables in each finite element cell. After that compute the uncounted dofs in each cell by looping over all cells, see Algorithm 1 for details.

Once the total number of global dofs is computed, a algorithm similar to that of Algorithm 1 can be used to number the global dofs. The indexed local dofs and the global dofs are shown in Figure 9.2. The same logic can be followed for all other mesh types to index local and global dofs.

Algorithm 1: Computing the number of global dofs.

```
n_global_dof=0
for i=0, i< n_cells do
    n_global_dof += n_interior_dof_in_cell_i

    if cell_i_fe_type==conforming then
        for j=0, j< n_vertices_in_cell_i do
            if vertex_j_clipboard==false then
                n_global_dof += 1
                vertex_j_clipboard = true

    for j=0, j< n_joints_in_cell_i do
        if joint_j_clipboard==false then
            if cell_i_fe_type==conforming then
                n_global_dof += n_joint_dof - n_vertices_in_joint
            else
                n_global_dof += n_joint_dof
            joint_j_clipboard = true
```

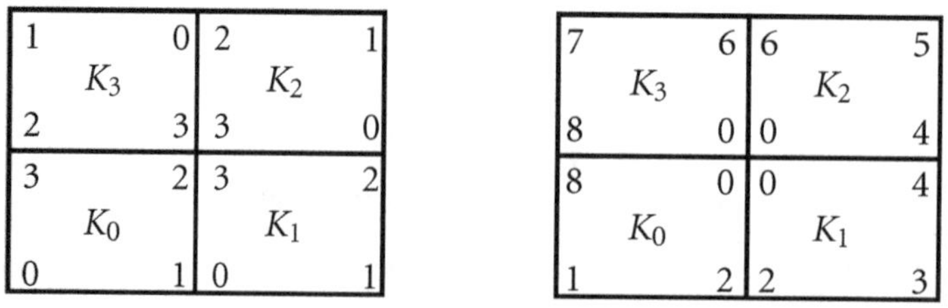

FIGURE 9.2 The finite element data structure of local (left) and global (right) dof numbering in a dof manager for Q_1 finite element on a quadrilateral mesh.

Next, the cell dof mapping can be stored in the dof manager using two one-dimensional arrays. For example, the cell dof mapping for the finite element data given in Figure 9.2 can be defined by

$$\begin{aligned}\texttt{cell\_dof\_ptr[]} &= \{0, 4, 8, 12, 16\},\\ \texttt{global\_dof[]} &= \{1, 2, 0, 8,\ \ 2, 3, 4, 0,\ \ 4, 5, 6, 0,\ \ 6, 7, 8, 0\}.\end{aligned}$$

Note that these arrays are defined in C++ style, that is, the numbering and the indices of the arrays start from 0. The first array, `cell_dof_ptr[]` stores the number of local dofs in each cell and the starting index of each cell in the `global_dof[]` array. The `global_dof[]` array stores the global dof of a local dof cellwise. For example, the number of local dofs in the cell K_1 is 4, that is, `cell_dof_ptr[2]` − `cell_dof_ptr[1]`, and its mapping to the global dofs starts at the position 4 , that is, at `global_dof[cell_dof_ptr[1]]` array. During computations, the global dofs of local dofs can be retrieved using Algorithm 2.

Algorithm 2: Retrieving global dofs of local dofs.

```
for i=0, i< n_cells do
    start = cell_dof_ptr[i]
    end = cell_dof_ptr[i+1]
    n_loc = end - start

    for j=start, j<end do
        dof = global_dof[j]
        ...
        ...
```

In addition to the cell dof mapping, joint dof mapping is needed to evaluate boundary and interface integrals. Analogue to the cell dof mapping, the local to global dof mapping for joints (edges in 2d and faces in 3d) can be constructed. For instance, the joint dof mapping for the finite element data given in Figure 9.2 can be defined by

$$\begin{aligned}\texttt{joint\_dof\_ptr[]} &= \{0, 2, 4, \ldots, 24\},\\ \texttt{global\_joint\_dof[]} &= \{1, 2,\ \ 2, 3,\ \ 3, 4, \ldots, 8, 1\}.\end{aligned}$$

9.2. NUMERICAL INTEGRATION

Numerical integration is one of the key components of finite element matrix assembling, where the integral over each cell in the finite element mesh needs to be evaluated. Numerical evaluation of a definite integral of a function (integrand) is called numerical quadrature. Though the name quadrature relates to the construction of a square with an area equal to that of a circle, numerical evaluation of multiple integrals (higher dimensions) is also called numerical quadrature. In addition, the term cubature is also used in the literature, see Cools (2003), to denote numerical evaluation of triple integrals. Nevertheless, we call the numerical evaluation of an integral over a cell $K_k \subset \mathbb{R}^d$, $d = 1,2,3$, $0 \leq k < n\_cells$, as numerical quadrature in this sequel. In this section, the standard quadrature formulas that are often used in finite element computations are provided without derivations, whereas the interested readers can refer the books on numerical analysis, see Gupta (2015) and Press et al. (2007) for more details.

In general, the integrals in finite element discrete forms are evaluated over a reference cell $\hat{K}$ instead over the original cell K_k using a suitable transformation, $F_{K_k} : \widehat{K} \to K_k$, that is,

$$\int_{K_k} f(x)\,\mathrm{d}x = \int_{\hat{K}} \hat{f}(\hat{x})\,\det DF_{K_k}(\hat{x})\,\mathrm{d}\hat{x},$$

see Chapter 3 for different transformation formulas and for integrands involving derivatives. Note that $\det DF_{K_k}$ will be a constant when the transformation is affine. In general, the numerical approximation of the above integral is obtained by a weighted sum of its function values, that is, using a quadrature

$$\int_{K_k} f(x)\,\mathrm{d}x = \int_{\hat{K}} \hat{f}(\hat{x})\,\det DF_{K_k}(\hat{x})\,\mathrm{d}\hat{x} \approx \sum_{q=0}^{NQ-1} w_q \hat{f}(\hat{x}_q)\,\det DF_{K_k}(\hat{x}_q),$$

where w_q are the weights of the quadrature, $\hat{x}_q$ the quadrature points and NQ the number of quadrature points. In the finite element context, the integrand $f(x)$ is defined by a combination of finite element basis functions $\{\phi_i(x)\}$, their derivatives and functions represent the data of the problem. For example, the time derivative in a scalar problem results in an integrand, $f(x) = \phi_i(x)\phi_j(x)$, and consequently it is evaluated by

$$\begin{aligned}\int_{K_k} \phi_i(x)\phi_j(x)\,\mathrm{d}x &= \int_{\hat{K}} \hat{\phi}_i(\hat{x})\hat{\phi}_j(\hat{x})\,\det DF_{K_k}(\hat{x})\,\mathrm{d}\hat{x} \\ &\approx \sum_{q=0}^{NQ-1} w_q \hat{\phi}_i(\hat{x}_q)\hat{\phi}_j(\hat{x}_q)\,\det DF_{K_k}(\hat{x}_q).\end{aligned}$$

Note that the reference basis functions, $\hat{\phi}_i(\hat{x}_q)$, and its derivatives do not vary in the integral over $\hat{K}$, whereas only $\det DF_{K_k}(\hat{x}_q)$ changes for different original cells K_k, $0 \leq k < n\_cells$, in the mesh. Therefore, in order to evaluate the integrals over all cells K_k, it is enough to evaluate $\det DF_{K_k}(\hat{x}_q)$ for each cell, whereas the values of basis functions and/or its derivatives at the quadrature points of a particular quadrature formula are evaluated only once. It is one of the advantages of using mapped finite elements.

Numerical quadrature for integrals over $[-1,1]^d$

The reference cell in 1d is the interval $\hat{K} = [-1, 1]$. The standard Newton–Cotes quadrature formulas such as Trapezoidal, Simpson's rules, midpoint rule, etc. or composite rules using Newton–Cotes quadrature can be used. The Newton–Cotes quadratures use equidistant quadrature points and are derived by interpolating polynomials. It is wellknown (Runge's phenomenon) that an oscillation occurs in the vicinity of the endpoints when polynomials of higher order over a set of equidistant interpolation points are used in polynomial interpolation. Therefore, methods such as Gaussian quadrature and Clenshaw–Curtis quadrature with unequally spaced abscissas are generally preferred. The Gaussian quadrature exactly integrates polynomials of order upto $2N+1$ with $NQ=N+1$, quadrature points, whereas the Clenshaw–Curtis quadrature exactly integrates polynomials of order up to N only. Even though there are advantages of using Clenshaw–Curtis quadratures for nonpolynomial integrands, the Gaussian quadrature is preferred in finite element computations as the finite element basis functions are polynomials in general. Hence, we discuss only the Gaussian quadrature here.

In general, the weights and quadrature points of the Gaussian quadratures are derived by using orthogonal polynomials, see Press et al. (2007). The Legendre polynomials are most commonly used, and the resulting quadrature is called Gauss–Legendre quadrature or simple Gaussian quadrature formula. Other orthogonal polynomials such as Chebyshev, Laguerre, Hermite and Jacobi can also be used, see Press et al. (2007). Moreover, Gaussian quadratures can also be extended with preassigned nodes, that is, fixing some quadrature points and determining the remaining quadrature points and the weights. The most common extensions of Gaussian quadratures are the Gauss–Radau quadrature, where one of the endpoints is a quadrature point, and Gauss–Lobatto quadrature, where both endpoints are in the list of quadrature points. The Gaussian quadratures for exact integral of polynomials of order up to 7 over the reference cell $\hat{K}=[-1,1]$ are given in Table 9.3. Note that the quadrature points of all quadrature are inside the reference cell $\hat{K}$.

An extension to multidimensional integrals is straightforward. Multidimensional quadrature formulas for integral over $\hat{K}=[-1,1]^d$, $d>1$ can be obtained by a tensor product of 1d

Table 9.3 Gauss–Legendre quadrature for $\hat{K}=[-1,1]$ which are exact for polynomials of order k.

NQ	k	$\hat{x}_i$	w_i
2	3	± 0.57735026919	1.00000000000
3	5	0	0.88888888888
		±0.77459666924	0.55555555555
4	7	±0.33998104358	0.65214515486
		±0.86113631159	0.34785484513
5	9	0	0.56888888888
		±0.53846931011	0.47862867049
		±0.90617984594	0.23692688506
6	11	±0.23861918608	0.46791393457
		±0.66120938647	0.36076157305
		±0.93246951420	0.17132449238
7	13	0	0.41795918367
		±0.40584515138	0.38183005051
		±0.74153118560	0.27970539149
		±0.94910791234	0.12948496617

formulas given in Table 9.3. For instance, two point Gaussian quadrature with nodes $-1/\sqrt{3}$ and $1/\sqrt{3}$ in 1d becomes four point quadrature with nodes $(-1/\sqrt{3}, -1/\sqrt{3})$, $(1/\sqrt{3}, -1/\sqrt{3})$, $(1/\sqrt{3}, 1/\sqrt{3})$ and $(-1/\sqrt{3}, 1/\sqrt{3})$ in 2d. It can be generalized to multidimensional cells, for example, in 3d we can evaluate

$$\begin{aligned}\int_{-1}^{1}\int_{-1}^{1}\int_{-1}^{1} f(x,y,z)\,\mathrm{d}x\,\mathrm{d}y\,\mathrm{d}z &\approx \int_{-1}^{1}\int_{-1}^{1}\left(\sum_{n=0}^{NQ-1} w_n f(x_n,y,z)\right)\mathrm{d}y\,\mathrm{d}z \\ &\approx \int_{-1}^{1}\left(\sum_{m=0}^{NQ-1}\sum_{n=0}^{NQ-1} w_m w_n f(x_n,y_m,z)\right)\mathrm{d}z \\ &\approx \sum_{\ell=0}^{NQ-1}\sum_{m=0}^{NQ-1}\sum_{n=0}^{NQ-1} w_\ell w_m w_n f(x_n,y_m,z_\ell).\end{aligned}$$

Numerical quadrature for simplex

The derivation of quadrature formulas for triangles and tetrahedrons is not straightforward as in the case of quadrilaterals or hexahedra, since a higher order nontensor quadrature formula for a simplex is difficult to construct. In particular, the construction of quadrature formulas for a simplex with nonnegative weights and quadrature points inside the simplex is very challenging.

The midpoint and the Gaussian quadrature for the reference triangle (0,0), (1,0) and (0,1) are presented in Table 9.4. Based on nonlinear least square approach, an array of quadrature formulas up to order 21 on triangles and up to order 14 on tetrahedra have been presented by Zhang et al. (2009). In addition, a collection of quadrature formulas for triangles and tetrahedron can also be found at *http://nines.cs.kuleuven.be/ecf*, see Cools (2003).

Table 9.4 Midpoint and Gaussian quadrature for the reference triangle which are exact for polynomials of order k.

NQ	k	$\hat{x}_i$	$\hat{y}_i$	w_i
3	2	0.5	0	0.16666666667
		0.5	0.5	0.16666666667
		0	0.5	0.16666666667
7	5	0.33333333333	0.33333333333	0.1125
		0.79742698535	0.10128650732	0.0629695902724
		0.10128650732	0.79742698535	0.0629695902724
		0.10128650732	0.10128650732	0.0629695902724
		0.05971587179	0.47014206411	0.0661970763943
		0.47014206411	0.05971587179	0.0661970763943
		0.47014206411	0.47014206411	0.0661970763943

Remark 9.1. In computations, the quadrature formulas need to be chosen optimally. During finite element matrix assembling, each cell uses the quadrature formula and in each cell the coefficients and the basis functions and its derivatives need to be evaluated at each quadrature point. Hence, the choice of a higher order quadrature formula results in higher computational costs. On the other hand, the use of a low order quadrature formula induces an additional error which might dominates the finite element discretization error. For more details on tailored use of quadrature formulas we refer to Section 4.5.3 of Grossmann et al. (2007).

9.3. Sparse matrix storage

Matrices generated from finite element discretizations are sparse due to the local support of the finite element basis functions. A matrix is said to be sparse if the number of nonzero entries in the matrix is much smaller than the total number of entries in the matrix. If the other extreme happens (number of nonzero entries is close to the total number of entries), the matrix is said to be dense. The fraction between the number of zero entries and the total number of entries in a matrix is called the sparsity of the matrix. Although the sparsity of the finite element matrices depends on many different facts, it will always be very large, and hence a sparse matrix storage is important for the efficiency of any finite element package.

There are many methods for storing a sparse matrix, for example, the compressed row storage, the compressed column storage, the block compressed row storage, the diagonal storage and the jagged diagonal storage to mention some of them, see Saad (2003). In this section, we explain the basic principles on how to store sparse matrices in an efficient and robust way.

Compressed row storage (CRS) or compressed sparse row (CSR) format

This is the most general and commonly used storage format for a sparse matrix. It does not require any assumption on the sparsity structure of the matrix. For the representation of a matrix in CSR format, a floating point array `val[]` and two integer arrays `col_ind[]`, `row_ptr[]` need to be created. The `val[]` array stores the nonzero entries of the matrix rowwise, whereas the column indices corresponding to nonzero entries are stored in the `col_ind[]` array, that is, if `val[k]` $= a_{\mathtt{ij}}$ then `col_ind[k]=j`. Further, the `row_ptr[]` array stores a list of indices (locations) in `val[]` and `col_ind[]` at which each row starts, that is, if `val[k]` $= a_{\mathtt{ij}}$ is a nonzero entry of the $\mathtt{i}^{th}$-row then `row_ptr[i]` $\leq$ `k` $\leq$ `row_ptr[i+1]`. To make things clear let us create the CSR format for the matrix

$$A = \begin{bmatrix} 4 & 5 & & & 3 & 2 \\ & 6 & & & 8 & \\ & 9 & 7 & 8 & & \\ & & & 9 & -2 & \\ 4 & 9 & & & 1 & \\ 2 & & & & 5 & 3 \end{bmatrix}.$$

The nonzero entries in A are stored rowwise in the `val[]` array. Further, the original column indices of all these nonzero entries are stored in `col_ind[]`, and the locations of `val[]` and `col_ind[]` at which each row starts are stored in `row_ptr[]`. Hence, we have

$$\texttt{val[]} = \{\underbrace{4,5,3,2,}_{\texttt{row 0}}\ \underbrace{6,8,}_{\texttt{row 1}}\ \underbrace{9,7,8,}_{\texttt{row 2}}\ \underbrace{9,-2,}_{\texttt{row 3}}\ \underbrace{4,9,1,}_{\texttt{row 4}}\ \underbrace{2,5,3}_{\texttt{row 5}}\},$$

$$\texttt{col\_ind[]} = \{\underbrace{0,1,4,5,}_{\texttt{row 0}}\ \underbrace{1,4,}_{\texttt{row 1}}\ \underbrace{1,2,3,}_{\texttt{row 2}}\ \underbrace{3,4,}_{\texttt{row 3}}\ \underbrace{0,1,4,}_{\texttt{row 4}}\ \underbrace{0,4,5}_{\texttt{row 5}}\},$$

$$\texttt{row\_ptr[]} = \{0,4,6,9,11,14,17\}.$$

Note that in the CSR format we do not need any specific sparsity pattern of the matrix and store only the nonzero entries of the matrix in the `val[]` array.

Compressed column storage (CCS) or compressed sparse column (CSC) format

This format for a sparse matrix is analogous to the CRS. Instead of the row traversal as in CSR format, the columns of the matrix is traversed and the row positions are stored in CSC format. As in the CSR format, a floating point array `val[]` and two integer arrays (`row_ind[], column_ptr[]`) are needed to generate the CSC format. The CSC format for the above matrix will be

$$\texttt{val[]} = \{4,4,2,\ 5,6,9,9,\ 7,\ 8,9,\ 3,8,-2,1,5,\ 2,3\},$$

$$\texttt{row\_ind[]} = \{0,4,5,\ 0,1,2,4,\ 2,\ 2,3,\ 0,1,3,4,5,\ 0,5\},$$

$$\texttt{column\_ptr[]} = \{0,3,7,8,10,15,17\}.$$

Note that the CSC format is the CSR format of the transposed matrix A^T.

Block compressed row storage (BCRS) format

This format can be used successfully when the nonzero entries in the sparse matrix are arranged in a regular pattern of dense squared blocks. A system matrix with such a block structure arises, for example, from the finite element discretization of vector-valued quantities like velocity and pressure in the 2d or 3d Navier–Stokes equations, see Section 8.8. The BCRS format becomes more efficient when the dimension of the blocks increases.

Let the dimension of each nonzero block in the matrix be n_b and the number of nonzero blocks be m. Then the row block dimension n_d can be defined as n/n_b, where n is number of rows in the sparse matrix. In order to represent a sparse block matrix in the BCRS format, three arrays are needed. A rectangular floating point array `val`$[m]$ $[n_b \times n_b]$ to store the nonzero blocks in rowwise, an integer array `col_ind`$[m]$ to store the original column index of the first entry in each nonzero block, and an integer array `row_ptr`$[n_d + 1]$ to store a list of indices (locations) in `val[]` and `col_ind[]` at which each block row starts. The BCRS format for the above matrix A will be

$$\texttt{val[][]} = \left\{ \begin{pmatrix} 4 & 5 \\ 0 & 6 \end{pmatrix}, \begin{pmatrix} 3 & 2 \\ 8 & 0 \end{pmatrix}, \begin{pmatrix} 9 & 7 \\ 0 & 0 \end{pmatrix}, \begin{pmatrix} 8 & 0 \\ 9 & -2 \end{pmatrix}, \begin{pmatrix} 4 & 9 \\ 2 & 0 \end{pmatrix}, \begin{pmatrix} 1 & 0 \\ 5 & 3 \end{pmatrix} \right\},$$

$$\texttt{col\_ind[]} = [0, 4, 1, 3, 0, 4],$$

$$\texttt{row\_ptr[]} = [0, 2, 4, 6].$$

In the matrix A, there are six blocks, and the starting column index of the first block is 0, the second block is 4, and so on. Further, there are three row blocks, that is, the block dimension $n_d = n/n_b = 6/2 = 3$. Note that, in contrast to the CRS and CCS format, a few zero entries are also stored in the BCRS format.

Matrix stencil of a finite element space

The finite element matrix stencil (sparsity pattern) depends on the choice of the finite element and on the underlying mesh. For allocating the right amount of storage for finite element sparse matrices, it is useful to know the matrix stencil. From the finite element and the finite element mesh we can determine the entries of the matrix which are potentially nonzero. Of course, we avoid to compute entries which are *a priori* zero (for example basis functions have no common support). For a better understanding, we consider the same example as in Figure 9.2, and construct a matrix stencil in CSR format. It is clear that the matrix resulting from this example will be a square matrix of size nine. However, the number of nonzero entries and the stencil of the matrix need to be determined.

In general, the number of nonzero entries, and the column and row indices of nonzero entries are needed to construct a sparse matrix stencil. In the finite element context, the number of nonzero entries in a row, i, is the number of dofs associated with the global dof, i. Further, the global dofs of the associated dofs are the column indices of the nonzero entries in row i. To collect these information, loop over all cells, and for each dof i in the cell append all dofs in the cell as associated dofs to a two-dimensional array `neib_dofs`$[i]$`[]`. For instance, let us consider the dof 0 in Figure 9.2. The set of all dofs $\{0, 1, 2, 8\}$, $\{3, 4\}$, $\{5, 6\}$ and $\{7\}$, from cells $K_0, \ldots, K_3$, respectively, are appended as associated dofs of dof 0. On the completion of the loop over all cells and for all dofs, we get

$$\texttt{neib\_dofs[0][]} = \{0, 1, \ldots, 7, 8\}, \qquad \texttt{neib\_dofs[1][]} = \{0, 1, 2, 8\},$$

$$\texttt{neib\_dofs[2][]} = \{0, 1, 2, 3, 4, 8\}, \qquad \texttt{neib\_dofs[3][]} = \{0, 2, 3, 4\},$$

$$\texttt{neib\_dofs[4][]} = \{0, 2, 3, 4, 5, 6\}, \qquad \texttt{neib\_dofs[5][]} = \{0, 4, 5, 6\},$$

$$\texttt{neib\_dofs[6][]} = \{0, 4, 5, 6, 7, 8\}, \qquad \texttt{neib\_dofs[7][]} = \{0, 6, 7, 8\},$$

$$\texttt{neib\_dofs[8][]} = \{0, 1, 2, 6, 7, 8\}.$$

The number of elements in `neib_dofs`$[i]$`[]` array is the number of nonzero entries in the row i, whereas the elements are the column indices of the nonzero entries in the row. Once these information are collected, the matrix stencil can be created in the CSR form as follows

$$\texttt{row\_ptr[]} = \{0, 9, 13, 19, 23, 29, 33, 39, 43, 49\},$$

$$\texttt{col\_ind[]} = \{0, 1, 2, 3, 4, 5, 6, 7, 8, \quad \ldots \quad 0, 1, 2, 6, 7, 8\}.$$

Here, the number of nonzero entries is 49, which is the last element in the `row_ptr[]` array. In object-oriented programming codes, a matrix class is constructed with the `col_ind[]` and the `row_ptr[]` arrays, and matrices are derived as an instance of the matrix class. For example, mass matrix, stiffness matrix, etc. are instances of the matrix class.

Retrieving a nonzero entry from a CSR format matrix

Assume that for a given pair of indices (i, j), the matrix entry a_{ij} is nonzero. The question is how to retrieve the value a_{ij} from the `val` array in the CSR format. Let the matrix be stored in the CSR format with the `val[]`, `col_ind[]` and `row_ptr[]` array. Then, the desirable operation is coded in Algorithm 3.

Algorithm 3: Retrieving $a_{ij} \neq 0$ in CSR format.

```
for (m= row_ptr[i]; m< row_ptr[i+1]; m++) do
    if (col_ind[m]== j) then
        a_ij = val[m]
        break
```

Matrix vector multiplication

The iterative solution of the algebraic system of equations is based on the repeated matrix vector multiplication. Therefore, it is important to be able to do this operation efficiently on the matrices stored in sparse format. Algorithm 4 shows how a sparse matrix vector multiplication can be coded. Note that only the necessary number of operations are performed.

Algorithm 4: Matrix vector multiplication in CSR format, $y = Ax$.

```
for (i= 0; i< N; i++) do
    y[i] = 0
    for (j= row_ptr[i]; j< row_ptr[i+1]; j++) do
        k=col_ind[j]
        y[i] += val_A[j]*x[k]
```

9.4. Assembling of system matrices and load vectors

Matrix assembling is one of the major steps in finite element computations. For the heat equation, considered in Chapter 6, for example, the matrix arising from the time derivative term is called mass matrix, whereas the matrix arising from the diffusion terms is called stiffness matrix. Further, the column matrix that includes the source term is called load vector. Finally, the system matrix is a combination of all these matrices.

Matrix assembling can be performed in two ways: (i) nodal based assembling, and (ii) cell based assembling. In the *nodal based assembling*, the non-zero entries a_{ij} are computed in one step by evaluating all contributions from all cells in which ϕ_i and ϕ_j have support. Thus, the integral over a same cell needs to be evaluated repeatedly for each a_{ij} associated with that cell. Even though the nodal based approach seems to be straightforward, it is computationally inefficient. Instead, the integral over a cell can be evaluated only once to compute and add the local contributions from the cell to all a_{ij} associated with it, and this approach is known as *cell based assembling*. This approach consists of two steps: first the local matrix of each cell is computed and then the local matrix is added to the global matrix using the local to global dof mapping defined in the dof manager. The system matrix assembling for scalar and vector problems are discussed in detail below.

Assembling the system matrix of a scalar problem

Let us consider the Poisson problem

$$-\Delta u = f \quad \text{in } \Omega, \qquad u = 0 \quad \text{on } \partial\Omega$$

as an example of the scalar problem. Let the finite dimensional space $V_h \subset H_0^1(\Omega)$ be spanned by the basis functions $\{\phi_i\}$, $i = 0, 1, \ldots, N-1$. Applying now the standard Galerkin finite element discretization to the Poisson problem, see Section 2.3, results in the system

$$\sum_{j=0}^{N-1} a_{ij}\, u_j = f_i, \quad i = 0, 1, \ldots, N-1,$$

where u_j, $j = 0, 1, \ldots, N-1$, are the unknown solution coefficients. Further, the nonzero entries a_{ij} and f_j of the stiffness matrix and the load vector, respectively, are given by

$$a_{ij} = \sum_{k=0}^{n\_cells-1} \int_{K_k} \nabla\phi_j \cdot \nabla\phi_i \, dx, \qquad f_i = \sum_{k=0}^{n\_cells-1} \int_{K_k} f\, \phi_i \, dx,$$

Now, these nonzero entries are computed by using numerical quadrature on the reference element as discussed before. Applying affine transformation to $\nabla\phi_i$ and ϕ_j, the above integrals can be transformed over the reference cell as

$$\int_{K_k} \nabla\phi_j(x) \cdot \nabla\phi_i(x)\, dx = \int_{\hat{K}} \left(\hat{\nabla}\hat{\phi}_j^T (B_{K_k}^T B_{K_k})^{-1} \hat{\nabla}\hat{\phi}_i \right) \det B_{K_k}\, d\hat{x},$$

$$\int_{K_k} f(x)\, \phi_i(x)\, dx = \int_{\hat{K}} \hat{f}(\hat{x})\, \hat{\phi}_i(\hat{x}) \det B_{K_k}\, d\hat{x},$$

for more details, see Section 3.4. Applying now the quadrature to get

$$a_{ij} \approx \sum_{k=0}^{n\_cells-1} \sum_{q=0}^{NQ-1} w_q \det B_{K_k} \left(\hat{\nabla}\hat{\phi}_j^T(\hat{x}_q)(B_{K_k}^T B_{K_k})^{-1} \hat{\nabla}\hat{\phi}_i(\hat{x}_q) \right),$$

$$f_i \approx \sum_{k=0}^{n\_cells-1} \sum_{q=0}^{NQ-1} w_q \det B_{K_k} \hat{f}(\hat{x}_q)\hat{\phi}_i(\hat{x}_q).$$

As discussed earlier, the cell-based assembling is more efficient, and an algorithm for assembling the above stiffness matrix and the load vector is given in Algorithm 5. In the algorithm, a *for* loop executes over all cells, and in each cell the local matrices are assembled and then the local matrix is added to the global matrix.

Assembling the system matrix of a vector-valued problem

Let us consider a 2d Oseen problem

$$-\nu\Delta\mathbf{u} + (\mathbf{b}\cdot\nabla)\mathbf{u} + \sigma\mathbf{u} + \nabla p = \mathbf{f}, \qquad \operatorname{div}\mathbf{u} = 0 \quad \text{in } \Omega, \qquad \mathbf{u} = 0 \quad \text{on } \partial\Omega, \tag{9.1}$$

as an example of a vector-valued problem. Denote the velocity and source vector components as $\mathbf{u} = (u^{\mathrm{I}}, u^{\mathrm{II}})^T$ and $\mathbf{f} = (f^{\mathrm{I}}, f^{\mathrm{II}})^T$. Let the finite dimensional spaces for velocity and pressure be $V_h \subset (H_0^1(\Omega))^2$ and $Q_h \subset L_0^2(\Omega)$, respectively. Now, following the standard Galerkin procedure, the discrete form of the Oseen problem (9.1) reads
Find $(\mathbf{u}_h, p_h) \in V_h \times Q_h$ such that

$$a_h(\mathbf{u}_h, \mathbf{v}_h) - b_h(p_h, \mathbf{v}_h) = (\mathbf{f}, \mathbf{v}_h), \qquad b_h(q_h, \mathbf{u}_h) = 0,$$

where

$$a_h(\mathbf{u}_h, \mathbf{v}_h) = \int_\Omega \nu\nabla\mathbf{u}_h : \nabla\mathbf{v}_h\, dx + \int_\Omega (\mathbf{b}\cdot\nabla)\mathbf{u}_h\cdot\mathbf{v}_h\, dx + \int_\Omega \sigma\,\mathbf{u}_h\cdot\mathbf{v}_h\, dx,$$

$$b_h(q_h, \mathbf{v}_h) = \int_\Omega q_h \operatorname{div}\mathbf{v}_h\, dx.$$

Let $\{\phi_i\}$, $i = 0,\ldots,N-1$, be a basis for each component of the vector space V_h, and $\{\psi_m\}$, $m = 0,\ldots,M-1$, be a basis of Q_h. Further, define

$$\mathbf{u}_h(x) = (u_h^{\mathrm{I}}, u_h^{\mathrm{II}})^T = \left(\sum_{j=0}^{N-1} u_j^{\mathrm{I}}\phi_j(x), \sum_{j=0}^{N-1} u_j^{\mathrm{II}}\phi_j(x) \right)^T, \qquad p_h(x) = \sum_{n=0}^{M-1} p_n\psi_n.$$

Note that the same set of basis functions is used for both velocity components. Using these discrete functions, the system matrix of the Oseen problem can be obtained as

$$\begin{bmatrix} A_{11} & A_{12} & B_1^T \\ A_{21} & A_{22} & B_2^T \\ B_1 & B_2 & 0 \end{bmatrix} \begin{bmatrix} U^{\mathrm{I}} \\ U^{\mathrm{II}} \\ P \end{bmatrix} = \begin{bmatrix} F^{\mathrm{I}} \\ F^{\mathrm{II}} \\ 0 \end{bmatrix}, \tag{9.2}$$

Algorithm 5: Assembling of the load vector and stiffness matrix of the Poisson problem.

```
for (k=0; k< n_cells; k++) do
    start = cell_dof_ptr[k]
    end = cell_dof_ptr[k+1]
    n_loc = end - start   // number of local DOFs
    loc_f[] = 0
    loc_a[] = 0
    /* assemble local matrix                                  */
    for (q=0; q<NQ; q++) do
        // loop over all quadrature points
        fact = w[q] * det B_{K_k} // weight * Jacobian
        for (i=0; i<n_loc; i++) do
            d = i * n_loc
            loc_f[i] += fact * f̂[q] * φ̂[q][i]   // load vector
            for (j=0; j<n_loc; j++) do
                /* local stiffness matrix                     */
                val = fact * (φ̂_x̂[q][i] * φ̂_x̂[q][j] + φ̂_ŷ[q][i] * φ̂_ŷ[q][j])
                loc_a[d+j] += val
    /* add local to global matrix                             */
    for (i=0; i<n_loc; i++) do
        d = i * n_loc
        I = global_dof[start+i]   // global test dof
        f[I] += loc_f[i]   // load vector
        for (j= 0; j< n_loc; j++) do
            J = global_dof[start+j]   // global ansatz dof
            for (m= row_ptr[I]; m< row_ptr[I+1]; m++) do
                if (col_ind[m]== J) then
                    a[m] += loc_a[d+j]   // stiffness matrix
                    break
```

where the components $U^{\mathrm{I}} = \mathrm{vec}(u^{\mathrm{I}}{}_j)$ and $U^{\mathrm{II}} = \mathrm{vec}(u^{\mathrm{II}}{}_j)$ are the vectorization of the unknown solution coefficients of the velocity, and $P = \mathrm{vec}(p_n)$ is the pressure. Further, the entries of the block matrices are given by

$$\{A_{11}\}_{i,j} = \sum_{k=0}^{n\_\text{cells}-1} \int_{K_k} \left\{ \nu \nabla \begin{pmatrix} \phi_j \\ 0 \end{pmatrix} : \nabla \begin{pmatrix} \phi_i \\ 0 \end{pmatrix} + \left[\begin{pmatrix} b_1 \\ b_2 \end{pmatrix} \cdot \nabla \begin{pmatrix} \phi_j \\ 0 \end{pmatrix} \right] \cdot \begin{pmatrix} \phi_i \\ 0 \end{pmatrix} + \sigma \begin{pmatrix} \phi_j \\ 0 \end{pmatrix} \cdot \begin{pmatrix} \phi_i \\ 0 \end{pmatrix} \right\} \mathrm{d}x$$

$$= \sum_{k=0}^{n\_\text{cells}-1} \int_{K_k} \left\{ \nu \left(\frac{\partial \phi_j}{\partial x_1} \frac{\partial \phi_i}{\partial x_1} + \frac{\partial \phi_j}{\partial x_2} \frac{\partial \phi_i}{\partial x_2} \right) + \left[b_1 \frac{\partial \phi_j}{\partial x_1} + b_2 \frac{\partial \phi_j}{\partial x_2} \right] \phi_i + \sigma \phi_j \phi_i \right\} \mathrm{d}x,$$

$$\{A_{12}\}_{i,j} = \sum_{k=0}^{n\_cells-1} \int_{K_k} \left\{ \nu \nabla \begin{pmatrix} 0 \\ \phi_j \end{pmatrix} : \nabla \begin{pmatrix} \phi_i \\ 0 \end{pmatrix} + \left[\begin{pmatrix} b_1 \\ b_2 \end{pmatrix} \cdot \nabla \begin{pmatrix} 0 \\ \phi_j \end{pmatrix} \right] \cdot \begin{pmatrix} \phi_i \\ 0 \end{pmatrix} + \sigma \begin{pmatrix} 0 \\ \phi_j \end{pmatrix} \cdot \begin{pmatrix} \phi_i \\ 0 \end{pmatrix} \right\} \mathrm{d}x,$$

$$\{A_{21}\}_{i,j} = \sum_{k=0}^{n\_cells-1} \int_{K_k} \left\{ \nu \nabla \begin{pmatrix} \phi_j \\ 0 \end{pmatrix} : \nabla \begin{pmatrix} 0 \\ \phi_i \end{pmatrix} + \left[\begin{pmatrix} b_1 \\ b_2 \end{pmatrix} \cdot \nabla \begin{pmatrix} \phi_j \\ 0 \end{pmatrix} \right] \cdot \begin{pmatrix} 0 \\ \phi_i \end{pmatrix} + \sigma \begin{pmatrix} \phi_j \\ 0 \end{pmatrix} \cdot \begin{pmatrix} 0 \\ \phi_i \end{pmatrix} \right\} \mathrm{d}x,$$

$$\{A_{22}\}_{i,j} = \sum_{k=0}^{n\_cells-1} \int_{K_k} \left\{ \nu \nabla \begin{pmatrix} 0 \\ \phi_j \end{pmatrix} : \nabla \begin{pmatrix} 0 \\ \phi_i \end{pmatrix} + \left[\begin{pmatrix} b_1 \\ b_2 \end{pmatrix} \cdot \nabla \begin{pmatrix} 0 \\ \phi_j \end{pmatrix} \right] \cdot \begin{pmatrix} 0 \\ \phi_i \end{pmatrix} + \sigma \begin{pmatrix} 0 \\ \phi_j \end{pmatrix} \cdot \begin{pmatrix} 0 \\ \phi_i \end{pmatrix} \right\} \mathrm{d}x,$$

$$= \sum_{k=0}^{n\_cells-1} \int_{K_k} \left\{ \nu \left(\frac{\partial \phi_j}{\partial x_1} \frac{\partial \phi_i}{\partial x_1} + \frac{\partial \phi_j}{\partial x_2} \frac{\partial \phi_i}{\partial x_2} \right) + \left[b_1 \frac{\partial \phi_j}{\partial x_1} + b_2 \frac{\partial \phi_j}{\partial x_2} \right] \phi_i + \sigma \phi_j \phi_i \right\} \mathrm{d}x,$$

$$\{B_1\}_{m,j} = \sum_{k=0}^{n\_cells-1} \int_{K_k} \nabla \cdot \begin{pmatrix} \phi_j \\ 0 \end{pmatrix} \psi_m \,\mathrm{d}x = \sum_{k=0}^{n\_cells-1} \int_{K_k} \frac{\partial \phi_j}{\partial x_1} \psi_m \,\mathrm{d}x,$$

$$\{B_2\}_{m,j} = \sum_{k=0}^{n\_cells-1} \int_{K_k} \nabla \cdot \begin{pmatrix} 0 \\ \phi_j \end{pmatrix} \psi_m \,\mathrm{d}x = \sum_{k=0}^{n\_cells-1} \int_{K_k} \frac{\partial \phi_j}{\partial x_2} \psi_m \,\mathrm{d}x,$$

$$\{F^{\mathrm{I}}\}_i = \sum_{k=0}^{n\_cells-1} \int_{K_k} \begin{pmatrix} f^{\mathrm{I}} \\ 0 \end{pmatrix} \cdot \begin{pmatrix} \phi_i \\ 0 \end{pmatrix} \mathrm{d}x = \sum_{k=0}^{n\_cells-1} \int_{K_k} f^{\mathrm{I}} \phi_i \,\mathrm{d}x,$$

$$\{F^{\mathrm{II}}\}_i = \sum_{k=0}^{n\_cells-1} \int_{K_k} \begin{pmatrix} 0 \\ f^{\mathrm{II}} \end{pmatrix} \cdot \begin{pmatrix} 0 \\ \phi_i \end{pmatrix} \mathrm{d}x = \sum_{k=0}^{n\_cells-1} \int_{K_k} f^{\mathrm{II}} \phi_i \,\mathrm{d}x.$$

It has to be noted that the above given off-diagonal block matrices, A_{12} and A_{21} will be zero. Further, the matrices A_{11} and A_{22} are equal, since the used basis functions for both velocity components are same. Since $A = A_{11} = A_{22}$ and $A_{12} = A_{21} = 0$, it is sufficient to assemble A, B_1, B_2, F^{I} and F^{II} and system (9.2) reduces to

$$\begin{bmatrix} A & 0 & B_1^T \\ 0 & A & B_2^T \\ B_1 & B_2 & 0 \end{bmatrix} \begin{bmatrix} U^{\mathrm{I}} \\ U^{\mathrm{II}} \\ P \end{bmatrix} = \begin{bmatrix} F^{\mathrm{I}} \\ F^{\mathrm{II}} \\ 0 \end{bmatrix}.$$

The algorithm for assembling the above matrices and the load vectors of the Oseen problem (9.1) is given in Algorithm 6.

It has to be noted that, these off-diagonal block matrices A_{12} and A_{21} will not be zero in the stress tensor form of the Oseen problem. Further, the diagonal block matrix A_{11} will be different from A_{22} in the stress tensor form. Therefore, all block matrices in (9.2) need to be assembled when the stress tensor form of the Stokes or Oseen or Navier–Stokes equations are considered.

Algorithm 6: Assembling of the load vector and the block matrices of the Oseen problem.

for *(k=0; k< n_cells; k++)* **do**
  n_u_loc = cell_u_dof_ptr$[k+1]$ - cell_u_dof_ptr$[k]$ `// local u DOFs`
  n_p_loc = cell_p_dof_ptr$[k+1]$ - cell_p_dof_ptr$[k]$ `// local p DOFs`
  $loc\_f1[\,] = 0;\ loc\_f2[\,] = 0;\ loc\_a[\,] = 0$
  $loc\_b1[\,] = 0;\ loc\_b2[\,] = 0$
  `/* assemble local matrix */`
  for *(q=0; q<NQ; q++)* **do**
    `// loop over all quadrature points`
    $fact = w[q] * \det DF_{K_k}[q]$ `// weight * Jacobian at quadrature`
    for *(i=0; i<n_u_loc; i++)* **do**
      $d = i * n\_u\_loc$
      $loc\_f1[i] \mathrel{+}= fact * f^{I}[q] * \hat{\phi}[q][i]$ `// load vector`
      $loc\_f2[i] \mathrel{+}= fact * f^{II}[q] * \hat{\phi}[q][i]$ `// load vector`
      for *(j=0; j<n_u_loc; j++)* **do**
        `/* local stiffness matrix */`
        $val = fact * \nu * \left(\hat{\phi}_{\hat{x}}[q][i] * \hat{\phi}_{\hat{x}}[q][j] + \hat{\phi}_{\hat{y}}[q][i] * \hat{\phi}_{\hat{y}}[q][j]\right)$
        $val \mathrel{+}= \left(b_1[q] * \hat{\phi}_{\hat{x}}[q][j] + b_2[q] * \hat{\phi}_{\hat{y}}[q][j]\right)\hat{\phi}[q][i]$
        $val \mathrel{+}= \sigma * \hat{\phi}[q][j] * \hat{\phi}[q][i]$
        $loc\_a[d+j] \mathrel{+}= val$
    for *(m=0; m<n_p_loc; m++)* **do**
      $d = m * n\_p\_loc$
      for *(j=0; j<n_u_loc; j++)* **do**
        $loc\_b1[d+j] = fact * \hat{\phi}_{\hat{x}}[q][j] * \hat{\psi}[q][m]$
        $loc\_b2[d+j] = fact * \hat{\phi}_{\hat{y}}[q][j] * \hat{\psi}[q][m]$
  `/* add local to global matrices */`
  ...

9.5. Inclusion of boundary conditions

Once the matrices in the finite element system are assembled, the boundary conditions need to be incorporated into the system before employing the algebraic solver. In this section, we discuss the inclusion of different boundary conditions in scalar and vector-valued problems (Oseen equations).

Inclusion of Dirichlet type boundary conditions

We here discuss the inclusion of inhomogeneous Dirichlet boundary condition in the 2d Oseen problem. Suppose k^{th} dof is a Dirichlet type and its value is nonzero, that is, $\mathbf{u}_k = \mathbf{g} := (g_1, g_2)^T$. Recall the definition of the discrete velocity

$$\mathbf{u}_h(x) = \left(u_h^{\mathrm{I}}, u_h^{\mathrm{II}}\right)^T = \left(\sum_{j=0}^{N-1} u_j^{\mathrm{I}}\phi_j(x), \sum_{j=0}^{N-1} u_j^{\mathrm{II}}\phi_j(x)\right)^T.$$

We now need to modify the system (9.2) in such a way that we should get $u_k^{\mathrm{I}} = g_1$ and $u_k^{\mathrm{II}} = g_2$ after solving the system (9.2). In general, Dirichlet type boundary conditions can be incorporated into the system in two ways. One approach is to modify the kth rows of both velocity components in the system (9.2) as follows:

$$\begin{array}{r} \\ k^{\text{th}}\text{ row of } u^{\mathrm{I}} \rightarrow \\ \\ \\ \\ k^{\text{th}}\text{ row of } u^{\mathrm{II}} \rightarrow \\ \\ \end{array}
\begin{bmatrix} \ddots & \vdots & \ddots & \vdots & \ddots & \vdots & \vdots \\ 0 & 1 & 0 & 0 & 0 & 0 & 0 \\ \ddots & \vdots & \ddots & \vdots & \ddots & \vdots & \vdots \\ & & & & & & \\ \vdots & \ddots & \vdots & \ddots & \vdots & \ddots & \vdots \\ 0 & 0 & 0 & 0 & 1 & 0 & 0 \\ \vdots & \ddots & \vdots & \ddots & \vdots & \ddots & \vdots \end{bmatrix}
\begin{bmatrix} \vdots \\ u_k^{\mathrm{I}} \\ \vdots \\ \\ \vdots \\ u_k^{\mathrm{II}} \\ \vdots \end{bmatrix}
=
\begin{bmatrix} \vdots \\ g_1 \\ \vdots \\ \\ \vdots \\ g_2 \\ \vdots \end{bmatrix}.$$

Here, for each velocity component, the diagonal entry of the respective k^{th} row is replaced by 1, and all other entries, including the entries of B_1^T and B_2^T, of the row are reset to zero. Further, the respective load vector row values are also replaced with the respective Dirichlet values.

Alternatively, the system matrix system (9.2) can be solved only for the non-Dirichlet dofs by taking the Dirichlet columns in the system matrix to the right side of the system. Suppose i^{th} non-Dirichlet dof is connected with the only Dirichlet dof k, then move the kth columns of both velocity component in the ith rows to the corresponding source terms that is,

$$\begin{aligned} f_i^{\mathrm{I}} &= f_i^{\mathrm{I}} - a_{ik}^{11} g_1 - a_{ik}^{12} g_2, \\ f_i^{\mathrm{II}} &= f_i^{\mathrm{II}} - a_{ik}^{21} g_1 - a_{ik}^{22} g_2 \end{aligned}$$

and drop the kth rows of both velocity components in the system. Further, if the kth columns in B_1 and B_2 exist, then these values should also be multiplied by g_1 and g_2, respectively, and moved to the right side of the system. Recall that a_{ik}^{12} and a_{ik}^{21} will be nonzero only in the deformation tensor form of the Oseen equations. The second approach has an advantage over the first, as the symmetry of the system, if exist, will be preserved, whereas the symmetry will be lost in the first approach.

Inclusion of the Robin type boundary conditions

We here discuss the inclusion of the Robin type boundary condition for a 2d scalar problem. The Robin type boundary condition (2.8)

$$\frac{\partial u}{\partial \mathbf{n}} = \sigma\,(u_\infty - u) \tag{9.3}$$

becomes homogeneous Neumann (do nothing) boundary condition when $\sigma = 0$, and becomes inhomogeneous Neumann (flux) boundary condition by considering $u = 0$ on the right side of (9.3).

The Robin type boundary condition introduces two boundary integral terms

$$\int_{\Gamma_R} \sigma\, u\, v\, ds \quad \text{and} \quad \int_{\Gamma_R} \sigma\, u_\infty\, v\, ds,$$

see (2.9), in the variational form. Both integrals need to be evaluated and then added to the system matrix and the load vector of the system, respectively. The algorithm for the inclusion of the Robin type boundary condition is presented in Algorithm 7. Note that the joint dof mapping array *joint_dof_ptr*[] need to be used to evaluate the boundary integrals.

Algorithm 7: Inclusion of Robin boundary condition in a scalar system.

```
for (j= 0; j< n_joints ; j++) do
    if (Joint [j]== (Robin OR Neumann) then
        loc_f[] = 0
        loc_a[] = 0
        n_loc = joint_dof_ptr[j+1] - joint_dof_ptr[j]
        /* boundary integral                                   */
        for (q=0; q<J_NQ; q++) do
            fact = w[q] * σ * |J_Jj[q]| // quadrature weight * Jacobian
            for (i=0; i< n_loc; i++) do
                d = i * n_loc
                val = fact * φ̂[q][i]
                loc_f[i] += u_∞[q] * val    // load vector
                for (j=0; j<n_loc; j++) do
                    loc_a[d + j] += val * φ̂[q][j]
    /* add local to global matrices                            */
    ...
```

Inclusion of the Navier-Slip boundary condition

We now discuss the inclusion of the Navier-Slip boundary condition for a 2d Oseen problem. Recall that the stress tensor form of the Oseen equations need to be used for models with the Navier-Slip boundary condition. Nevertheless, the pressure space Q is still $L_0^2(\Omega)$, for a detailed discussion on the choice of the pressure space see Section (8.1). The Navier-Slip boundary condition (8.8) on the solid boundary Γ_S in 2d becomes

$$\mathbf{u}\cdot\mathbf{n}=0, \qquad \beta\,\mathbf{u}\cdot\boldsymbol{\tau}+\boldsymbol{\tau}\cdot\mathbb{S}(\mathbf{u},p)\cdot\mathbf{n}=0,$$

where $\boldsymbol{\tau}=(\tau_1,\tau_2)^T$ and $\mathbf{n}=(n_1,n_2)=(\tau_2,-\tau_1)^T$ be the tangential and normal vectors, respectively, on the boundary Γ_S. The Navier-Slip boundary condition introduces additional boundary integral term

$$\beta\int_{\Gamma_S}\mathbf{u}\cdot\boldsymbol{\tau}\,\mathbf{v}\cdot\boldsymbol{\tau}\,\mathrm{d}s$$

in the variational form of the Navier–Stokes equations, see Section 8.8 for the derivation. Further, the no penetration condition $\mathbf{u}\cdot\mathbf{n}=0$ has to be incorporated in the velocity test space as in the Dirichlet boundary case.

Let us impose the Navier-Slip boundary condition on the kth dof of the velocity on Γ_S. Let τ^k and n^k be the tangential and normal vectors at the kth dof on Γ_S. Assume that j^{th} dof on Γ_S is connected with k^{th} dof. The corresponding $(k,j)^{\text{th}}$ matrix entries for the boundary integral are

$$S_{k,j}^{11}=\beta\int_{\Gamma_S}\phi_j\tau_1\,\phi_k\tau_1\,\mathrm{d}s, \qquad S_{k,j}^{12}=\beta\int_{\Gamma_S}\phi_j\tau_2\,\phi_k\tau_1\,\mathrm{d}s,$$
$$S_{k,j}^{21}=\beta\int_{\Gamma_S}\phi_j\tau_1\,\phi_k\tau_2\,\mathrm{d}s, \qquad S_{k,j}^{22}=\beta\int_{\Gamma_S}\phi_j\tau_2\,\phi_k\tau_2\,\mathrm{d}s.$$

The values are added in the system as follows:

$$\begin{array}{r} \\ k^{\text{th}}\text{ row of }u^{\text{I}}\rightarrow \\ \\ \\ \\ k^{\text{th}}\text{ row of }u^{\text{II}}\rightarrow \\ \\ \end{array}
\begin{bmatrix} \ddots & \vdots & \ddots & \ddots & \vdots & \ddots & \vdots \\ \cdots & c_{k,j}^{11} & \cdots & \cdots & c_{k,j}^{12} & \cdots & d_{k,m} \\ \ddots & \vdots & \ddots & \ddots & \vdots & \ddots & \vdots \\ & & & & & & \\ \ddots & \vdots & \ddots & \ddots & \vdots & \ddots & \vdots \\ 0 & n_1^k & 0 & 0 & n_2^k & 0 & 0 \\ \ddots & \vdots & \ddots & \ddots & \vdots & \ddots & \vdots \end{bmatrix}
\begin{bmatrix} \vdots \\ u_k^{\text{I}} \\ \vdots \\ \\ \vdots \\ u_k^{\text{II}} \\ \vdots \end{bmatrix}
=
\begin{bmatrix} \vdots \\ g_k^{\text{I}} \\ \vdots \\ \\ \vdots \\ 0 \\ \vdots \end{bmatrix},$$

where

$$
\begin{aligned}
c^{11}_{k,j} &= \{A_{11}\}_{k,j}\,\tau_1^k + \{A_{21}\}_{k,j}\,\tau_2^k + S^{11}_{k,j} + S^{21}_{k,j},\\
c^{12}_{k,j} &= \{A_{12}\}_{k,j}\,\tau_1^k + \{A_{22}\}_{k,j}\,\tau_2^k + S^{12}_{k,j} + S^{22}_{k,j},\\
d_{k,m} &= \left\{B_1^T\right\}_{k,m}\tau_1^k + \left\{B_2^T\right\}_{k,m}\tau_2^k,\\
g_k^{\mathrm{I}} &= f_k^{\mathrm{I}}\tau_1^k + f_k^{\mathrm{II}}\tau_2^k
\end{aligned}
$$

for all j and m in the kth row. In general, the kth row of the first velocity component has to be multiplied by τ_1^k, and then the kth row of the second velocity component has to be added after multiplying it by τ_2^k to the kth row of the first component. Finally, the diagonal entries of the kth row in A_{12} and in A_{22} have to replaced by n_1^k and n_2^k, respectively, and reset all other entries in the row to zero. Finally, set $f_k^{\mathrm{II}} = 0$. Analogue to the 2d case, the Navier-slip condition in 3d can be implemented in a similar way.

Handling $L_0^2(\Omega)$ pressure space

The choice of the pressure space depends on the boundary condition. For Navier–Stokes with Dirichlet or Navier-slip boundary, the pressure space has to be $L_0^2(\Omega)$, for a detailed discussion see Section 8.8. In this case, the mean value of the pressure has to be eliminated from the computed pressure. It can be achieved by defining the discrete pressure, $p_{h,0} \in L_0^2(\Omega)$ as

$$
p_{h,0}(x) = \sum_{n=0}^{M-1} (p_n - \bar{p})\psi_n, \qquad \bar{p} = \frac{\int_\Omega p_h \,\mathrm{d}x}{\int_\Omega 1 \,\mathrm{d}x},
$$

where $p_h(x)$ is the computed discrete pressure. It is obvious to verify that

$$
\int_\Omega p_{h,0}(x)\,\mathrm{d}x = 0.
$$

Alternatively, the nonuniqueness of the pressure upto an additive constant can also be handled by imposing a zero value to one of the pressure dofs, see Chapter 5.2.5 in Boffi et al. (2013). In practice, the value of first dof of the pressure is set to zero. This approach is efficient and much easier in computations.

Evaluating integrals with surface gradient

Navier–Stokes equations with free surface boundaries will have surface integrals with surface gradient ∇_Γ, for example, see Section 8.8 where we have

$$
\int_{\Gamma_F} \sigma \mathbb{P} : \nabla_\Gamma \mathbf{v}\,\mathrm{d}s, \qquad \mathbb{P} := \mathbb{I} - \mathbf{n}\otimes\mathbf{n}.
$$

Let us consider a 2d case, and let $\mathbf{n} = (n_1, n_2)^T$ be the outward normal on Γ_{F}. Then we have

$$
\mathbb{P} := \mathbb{I} - \mathbf{n}\otimes\mathbf{n} = \begin{bmatrix} 1 - n_1 n_1 & -n_1 n_2 \\ -n_1 n_2 & 1 - n_2 n_2 \end{bmatrix}.
$$

The surface gradient of the velocity test function is defined as

$$\nabla_\Gamma \mathbf{v} := \nabla \mathbf{v} - (\mathbf{n} \cdot \nabla \mathbf{v})\mathbf{n}.$$

Further, for a vector basis function $\mathbf{v} = \phi_i$, we have

$$\nabla_\Gamma \phi_i = \begin{bmatrix} \dfrac{\partial \phi_i}{\partial x_1} - n_1 d & \dfrac{\partial \phi_i}{\partial x_1} - n_1 d \\ \dfrac{\partial \phi_i}{\partial x_2} - n_2 d & \dfrac{\partial \phi_i}{\partial x_2} - n_2 d \end{bmatrix}, \qquad d = n_1 \frac{\partial \phi_i}{\partial x_1} + n_2 \frac{\partial \phi_i}{\partial x_2}.$$

Hence, the additional entries for the load vector due to the force balancing boundary condition on the free surface Γ_F are

$$\{F^{\mathrm{I}}\}_i += \int_{\Gamma_F} \sigma \left((1 - n_1 n_1) \left(\frac{\partial \phi_i}{\partial x_1} - n_1 d \right) - n_1 n_2 \left(\frac{\partial \phi_i}{\partial x_2} - n_2 d \right) \right) \mathrm{d}s,$$

$$\{F^{\mathrm{II}}\}_i += \int_{\Gamma_F} \sigma \left(-n_1 n_2 \left(\frac{\partial \phi_i}{\partial x_1} - n_1 d \right) + (1 - n_2 n_2) \left(\frac{\partial \phi_i}{\partial x_2} - n_2 d \right) \right) \mathrm{d}s.$$

In the case of semiimplicit treatment of the curvature on the free surface, an additional bilinear form

$$\int_{\Gamma_F} \sigma \nabla_\Gamma \mathbf{u} : \nabla_\Gamma \mathbf{v} \, \mathrm{d}s$$

has to be added to the variational form, see Section 8.8. This free surface integral generates additional entries to the diagonal block matrices, A_{11} and A_{22}, of the Navier–Stokes system.

9.6. Solution of the algebraic systems

The solution of a large, sparse linear algebraic system is itself a major research area in numerical analysis, and this topic has evolved in all directions: theory, algorithms and parallel implementations, see Hackbusch (1985; 1994) and Saad (2003). This section provides a brief overview of the solvers for the algebraic system that arise from finite element discretizations, rather than the theory and algorithms of the solvers. Nevertheless, details of the existing methods and solvers for sparse linear systems are discussed briefly.

The size and the sparsity pattern of a system matrix are determined by the choice of the mesh, the finite element discretization and the finite elements. Since the support of finite basis functions increases with an increase in the dimension of the problem, the sparsity of the system matrix reduces with increase in the dimension of the model. Similarly, an increase in the order of finite elements will decrease the sparsity of the system matrix. Hence, not only the size of the system but also the order of finite elements and the dimension of the model problem influence the computational complexity associated with the solution of the system matrix. In addition, the choice of finite elements (continuous, nonconforming, discontinuous) influence the sparsity of the matrix. Moreover, the nonconforming and discontinuous elements are more efficient for parallel computations as the matrix will be more sparse due to less connectivity, and eventually less communication is needed in comparison with the conforming finite elements. Irrespective of the sparsity of the algebraic system, the solution can be obtained by using either a direct solver or an iterative solver.

Direct solvers

The solution of an algebraic system can be obtained using the Gauss elimination method or one of its variants. In the Gauss elimination method, a sequence of row operations are performed on the algebraic system to transform its system matrix into an equivalent upper triangular matrix. Finally, the solution can be obtained by backward substitution with the computational complexity of $\mathcal{O}(n^3)$, where n is the number of unknowns. Even though the Gauss elimination is straightforward to implement, LU decomposition is preferred in direct solvers. In LU decomposition, the system matrix is factorized into lower (L) and upper (U) triangular matrices, and the solution is obtained by solving two triangular systems

$$LY = b, \qquad UX = Y.$$

Though the computational complexity of a LU decomposition and the Gauss elimination is same, there are advantages in using LU decomposition. For instance, the same LU decomposition can be used for different or several right hand sides, when the system matrix remains same for all right hand sides. Nevertheless, a proper ordering or permutations are necessary for an efficient and robust LU decomposition with minimized fill-in, especially when the system matrix is sparse.

Another variant of the Gauss elimination is the frontal solvers proposed by Irons (1970), which is extensively used for solving sparse linear systems arising from finite element discretization. Multifrontal solver is an improved variant of the frontal solvers proposed by Duff and Reid (1983). Some of the popular direct solvers are SuperLU (Demmel et al., 1999), UMFPACK (Davis and Duff, 1997; 1999), PARDISO (Schenk et al., 2000), MUMPS (Amestoy et al., 2001; 2006), PaStiX (Hénon et al., 1999). The direct solvers are preferred for small and reasonably large systems due to its robustness and accuracy. However, the huge memory (RAM) requirement due to fill-in during the LU decomposition is the main limitation of direct solvers for large systems, especially systems arising from 3d models in practical applications.

Iterative solvers

In general, iterative solvers are preferred over the direct solvers for large sparse linear systems due to the memory constraint associated with the direct solvers. In particular, iterative solvers are preferred in massively parallel computing with distributed memory, as the memory requirement in each compute node depends only on the size of the partitioned subsystem. In this section, we briefly present the basics of different iterative solvers, for more details, refer to Saad (2003). Let

$$Au = b$$

be a linear algebraic system with n unknowns. Decomposing the matrix A into M and N such that $A = M - N$, where M is a nonsingular matrix, the above system can be reformulated as

$$u = M^{-1}Nu + M^{-1}b.$$

On the application of Banach fixed point Theorem 1.3, a fixed point iteration

$$u^m = Su^{m-1} + M^{-1}b, \quad m = 1, 2, \ldots \tag{9.4}$$

with an initial guess u^0 can be constructed for the solution of the above system. Here, $S = M^{-1}N$ is an iteration matrix. Moreover, the iteration (9.4) can be classified into stationary and nonstationary methods based on the dependency of S and $M^{-1}b$ on the iteration number. Neither S nor $M^{-1}b$ depend on the iteration number m in stationary iteration methods, for example, Jacobi, Gauss–Seidel, Successive over relaxation (SOR), etc. On contrary, either S or $M^{-1}b$ or both depend on the iteration number in nonstationary methods, for example, Conjugate gradient (CG), Generalized minimum residual (GMRES), BiCG, etc. The iteration (9.4) can also be damped at each iteration step m, that is,

$$u^* = Su^{m-1} + M^{-1}b, \qquad u^m = \omega u^* + (1-\omega)u^{m-1}, \quad \omega \in \mathbb{R}^+.$$

Combining both steps together to get

$$u^m = [\omega S + (1-\omega)\mathbb{I}]\, u^{m-1} + \omega M^{-1}b,$$

where $\mathbb{I}$ is the identity matrix.

To obtain different iterative schemes, split $A = D - L - U$, where D, L and U are diagonal, lower diagonal and upper diagonal matrices, respectively. Now by choosing $M = D$ and $N = -L - U$, the Jacobi iterative scheme can be obtained. Similarly, the Gauss–Seidel scheme can be obtained by choosing $M = D - L$ and $N = -U$. In addition to these methods, projection-based iterate schemes can also be developed. The main idea in the projection-based technique is to seek a solution $\tilde{u}$ to $Au = b$ by imposing $\tilde{u} \in \mathcal{K}$ (search space) such that $(b - Au) \perp \mathcal{L}$ (subspace of constraints), where $\mathcal{K}$ and $\mathcal{L}$ are m-dimensional subspaces of $\mathbb{R}^n$. Different combinations of these subspaces $\mathcal{K}$ and $\mathcal{L}$ lead to nonstationary iterative schemes such as Full orthogonalization method (FOM), GMRES, CG, etc.

In general, these iterative schemes work well when the condition number of the matrix is small. Nevertheless, a preconditioner has to be used for matrices with large condition number. Further, the standard iterative schemes are able to damp highly oscillating error modes in the system very quickly, whereas these schemes will be very slow in damping smooth error modes. Nevertheless, the smooth error mode on a given grid is in general more oscillatory on a coarser grid. It is the key idea to develop a multigrid method. Let us explain the general strategy in the two-level multigrid algorithm. Let Ω_{2h} be the coarse mesh, and Ω_h be the fine mesh obtained by uniformly refining Ω_{2h}. Further, denote the restriction operator from Ω_h to Ω_{2h} by I_h^{2h}, and the prolongation operator from Ω_{2h} to Ω_h by I_{2h}^h. Then, the general steps in two-level algorithm are:

- Step 1 (Pre-smoothing): Apply the smoother γ_1 times on the system $A_h u_h = b_h$ with an initial guess u_0.

- Step 2 (Coarse grid correction): Calculate the residual $r_h = b_h - A_h u_h$, and restrict it to Ω_{2h}, that is, $b_{2h} = I_h^{2h} r_h$, and solve the coarse grid problem $A_h e_{2h} = b_{2h}$.
- Step 3 (Prolongation): Prolongate e_{2h}, that is, obtain $e_h = I_{2h}^h e_{2h}$, and then update the solution u_h with e_h and denote $v_h = u_h + e_h$.
- Step 4 (Post-smoothing): Apply the smoother γ_2 times on the system $A_h u_h = b_h$ with the initial guess v_h.

This approach is called a coarse grid correction scheme or two-level method. Note that this scheme raises another question that how to solve the coarse grid problem $A_h e_{2h} = b_{2h}$. An idea is to use the two-level method for the coarse grid problem, and recursive application of two-level method to coarse grid problem leads to the multigrid method. In particular, the above scheme leads to a geometric multigrid method. The coarse system matrices in the multigrid algorithm can also be obtained by coarsening the finer matrix, A_h, without using hierarchy of meshes, and this approach is classified as algebraic multigrid method. For more details on multigrid method, refer to Hackbusch (1985) and Shaïdurov (1995).

9.7. Object-oriented C++ programming

We conclude this chapter by recalling the basic concepts of the object-oriented programming in the context of finite element algorithm implementations. For more details, the readers are suggested to refer the standard text books on the object-oriented C++ programming.

Class and object in 1d finite element

The key components of an object-oriented package are the main program and its associated classes and objects. To understand the object-oriented implementation of finite element algorithms, we start with a very simple example of storing a set of vertices in a one-dimensional finite element mesh. Since the object-oriented concepts for two- and three-dimensional finite element algorithms are similar to that of one-dimensional concepts, it is sufficient to start with the one-dimensional example. Further, the beginners can also easily understand the object-oriented concepts using this example.

Let us first consider a standard C++ program without object-oriented concepts. Suppose the mesh contains N vertices, then a floating point type array of dimension N is needed to store the positions of N vertices. Further, the array has to be initialized with the position values. These operations are performed in the Main.C code given in Table 9.5. In Main.C, the variable x is a pointer that stores the address of an array of float type. Then, the memory for N float values are dynamically allocated using the keyword **new**. Finally, the array is initialized with the position value of the vertices.

Let us now consider the object-oriented code, MainOOP.C in Table 9.5, that performs the same operation. Instead of a pointer of float type, a double pointer of TVertex type is declared. Then, a memory block of size N that stores the address of an array of TVertex type is dynamically

Table 9.5 A main program without (Main.C) and with (MainOOP.C) using the object-oriented concepts for storing a set of one-dimensional vertices.

Main.C

```
int main()
{
 int i, N=11;
 float h, *x;
 x = new float[N];

 h = 1.0/(N-1.);
 for(i=0;i<N;i++)
   x[i] = i*h;

 return 0;
}
```

MainOOP.C

```
#include "Vertex.h"

int main()
{
 int i, N=10;
 float h;
 TVertex **v;
 v = new TVertex *[N+1];

 h = 1.0/N;
 for(i=0;i<N+1;i++)
   v[i] = new TVertex(i*h);

 return 0;
}
```

allocated using the keyword **new**. After that N instances of the TVertex type are created in the **for** loop using the constructor of TVertex.

As we can observe, the class (TVertex) and objects ($V[i]$) are already introduced in the code MainOOP.C, and it necessitates to understand more about the class and objects. Further, we discuss the object-oriented concepts such as inheritance, polymorphism, virtual member, etc.

Definition 9.1. Class is a collection of data and member functions that provides an implementation for the used data structure and operations. The member functions are often called methods and it perform operations using the data of the class. For example, TVertex defined in Table 9.6 is a class.

Definition 9.2. Object is an instance of a class and there can be more than one instance of the class. Each object can uniquely be identified by its name. Further, each object has its own memory for their respective data in the class. For example, the instance of the class, v[i] created in the **for** loop of MainOOP.C is an object of the class TVertex.

We now define the TVertex class that is used in the main program MainOOP.C given in Table 9.5. The basic stencil or blue print that represents a vertex must have a variable to store the position of the vertex. Further, the variable needs to be initialized, and the stored value should be able to access from outside of the class. Note that, this stencil and the operations are common for all vertices in the mesh. The stencil and the functions that perform all these operations are combined to define the class TVertex, see Table 9.6. The class named TVertex is defined in the header file Vertex.h, whereas operations of its methods are defined in their corresponding source file Vertex.C. We can include the operations of its methods in the class

Table 9.6 The header and its source files of the TVertex class.

Vertex.h (header file)

```
class TVertex
{
 protected:
  float x;

 public:
  TVertex(float x_);

  ~TVertex();

  void ADDdx(float dx);

  float GetX();
};
```

Vertex.C (source file)

```
#include "Vertex.h"

TVertex::TVertex(float x_)
{x=x_;}

TVertex::~TVertex()
{ }

void TVertex::ADDdx(float dx)
{x +=dx;}

float TVertex::GetX()
{return x;}
```

definition itself. However, it is advisable to write a separate source files for the class in order to maintain a structure and to exploit the advantages of the object-oriented concepts.

We now briefly discuss the structure and the operations in the header file Vertex.h. The class TVertex is defined with the keyword **class**. The class environment is separated into two parts, **protected** and **public**. As the names suggest, the variables and methods under the **protected** environment can only be accessed by the member functions of the class and its derived/child class, whereas the methods under the **public** environment can also be accessed from outside of the class. In addition, a **private** environment in the class is also possible. Moreover, the member functions consist at least a class constructor and one class destructor apart from other methods.

A *class constructor* is a special member function of a class with the same name as the class. It is called to create a new object of the class, and it does not return any type of values including void. Moreover, it takes arguments as other member functions. Note that a class can have more than one constructor with different arguments. For example, the member function '`TVertex(float x_);`' in the class TVertex is one of its constructor. It takes one argument `x_` and stores it in the class variable `x`.

A *class destructor* is a special member function of a class with the same name as the class prefixed with a tilde (~). It can be called from outside of the class whenever an object of the class needs to be deleted, and it neither takes arguments nor return a value. The destructor must be called whenever an object goes out of scope otherwise the out of scope object cause a memory leak, especially memory blocks are allocated by a constructor of the class. For example, '`~TVertex();`' in the class TVertex is its destructor. Since no memory is allocated by the class constructor, nothing to deallocate in the destructor.

In addition to the TVertex class, new classes for edges (TEdge), joints (TJoint) and cells (TBaseCell) are necessary, see Figure 9.1. Even though the edges and joints (faces) are distinct

Table 9.7 The definitions of TJoint, TBaseCell and TMesh classes.

```
class TJoint
{
 protected:
  TVertex *v;

  TBaseCell *neib0, *neib1;

 public:
  TJoint(TVertex *v_);

  ~TJoint();

  void SetNeib(int i, TBaseCell *neib);
};
```

```
class TBaseCell
{
 protected:
  TVertex *v0, *v1;

  TJoint *j0, *j1;

 public:
  TBaseCell(TVertex *v0_, TVertex *v1_);

  ~TBaseCell();

  void SetJoint(int i, TJoint *j);
};
```

```
class TMesh
{
 protected:
  int N_Cells;

  TBaseCell **Cells;

 public:
  TMesh(int N, TBaseCell **cells_);

  ~TMesh();
};
```

in three dimensions, these terminologies are same in one and two dimensions. Thus, the new classes for joints (TJoint) and cells (TBaseCell) are defined here, see Table 9.7. The basic stencil of the TJoint class in one dimension should have one pointer variable of type TVertex and two

Table 9.8 The main program to store an one-dimensional mesh in object-oriented concepts.

```
int main()
{
 int i, N=10;
 float h;
 TVertex **v;
 TJoint **Joints;
 TBaseCell **Cells;
 TMesh *Domain;

 v = new TVertex *[N+1];
 Joints = new TJoint *[N+1];
 Cells = new TBaseCell *[N];

 h = 1.0/N;
 for(i=0;i<N+1;i++)
  {
   v[i] = new TVertex(i*h);
   Joints[i] = new TJoint(v[i]);
  }

 for(i=0;i<N;i++)
  {
   Cells[i] = new TBaseCell(v[i], v[i+1]);
   Cells[i]->SetJoint(0, Joints[i]);
   Cells[i]->SetJoint(1, Joints[i+1]);
  }

  Joints[0]->SetNeib(1, Cells[0]);
  for(i=1;i<N;i++)
   {
    Joints[i]->SetNeib(0, Cells[i-1]);
    Joints[i]->SetNeib(1, Cells[i]);
   }
  Joints[N]->SetNeib(0, Cells[N-1]);

  Domain = new TMesh(N, Cells);

 return 0;
}
```

pointer variables of type TBaseCell to store the information of the vertex and its neighbouring cells, respectively, of the joint. Similarly, the basic stencil of the TBaseCell class should have two pointer variables of type TVertex and two pointer variables of type TJoint to store the information of the vertices and the joints, respectively, in the cells. Finally, a class that stores all these information and is able to produce output of the cells needs to be defined, and it is

Table 9.9 Example of an inheritance.

```
class TVertex2D: public TVertex
{
 protected:
  float y;

 public:
  TVertex(float x_, float y_);

  ...
};
```

defined as TMesh. Using all these classes, the mesh can be stored with object-oriented concepts, and the corresponding main program, MainMesh.C is given in Table 9.8.

The above defined classes can easily be extended to 2d and 3d finite element algorithms. To handle high dimensional meshes, we can make use of some of the features such as inheritance, virtual function, friend, etc. of the classes.

Inheritance is a process of forming a new class from an existing base class. The new class is called a derived or child class whereas the base class is known as a parent or super class. Inheritance helps in reducing the overall code size of the program by reusing the data and methods defined in the base class. Further, a pointer to a derived class is type-compatible with a pointer to its base class. The application of this feature is called polymorphism. Although the two- and three-dimensional vertices can be handled with the TVertex class using *#ifdef ... #endif* conditions, TVertex2d class is presented in Table 9.9 to illustrate the inheritance.

A virtual function is a member function of a base class that can be redefined in the derived/child classes of the base class. A member function in the base class can be declared as a virtual function by preceding the function with the keyword 'virtual'. Suppose TTriangle and TQuadrilateral are the two classes derived from the base class TBaseCell. Then, a virtual function GetArea() defined in the TBaseCell can be redefined in the derived classes TTriangle and TQuadrilateral with respective to their area calculation.

A friend function is not a member function of the class but it is defined inside the class with the keyword **friend**. Further, the friend function has the right to access all private and protected members of the class in which it is defined. Even though the friend function functionality seems to have some advantage, it is better to avoid many friend functions in a large scientific code, since it breaks the data hiding principle and it is difficult to handle too many functions. In addition to these definitions and concepts, readers are suggested to understand additional functionalities such as abstract classes, friend function or classes, public and private permissions, etc. in the object-oriented programming.

BIBLIOGRAPHY

Adams, Robert A. 1975. *Sobolev spaces.* Pure and Applied Mathematics, vol. 65. New York-London: Academic Press [A subsidiary of Harcourt Brace Jovanovich, Publishers].

Ainsworth, Mark, and J. Tinsley Oden. 2000. *A Posteriori Error Estimation in Finite Element Analysis.* Pure and Applied Mathematics (New York). New York: Wiley-Interscience [John Wiley & Sons].

Amestoy, Patrick R., Iain S. Duff, Jean-Yves L'Excellent, and Jacko Koster. 2001. 'A Fully Asynchronous Multifrontal Solver Using Distributed Dynamic Scheduling'. *SIAM J Matrix Anal Appl* 23 (1): 15–41.

Amestoy, Patrick R., Abdou Guermouche, Jean-Yves L'Excellent, and Stéphane Pralet. 2006. 'Hybrid scheduling for the parallel solution of linear systems'. *Parallel Computing* 32 (2): 136–56.

Arndt, Daniel. 2013. 'Augmented Taylor–Hood elements for incompressible flow'. MPhil thesis, University Göttingen.

Arnold, Douglas N. 1993. 'On nonconforming linear-constant elements for some variants of the Stokes equations'. *Istit Lombardo Accad Sci Lett Rend A* 127 (1): 83–93 (1994).

Arnold, Douglas N., and Franco Brezzi. 1993. 'Some New Elements for the Reissner–Mindlin Plate Model'. *Boundary value problems for partial differential equations and applications.* RMA Res. Notes Appl. Math., vol. 29, 287–92. Paris: Masson.

Arnold, Douglas N., Franco Brezzi, and Michel Fortin. 1984. 'A Stable Finite Element for the Stokes Equations'. *Calcolo* 21 (4): 337–44 (1985).

Arnold, Douglas N., Daniele Boffi, Richard S. Falk, and Lucia Gastaldi. 2001. 'Finite Element Approximation on Quadrilateral Meshes'. *Comm Numer Methods Engrg* 17 (11): 805–12.

Arnold, Douglas N., Daniele Boffi, and Richard S. Falk. 2002. 'Approximation by Quadrilateral Finite Elements'. *Math Comp* 71 (239): 909–22 (electronic).

Bathe, Klaus-Jürgen, and Eduardo N. Dvorkin. 1985. 'A Four-node Plate Bending Element Based on Mindlin/Reissner Plate Theory and a Mixed Interpolation'. *International Journal for Numerical Methods in Engineering* 21 (2): 367–83.

Becker, Roland, and Malte Braack. 2001. 'A Finite Element Pressure Gradient Stabilization for the Stokes Equations Based on Local Projections'. *Calcolo* 38 (4): 173–99.

Benzi, Michele, Gene H. Golub, and Jörg Liesen. 2005. 'Numerical Solution of Saddle Point Problems'. *Acta Numer* 14: 1–137.

Bercovier, Michel, and O. A. Pironneau. 1979. 'Error Estimates for Finite Element Method Solution of the Stokes Problem in the Primitive Variables'. *Numer Math* 33 (2): 211–24.

Bernardi, Christine. 1989. 'Optimal Finite-Element Interpolation on Curved Domains'. *SIAM J Numer Anal* 26 (5): 1212–40.

Bernardi, Christine, and Geneviève Raugel. 1985. 'Analysis of Some Finite Elements for the Stokes Problem'. *Math Comp* 44 (169): 71–9.

Boffi, Daniele, and Lucia Gastaldi. 2002. 'On the Quadrilateral Q_2-P_1 Element for the Stokes Problem'. *Internat J Numer Methods Fluids* 39 (11): 1001–11.

Boffi, Daniele, Nicola Cavallini, Francesca Gardini, and Lucia Gastaldi. 2012. 'Local Mass Conservation of Stokes Finite Elements'. *J Sci Comput* 52 (2): 383–400.

Boffi, Daniele, Franco Brezzi, and Michel Fortin. 2013. *Mixed Finite Element Methods and Applications*. Springer Series in Computational Mathematics, vol. 44. Heidelberg: Springer.

Boland, J. M., and Roy Nicolaides. 1984. 'On the Stability of Bilinear–Constant Velocity–Pressure Finite Elements'. *Numer Math* 44 (2): 219–22.

Braess, Dietrich. 2007. *Finite Elements: Theory, Fast Solvers, and Applications in Elasticity Theory*, Third edn. Trans. Larry L. Schumaker. Cambridge: Cambridge University Press.

Brenner, Susanne C., and L. Ridgway Scott. 2008. *The Mathematical Theory of Finite Element Methods: Texts in Applied Mathematics*, Third edn. vol. 15. New York: Springer.

Brezzi, Franco, and R. S. Falk. 1991. 'Stability of Higher-order Hood-Taylor Methods'. *SIAM J Numer Anal* 28 (3): 581–90.

Bristeau, Marie-Odile, Roland Glowinski, and Jacques Periaux. 1987. 'Numerical Methods for the Navier–Stokes Equations. Application to the Simulation of Compressible and Incompressible Flows'. *Comp Phys* 6: 73–188.

Bungartz, Hans-Joachim, and Michael Griebel. 2004. 'Sparse grids'. *Acta Numerica* 13: 1–123.

Ciarlet, Philippe G. 2002. *The Finite Element Method for Elliptic Problems*. Classics in Applied Mathematics, vol. 40. Philadelphia, PA. Society for Industrial and Applied Mathematics (SIAM), Reprint of the 1978 original [North-Holland, Amsterdam; MR0520174 (58 #25001)].

Cools, Ronald. 2003. 'An Encyclopaedia of Cubature Formulas'. *Journal of Complexity* 19 (3): 445–53.

Crouzeix, Michel, and Richard S. Falk. 1989. 'Nonconforming Finite Elements for the Stokes Problem'. *Math Comp* 52 (186): 437–56.

Crouzeix, Michel, and Pierre-Arnaud Raviart. 1973. 'Conforming and Nonconforming Finite Element Methods for Solving the Stationary Stokes Equations I'. *RAIRO Anal Numér* 7: 33–76.

Davis, Timothy A., and Iain S. Duff. 1997. 'An Unsymmetric-pattern Multifrontal Method for Sparse *LU* Factorization'. *SIAM J Matrix Anal Appl* 18 (1): 140–58.

———. 1999. 'A Combined Unifrontal/Multifrontal Method for Unsymmetric Sparse Matrices'. *ACM Trans Math Software* 25 (1): 1–20.

Demmel, James W., Stanley C. Eisenstat, John R. Gilbert, Xiaoye S. Li, and Joseph W. H. Liu. 1999. 'A Supernodal Approach to Sparse Partial Pivoting'. *SIAM J Matrix Anal Appl* 20 (3): 720–55.

Duff, Iain S., and John Reid. 1983. 'The Multifrontal Solution of Indefinite Sparse Symmetric Linear'. *ACM Trans Math Softw* 9 (3): 302–25.

Dziuk, Gerhard, and Charles M. Elliott. 2013. 'Finite Element Methods for Surface PDEs'. *Acta Numer* 22: 289–396.

Ern, Alexandre, and Jean-Luc Guermond. 2004. *Theory and Practice of Finite Elements*. Applied Mathematical Sciences, vol. 159. New York: Springer-Verlag.

Fortin, Michel. 1972. 'Calcul numérique des écoulements des fluides de Bingham et des fluides newtoniens incompressibles par la méthode des élements finis'. PhD thesis, Université de Paris VI.

———. 1977. 'An Analysis of the Convergence of Mixed Finite Element Methods'. *RAIRO Anal Numér* 11 (4): 341–54, iii.

———. 1981. 'Old and New Finite Elements for Incompressible Flows'. *Internat J Numer Methods Fluids* 1 (4): 347–64.

Fortin, Michel, and M. Soulie. 1983. 'A Non-conforming Piecewise Quadratic Finite Element on Triangles'. *Int J Numer Meth Engrg* 19 (4): 505–20.

Gajewski, Herbert, Konrad Gröger, and Klaus Zacharias. 1974. *Nichtlineare Operatorgleichungen und Operatordifferentialgleichungen.* Berlin: Akademie-Verlag. Mathematische Lehrbücher und Monographien, II. Abteilung, Mathematische Monographien, Band 38.

Ganesan, Sashikumaar. 2015. 'Simulations of Impinging Droplets with Surfactant-dependent Dynamic Contact Angle'. *J Comput Phys* 301: 178–200.

Ganesan, Sashikumaar, and Lutz Tobiska. 2009. 'A Coupled Arbitrary Lagrangian–Eulerian and Lagrangian Method for Computation of Free Surface Flows with Insoluble Surfactants'. *J Comput Phys* 228 (8): 2859–73.

———. 2012. 'Arbitrary Lagrangian–Eulerian Finite-Element Method for Computation of Two-phase Flows with Soluble Surfactants'. *J Comput Phys* 231 (9): 3685–702.

———. 2013. 'Operator-splitting Finite Element Algorithms for Computations of High-dimensional Parabolic Problems'. *Appl Math Comp* 219: 6182–96.

Ganesan, Sashikumaar, Gunar Matthies, and Lutz Tobiska. 2008. 'Local Projection Stabilization of Equal Order Interpolation Applied to the Stokes Problem'. *Math Comput* 77: 2039–60.

Ganesan, Sashikumaar, Andreas Hahn, Kristin Simon, and Lutz Tobiska. 2016a. 'Finite Element Computations for Dynamic Liquid-fluid Interfaces'. In *Computational Methods for Complex Liquid–Fluid Interfaces.* Progress in Colloid and Interface Science, edited by Rahni, M. T., Karbaschi, M., and Miller, R., vol. 5, 331–51. CRC Press Taylor & Francis Group.

Ganesan, Sashikumaar, Volker John, Gunar Matthies, Meesala Raviteja, Abdus Shamim, and Ulrich Wilbrandt. 2016b. 'An Object Oriented Parallel Finite Element Scheme for Computations of PDEs: Design and Implementation'. 2016 IEEE 23rd International Conference on High Performance Computing Workshops (HiPCW), 00, 106–15.

Geuzaine, Christophe, and Jean-Francois Remacle. 2009. 'Gmsh: A Three-dimensional Finite Element Mesh Generator with Built-in Pre- and Post-processing Facilities'. *Int J Numer Meth Engg* 79 (11): 1309–31.

Girault, Vivette, and Pierre-Arnaud Raviart. 1986. *Finite Element Methods for Navier–Stokes Equations.* Springer Series in Computational Mathematics Theory and Algorithms, vol. 5. Berlin: Springer-Verlag. .

Grisvard, Pierre. 1992. *Singularities in Boundary Value Problems.* Recherches en Mathématiques Appliquées [Research in Applied Mathematics], vol. 22. Paris: Masson, Berlin: Springer-Verlag.

Grossmann, Christian, Hans-Gorg Roos, and Martin Stynes. 2007. *Numerical Treatment of Partial Differential Equations.* Berlin, Heidelberg: Springer.

Gupta, Radhey S. 2015. *Elements of Numerical Analysis.* Second edn. Cambridge University Press.

Hackbusch, Wolfgang. 1985. *Multigrid Methods and Applications.* Springer Series in Computational Mathematics, vol. 4. Berlin: Springer-Verlag.

———. 1994. *Iterative Solution of Large Sparse Systems of Equations.* Applied Mathematical Sciences, vol. 95. New York: Springer-Verlag. Trans. and revised from the 1991 German original.

Hairer, Ernst, and Gerhard Wanner. 1996. *Solving Ordinary Differential Equations II.* Springer Series in Computational Mathematics, vol. 14. Berlin: Springer-Verlag.

Han, Hou De. 1984. 'Nonconforming Elements in the Mixed Finite Element Method'. *J Comput Math* 2 (3): 223–33.

Hansbo, Peter, and Mats G. Larson. 2003. 'A P^2-continuous, P^1-discontinuous Finite Element Method for the Mindlin–Reissner Plate Model'. *Numerical Mathematics and Advanced Applications*, 765–74. Milan: Springer Italia.

———. 2014. 'Locking Free Quadrilateral Continuous/discontinuous Finite Element Methods for the Reissner–Mindlin Plate'. *Comput Methods Appl Mech Engrg* 269: 381–93.

Hansbo, Peter, David Heintz, and Mats G. Larson. 2011. 'A Finite Element Method with Discontinuous Rotations for the Mindlin–Reissner Plate Model'. *Comput Methods Appl Mech Engrg* 200 (5–8): 638–48.

Hénon, Pascal, Pierre Ramet, and Jean Roman. 1999. 'A Mapping and Scheduling Algorithm for Parallel Sparse Fan-in Numerical Factorization'. In *Euro-Par'99 Parallel Processing: 5th International Euro-Par Conference Toulouse, France, August 31–September 3, 1999 Proceedings*, 1059–67. Berlin, Heidelberg: Springer.

Hughes, Thomas J. R., Leopoldo P. Franca, and Marc Balestra. 1986. 'A New Finite Element Formulation for Computational Fluid Dynamics. V: Circumventing the Babuška–Brezzi Condition: A Stable Petrov–Galerkin Formulation of the Stokes Problem Accomodating Equal-order Interpolations'. *Comput Methods Appl Mech Eng* 59: 85–99.

Irons, Bruce M. 1970. 'A Frontal Solution Scheme for Finite Element Analysis'. *Int J Numer Methods Eng* 2: 5–32.

Jiang, Jin Sheng, and Cheng Xiao Liang. 1997. 'Stability of Locally Mass-conserving Higher Order Taylor-Hood Elements'. *J Sys Sci Math Scis* 17 (3): 193–97.

John, Volker. 2016. *Finite Element Methods for Incompressible Flow Problems.* Springer International Publishing.

John, Volker, and Matthies Gunar. 2004. 'MooNMD - a Program Package Based on Mapped Finite Element Methods'. *Comput Visual Sci* 6 (2–3): 163–70.

Lascaux, Patrick, and Lesaint Pierre. 1975. 'Some Nonconforming Finite Elements for the Plate Bending Problem'. *Rev Française Automat Informat.Recherche Operationnelle Sér Rouge Anal Numér* 9 (R-1): 9–53.

Linke, Alexander. 2014. 'On the Role of the Helmholtz Decomposition in Mixed Methods for Incompressible Flows and a New Variational Crime'. *Comput Methods Appl Mech Engrg* 268: 782–800.

Linke, Alexander, Matthies Gunar, and Tobiska Lutz. 2016. 'Robust Arbitrary Order Mixed Finite Element Methods for the Incompressible Stokes Equations with Pressure Independent Velocity Errors'. *ESAIM Math Model Numer Anal* 50 (1): 289–309.

Lions, Jacques-Louis, and Magenes Enrico. 1968. *Problèmes aux limites non homogènes et applications. Vol. 1.* Travaux et Recherches Mathématiques, No. 17. Paris: Dunod.

Mansfield, Lois. 1982. 'Finite Element Subspaces with Optimal Rates of Convergence for the Stationary Stokes Problem'. *ESAIM: Mathematical Modelling and Numerical Analysis* 16 (1): 49–66.

Matthies, Gunar. 2001. 'Mapped Finite Elements on Hexahedra. Necessary and Sufficient Conditions for Optimal Interpolation Errors'. *Numer Algorithms* 27 (4): 317–27.

———. 2007. 'Inf-sup Stable Nonconforming Finite Elements of Higher Order on Quadrilaterals and Hexahedra'. *ESAIM: Mathematical Modelling and Numerical Analysis* 41 (9): 855–74.

Matthies, Gunar, and Tobiska Lutz. 2002. 'The Inf-sup Condition for the Mapped Q_k-P_{k-1}^{disc} Element in Arbitrary Space Dimensions'. *Computing* 69 (2): 119–39.

———. 2005. 'Inf-sup Stable Non-conforming Finite Elements of Arbitrary Order on Triangles'. *Numerische Mathematik* 102 (2): 293–309.

Matthies, Gunar, Skrzypacz Piotr, and Tobiska Lutz. 2007. 'A Unified Convergence Analysis for Local Projection Stabilisations Applied to the Oseen Problem'. *M2AN Math Model Numer Anal* 41 (4): 713–42.

McLachlan, Robert I., and Quispel G. Reinout W. 2002. 'Splitting Methods'. *Acta Numerica* 11 (1): 341–434.

Ming, Wang, and Xu Jinchao. 2007. 'Nonconforming Tetrahedral Finite Elements for Fourth Order Elliptic Equations'. *Math Comp* 76 (257): 1–18 (electronic).

Press, William H., Saul A. Teukolsky, William T. Vetterling, and Brian P. Flannery. 2007. *Numerical Recipes.* Third edn. Cambridge University Press. The Art of Scientific Computing.

Rannacher, Rolf, and Turek Stefan. 1992. 'Simple Nonconforming Quadrilateral Stokes Element'. *Numer Meth Part Diff Equ* 8: 97–111.

Saad, Yousef. 2003. *Iterative methods for sparse linear systems.* Second edn. Society for Industrial and Applied Mathematics, Philadelphia, PA.

Schenk, Olaf, K. Gärtner, and Wolfgang Fichtner. 2000. 'Efficient Sparse LU Factorization with Left-Right Looking Strategy on Shared Memory Multiprocessors'. *BIT Numerical Mathematics* 40 (1): 158–76.

Schieweck, Friedhelm. 2010. 'A-stable Discontinuous Galerkin–Petrov Time Discretization of Higher Order'. *Journal of Numerical Mathematics* 18 (1): 25–57.

Scott, L. Ridgway, and Michael Vogelius. 1985a. 'Conforming Finite Element Methods for Incompressible and Nearly Incompressible Continua'. In *Large-scale computations in fluid mechanics, Part 2 (La Jolla, Calif., 1983).* Lectures in Appl. Math., vol. 22. 221–44. Providence, RI: American Mathematical Society.

———. 1985b. 'Norm Estimates for a Maximal Right Inverse of the Divergence Operator in Spaces of Piecewise Polynomials'. *RAIRO, Modélisation Math Anal Numér* 19: 111–43.

Scott, L. Ridgway, and Shangyou Zhang. 1990. 'Finite Element Interpolation of Nonsmooth Functions Satisfying Boundary Conditions'. *Math Comp* 54 (190): 483–93.

Shaĭdurov, Vladimir V. 1995. *Multigrid Methods for Finite Elements.* Mathematics and its Applications, vol. 318. Dordrecht: Kluwer Academic Publishers Group. Trans. from the 1989 Russian original by N. B. Urusova and revised by the author.

Shewchuk, Jonathan Richard. 1996. 'Triangle: Engineering a 2D Quality Mesh Generator and Delaunay Triangulator'. 203–22 In *Applied Computational Geometry: Towards Geometric Engineering.* edited by Lin, M. C., and Manocha, D., Lecture Notes in Computer Science. Berlin: Springer-Verlag.

Si, Hang. 2015. 'TetGen, a Delaunay-based Quality Tetrahedral Mesh Generator'. *ACM Trans Mathematical Software* 41 (2): 11.

Stenberg, Rolf. 1984. 'Analysis of Mixed Finite Elements Methods for the Stokes Problem: A Unified Approach'. *Math Comp* 42 (165): 9–23.

Thatcher, Ronald W. 1990. 'Locally Mass-conserving Taylor–Hood Elements for Two- and Three-dimensional Flow'. *Int J Numer Methods Fluids* 11 (3): 341–53.

Thomée, Vidar. 2006. *Galerkin Finite Element Methods for Parabolic Problems.* Second edn. Springer Series in Computational Mathematics, vol. 25. Berlin: Springer-Verlag.

Toulorge, Thomas, Christophe Geuzaine, Jean-Francois Remacle, and Jonathan Lambrechts. 2013. 'Robust Untangling of Curvilinear Meshes'. *J Comput Phys* 254: 8–26.

Turek, Stefan. 1994. 'Multigrid Techniques for Simple Discretely Divergence-free Finite Element Spaces'. In *Multigrid Methods, IV (Amsterdam, 1993).* Internat. Ser. Numer. Math., vol. 116, 321–32. Basel: Birkhäuser.

Verfürth, Rüdiger. 1984. 'Error Estimates for a Mixed Finite Element Approximation of the Stokes Equations'. *ESAIM: Mathematical Modelling and Numerical Analysis* 18 (2): 175–82.

———. 2013. *A Posteriori Error Estimation Techniques for Finite Element Methods.* Numerical Mathematics and Scientific Computation. Oxford: Oxford University Press.

Walkington, Noel J. 2014. 'A C^1 Tetrahedral Finite Element Without Edge Degrees of Freedom'. *SIAM J Numer Anal* 52 (1): 330–42.

Wilbrandt, Ulrich, Clemens Bartsely, Naveed Ahmed, Najib Alia, Felix Anker, Laura Blank, Alfonso Caiazzo, Sashikumaar Ganesan, Swetlana Giere, Gunar Matthies, Raviteja Meesala, Abdus Shamim, Jagannath Venkatesan, and Volker John. 2017. 'ParMoon–A Modernized Program Package Based on Mapped Finite Elements'. *Computers and Mathematics with Applications.* doi:10.1016/j.camwar2016.12.020.

Zeidler, Eberhard. 1990. *Nonlinear Functional Analysis and its Applications. II/A.* Linear Monotone Operators. New York: Springer-Verlag. Trans. from the German by the author and Leo F. Boron.

Ženíšek, Alexander. 1974a. 'A General Theorem on Triangular Finite $C^{(m)}$-elements'. *Rev Française Automat Informat Recherche Opérationnelle Sér Rouge* 8 (R-2): 119–27.

———. 1974b. 'Tetrahedral Finite $C^{(m)}$-elements'. In *Proceedings of the Third Conference on Basic Problems of Numerical Mathematics (Prague, 1973)*, vol. 15, 189–193.

Zhang, Shangyou. 2005. 'A New Family of Stable Mixed Finite Elements for the 3D Stokes Equations'. *Mathematics of Computation* 74 (250): 543–54.

———. 2016. 'Stable Finite Element Pair for Stokes Problem and Discrete Stokes Complex on Quadrilateral Grids'. *Numer Math* 133: 371–408.

Zhang, Linbo, Tao Cui, and Hui Liu. 2009. 'Symmetric Quadrature Rules on Triangles and Tetrahedra'. *Journal of Computational Mathematics* 17 (1): 89–96.

INDEX

For EU product safety concerns, contact us at Calle de José Abascal, 56–1°,
28003 Madrid, Spain or eugpsr@cambridge.org.

www.ingramcontent.com/pod-product-compliance
Lightning Source LLC
LaVergne TN
LVHW081517060526
838200LV00005B/201

* 9 7 8 1 1 0 8 4 1 5 7 0 5 *